GTPases

EDITED BY

Alan Hall

*Medical Research Council Laboratory for Molecular
Cell Biology and Department
of Biochemistry University College London*

OXFORD
UNIVERSITY PRESS

OXFORD

UNIVERSITY PRESS

Great Clarendon Street, Oxford OX2 6DP

Oxford University Press is a department of the University of Oxford
and furthers the University's aim of excellence in research, scholarship,
and education by publishing worldwide in

Oxford New York

Athens Auckland Bangkok Bogotá Buenos Aires Calcutta
Cape Town Chennai Dar es Salaam Delhi Florence Hong Kong Istanbul
Karachi Kuala Lumpur Madrid Melbourne Mexico City Mumbai
Nairobi Paris São Paulo Singapore Taipei Tokyo Toronto Warsaw

with associated companies in Berlin Ibadan

Oxford is a registered trade mark of Oxford University Press
in the UK and in certain other countries

Published in the United States
by Oxford University Press Inc., New York

© Oxford University Press, 2000

A catalogue record for this book is available from the British Library

Library of Congress Cataloging in Publication Data

GTPases / [edited by] Alan Hall.
p. cm.—(Frontiers in molecular biology ; 24)
Includes bibliographical references and index.
1. Guanosine triphosphatase. 2. Guanosine triphosphatase—Biotechnology. I. Hall, A.
(Alan) II. Series.
QP609.G83 G77 2000   571.6–dc21   99-046596
ISBN 0-19-963745-8 (hbk.)
ISBN 0-19-963744-X (pbk.)

Typeset by Footnote Graphics Warminster, Wilts

Printed by the Bath Press, Avon

GTPases

Book are to b

Frontiers in Molecular Biology

B. D. Hames

*Department of Biochemistry
and Molecular Biology
University of Leeds, Leeds LS2 9JT, UK*

D. M. Glover

*Cancer Research Laboratories,
Department of Anatomy and Physiology,
University of Dundee, Dundee DD1 4HN, UK*

TITLES IN THE SERIES

Preface

All cells need to be able to regulate multiple biochemical pathways within a complex environment and both prokaryotes and eukaryotes have devised a multitude of ingenious ways to deal with this problem. The most common mechanism for controlling biochemical pathways involves protein phosphorylation and dephosphorylation, but it is now clear that cells also make use of a wide repertoire of dedicated regulatory proteins called GTP-binding proteins or GTPases. These proteins act as binary molecular switches, interconverting between an active (GTP-bound) and an inactive (GDP-bound) conformation. In the active conformation, the protein can interact with one or more cellular targets to promote a response, while hydrolysis of GTP to GDP regenerates the resting state. This simple idea has been modified and perfected throughout evolution to the point where members of the GTPase superfamily are able to regulate highly sophisticated and complex cellular processes.

This book covers all the main families of GTPases currently under investigation. Chapters 1 and 2 describe the most widely studied GTPase family, the heterotrimeric G proteins, which were discovered back in 1980 and for which Rodbell and Gilman received the 1994 Nobel prize. G proteins are found in all eukaryotic cells and their role is to couple activated heptahelical, transmembrane receptors to cellular responses. The *Caenorhabditis elegans* sequencing project has revealed that 5% of the worm's genes (almost 1000) encode heptahelical receptors and their ability to influence cell behaviour in response to ligand activation is funnelled through 14 distinct G proteins. G proteins classically control the levels of second messengers such as cAMP, cGMP, inositol 1,4,5-triphosphate (IP$_3$) and Ca^{2+}, although more recently it has become clear that they can regulate many other cellular pathways, including those that are directly under the control of the next most studied family of regulatory GTP-binding proteins, the small GTPases.

The *Ras* gene caused a great deal of excitement in the early 1980s, as it was the first human oncogene to be characterized. Subsequent analysis of the Ras protein showed that it behaved as a monomeric, regulatory GTP-binding protein and that it belonged to a large family of related proteins (now totalling more than 60 in humans) referred to as 'small GTPases'. Each of the five major classes of small GTPases, Ras, Rho, Rab, Arf and Ran, has taken on a life of its own and these are covered separately in Chapters 3-7 of this book. The members of the Ras and Rho families that have been analysed so far regulate intracellular signal transduction pathways in response to extracellular signals. The most famous example being growth factor induced activation of Ras followed by Ras.GTP-induced activation of a mitogen-activated protein (MAP) kinase cascade, leading to gene transcription and entry into the cell cycle. In fact many fundamental aspects of cell behaviour are controlled by Ras and Rho GTPases, including decisions to proliferate or differentiate, cell polarity and, through modifications to the actin cytoskeleton, cell shape and cell movement. The Rab and Arf families, on the other hand, play specialized roles in promoting steps involved in intracellular transport. The mammalian Rab family alone currently contains over 30 members, while even the lowly yeast *Saccharomyces cerevisiae* has 11 Rabs. Vesicular trafficking is clearly a fundamental requirement of all eukaryotic cells and the Rab and Arf family proteins must have evolved early to deal with the sorting and recognition problems involved. Finally, Ran, which is by far the most abundant GTPase, comprising 0.4% of total cellular protein in human cells, is in a class of its own; it regulates both the import and export of proteins and RNA through the nuclear membrane and, not surprisingly therefore, is highly conserved in all eukaryotes.

Two further families of GTPases are discussed in Chapter 8. The first includes the signal recognition particle (SRP) and the SRP receptor, involved in targetting secreted and membrane proteins to the endoplasmic reticulum or to the plasma membrane of eukaryotes and prokaryotes, respectively. The SRP and the SRP receptor protein complexes each contain a GTPase and these two molecular switches co-ordinate the assembly and the unidirectionality of the protein targetting reaction. The second family includes GTPases involved in protein synthesis—the elongation and trans-location factors. EF-Tu, the elongation factor from *Escherichia coli*, has been studied biochemically since the early 1970s. In its GTP-bound conformation, EF-Tu transports aminoacylated tRNA to the ribosome and, provided there is a correct match between codon and anticodon, EF-Tu then dissociates in response to ribosome-induced GTP hydrolysis.

In 1989, the predicted conformational change associated with GTPases was finally visualized directly, with the X-ray structure determination of Ras in both its GDP- and GTP-bound states. As discussed in Chapter 9, this has been followed by a huge amount of structural analysis that now encompasses all families of GTPases, including the heterotrimeric G proteins, as well as many of the proteins with which they interact. As a consequence, the molecular details underlying GTPase target recognition are being clarified.

The final chapter describes the numerous bacterial toxins for which GTPases are the cellular targets, including the elongation factors (diphtheria toxin), hetero-trimeric G proteins (pertussis and cholera toxins) and small GTPases (clostridial toxins). The diverse clinical effects of these toxins provide a clear demonstration of the importance of GTPases in a wide variety of physiological settings. In addition to providing important insights into mechanisms of bacterial pathogenesis, this work is also providing new experimental reagents with which to probe cellular aspects of GTPase function.

My hope in bringing together all GTPase families under the cover of one book was to provide a better and more coherent overview of the versatility and diversity of this class of regulatory molecule. The analysis of GTPase function has proven to be an enormously fruitful way in which to probe the biochemical pathways associated with many fundamental processes in cell biology and physiology. There is no reason to believe that this will not continue to be the case in the foreseeable future and I hope this book will provide a reference point for those venturing into this area.

London
August 1999

A.H.

Contents

2   G proteins II: G_q, G_{12}, and G_z 35

JENNIFER L. GLICK, THOMAS E. MEIGS, AND PATRICK J. CASEY

3   Ras   67

JOHANNES L. BOS

4 Rho

ANNE J. RIDLEY

5 Rab   137

RUTH N. COLLINS AND PATRICK BRENNWALD

6 Arf

MICHAEL G. ROTH

10 GTPases targetted by bacterial toxins   311

KLAUS AKTORIES, GUDULA SCHMIDT, AND FRED HOFMANN

Contributors

KLAUS AKTORIES
Institut für Pharmakologie und Toxikologie, Albert-Ludwigs-Universität Freiburg, Hermann Herder Strasse 5, D-79104 Freiburg,Germany.

JOHANNES L. BOS
Laboratory for Physiological Chemistry and Centre for Biomedical Genetics, Universiteitsweg 100, 3584 CG Utrecht, The Netherlands.

PATRICK J. BRENNWALD
Department of Cell Biology, Cornell University Weill Medical College, 1300 York Avenue, New York, NY 10021, USA.

AMY BROWNAWELL
Markey Center for Cell Signaling, University of Virginia, Box 577 HSC, Charlottesville, VA 22908, USA.

PATRICK J. CASEY
Department of Pharmacology and Cancer Biology, Duke University Medical Center, Research Drive, C303 LSRC, Durham, NC 27710-3686, USA.

RUTH N. COLLINS
Department of Molecular Medicine, Veterinary Medical College, Cornell University, Ithaca, NY 14853, USA.

DOUGLAS M. FREYMANN
Department of Molecular Pharmacology and Biological Chemistry, Northwestern University Medical School, Chicago, Illinois 60611, USA.

JENNIFER L. GLICK
Department of Pharmacology and Cancer Biology, Duke University Medical Center, Research Drive, C303 LSRC, Durham, NC 27710-3686, USA.

ALAN HALL
MRC Laboratory for Molecular Cell Biology, University College London, Gower Street, London WC1E 6BT, UK.

FRED HOFMANN
Institut für Pharmakologie und Toxikologie, Albert-Ludwigs-Universität Freiburg, Hermann Herder Strasse 5, D-79104 Freiburg, Germany.

IAN MACARA
Markey Center for Cell Signaling, University of Virginia, Box 577 HSC, Charlottesville, VA 22908, USA.

THOMAS E. MEIGS

Department of Pharmacology and Cancer Biology, Duke University Medical Center, Research Drive, C303 LSRC, Durham, NC 27710-3686, USA.

SUSANNE M. MUMBY

Department of Pharmacology, University of Texas Southwestern Medical Center at Dallas, 5323 Harry Hines Blvd., Dallas, TX 95235-9041, USA.

ANNE J. RIDLEY

Ludwig Institute for Cancer Research, University College Branch, 91 Riding House Street, London W1P 8BT; and Department of Biochemistry and Molecular Biology, University College London, Gower Street, London WC1E 6BT, UK.

MICHAEL G. ROTH

Department of Biochemistry, University of Texas Southwestern Medical Center at Dallas, 5323 Harry Hines Blvd., Dallas, TX 75235-9038, USA.

GUDULA SCHMIDT

Institut für Pharmakologie und Toxikologie, Albert-Ludwigs-Universität Freiburg, Hermann Herder Strasse 5, D-79104 Freiburg, Germany.

PETER WALTER

Howard Hughes Medical Institute and Dept Biochemistry and Biophysics, University of California-San Francisco, 513 Parnassus Ave - S964, San Francisco, CA 94143-0448, USA.

KATIE WELCH

Markey Center for Cell Signaling, University of Virginia, Box 577 HSC, Charlottesville, VA 22908, USA.

ALFRED WITTINGHOFER

Max-Planck-Institut fur molekular Physiologie, Abteilung Strukturelle Biologie, Otto-Hahn-Str. 11, 44227 Dortmund, Germany.

Abbreviations

AC	adenylyl cyclase
AHO	Albright hereditary osteodystrophy
AP	adaptor protein
ARF	ADP-ribosylating factor
ARL	Arf-like
ASK1	apoptosis signal-regulating kinase 1
ATP	adenosine triphosphate
BIB	basic importin-β binding
Btk	Bruton's tyrosine kinase
cAMP	cyclic adenosine monophosphate
CCK	cholecysto kinin
Cdk5	cyclin-dependent kinase 5
cDNA	complentary DNA
CDR	complementarity determining region
cGMP	cyclic guanosine monophosphate
CNF	cytotoxic necrotizing factor
COP	coat protein
CRIB	Cdc42/Rac-interactive binding (motif)
DAG	diacylglycerol
DH	Dbl homology
DNT	dermonecrotic toxin
EDIN	epidermal differentiation inhibitor
EDTA	ethylenediaminetetraacetic acid
EF	elongation factor
EGF	epidermal growth factor
Epac	exchange protein activated by cAMP
EPR	electron paramagnetic resonance
ER	endoplasmic reticulum
ERK	extracellular-signal-regulated kinase
ERM	ezrin, moesin, radixin
ESEEM	electron spin-echo envelope modulation
fMLP	N-formylmethionyl-leucyl-phenylalanine
GAP	GTPase-activating protein
GC	germinal centre
GDF	GDI displacement factor
GDI	guanine nucleotide dissociation inhibitor
GDP	guanosine diphosphate
GDS	guanine nucleotide dissociation stimulator
GEF	guanine nucleotide exchange factor

GFAP	glial fibrillary acidic protein
GFP	green fluorescent protein
GGTase	geranylgeranyl transferase
GH	growth hormone
GPCR	G-protein-coupled receptor
GRK	G-protein-coupled receptor kinase
GTP	guanosine triphosphate
GTPase	guanosine triphosphatase
HGF	hepatocyte growth factor
5HT	5-hydroxytryptamine (serotonin)
IBD	I-box domain
ICAM	Intercellular, calcium-dependent, adhesion molecule
IL-2	interleukin-2
IP$_3$	inositol 1,4,5-triphosphate
IP$_4$	inositol 1,3,4,5-tetraphosphate
JNK	c-*Jun* N-terminal kinase
LPA	lysophosphatidic acid
MAP	mitogen-activated protein
MAPK	mitogen-activated protein kinase
MAPKKK	MAP kinase kinase kinase
MLC	myosin light chain
mRNA	messenger RNA
NAD	nicotinamide adenine dinucleotide
NADPH	reduced nicotinamide adenine dinucleotide phosphate
NES	nuclear export signal
NF	nuclear factor
NGF	nerve growth factor
NHE	Na$^+$/H$^+$ exchanger
NK	natural killer
NLS	nuclear localization signal
NMR	nuclear magnetic resonance
NPC	nuclear pore complex
PA	phosphatidic acid
PAH	phosphatidate hydrolase
PAK	p21-activated protein kinase
PC	phosphatidylcholine
PDE	phosphodiesterase
PDGF	platelet-derived growth factor
PEPCK	phosphoenolpyruvate carboxykinase
PH	pleckstrin homology
PHP	pseudohypoparathyroidism
PI	phosphatidylinositol
PI3K	phosphoinositide 3-kinase
PI3P	phosphotidylinositol 3-phosphate

PI(3)K	phosphatidylinositol-3 hydroxykinase
PI3,4P$_2$	phosphatidylinositol 3,4-bisphosphate
PI3,4,5P$_3$	phosphatidylinositol 3,4,5-trisphosphate
PI4P	phosphoinositol 4-phosphate
PI4,5P$_2$	phosphoinositol 4,5-bisphosphate
PI5K	phosphoinositide 5-kinase
PIP$_2$	phosphatidylinositol 4,5-bisphosphate
PIP5-K	phosphoinositol (4) phosphate 5-kinase
PI-PLC	phosphoinositide-specific phospholipase C
PKA	protein kinase A
PKB	protein kinase B
PKC	protein kinase C
PLA$_2$	phospholipase A$_2$
PLC-β	phosphoinositide-specific phospholipase C-β
PLD	phospholipase D
PRK	PKC-related kinase
PTB	phosphotyrosine binding
PTX	pertussis toxin
Ran	Ras-related protein in the nucleus
RBD	Ras-binding domain
RCC	regulator of chromosome condensation
REM	Rho effector motif
REP	Rab escort protein
RF	release factor
RGS	regulators of G-protein signalling
RKH	Rok-kinectin homology
ROK	Rho kinase
rRNA	ribosomal RNA
SCC	superior colliculus
SDS	sodium dodecyl sulphate
SF	scatter factor
Sos	son of sevenless
SR	SRP receptor
SRF	serum response factor
SRP	signal recognition particle
SUMO	small ubiquitin-related modifier
TGF	transforming growth factor
TGN	trans-Golgi network
TNF	tumour necrosis factor
TPA	12-O-tetradecanoylphorbol-13-acetate
tRNA	transfer RNA
UDP	uridine diphosphate
WAS	Wiskott-Aldrich syndrome
XTP	xanthosine triphosphate

1 | G proteins I: G_s and G_i

SUSANNE M. MUMBY

1. Introduction: from pheromones to photons

Cells respond with remarkable speed and specificity to the many environmental cues that they encounter. Signal transduction mediated by heterotrimeric G proteins is a common strategy among eukaryotes to relay extracellular information to the interior of the cell. This strategy is employed in physiological functions ranging from mating in yeast to vision in humans. External primary messengers, such as light, odorants, hormones, chemoattractants, and neurotransmitters, stimulate antennae-like receptors that activate G proteins. The activity of second-messenger-producing effectors are, in turn, modulated by the activated G proteins. A relay diagram of G-protein signalling is shown in Fig. 1. The specific examples shown, hormone-sensitive adenylyl cyclase and visual transduction (Fig. 1, lines a–c), are the most extensively studied systems, on which much of our current knowledge of G-protein-mediated signal transduction is based.

Signal transduction is initiated at the plasma membrane by G-protein-coupled receptors, characterized by amino-acid sequences predicted to form seven membrane spans (Fig. 2). Extracellular-facing aspects of the receptor bind agonist and intracellular portions thus are enabled to interact fruitfully with a cognate G protein positioned at the inner face of the plasma membrane. G proteins are composed of α, β, and γ subunits whose deduced amino-acid sequences lack clear hydrophobic stretches that would be anticipated to span the bilayer. Instead, G proteins are modified by lipids that facilitate their interaction with proteins and/or lipids comprising the plasma membrane. Effectors, including adenylyl cyclase and ion channels, span the membrane (Fig. 2) but others do not. Light-activated cGMP phosphodiesterase (PDE), for example, is found in cytosolic and membranous fractions of retinal cells.

1.1 In the beginning . . .

Appreciation of G-protein-mediated signal transduction began with the character-ization of hormone-sensitive adenylyl cyclase. Rodbell first promulgated the notion of a transducer acting as an intermediary between receptors and adenylyl cyclase. He and his colleagues established that receptor agonist was not sufficient to activate the

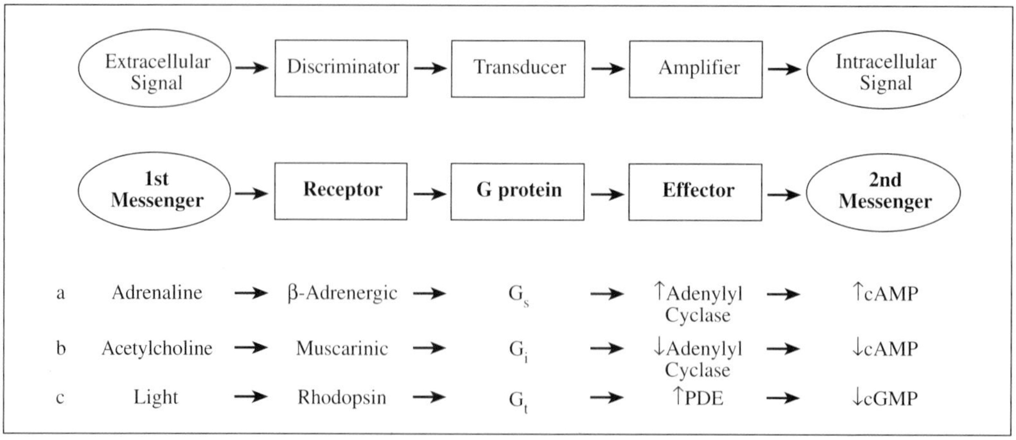

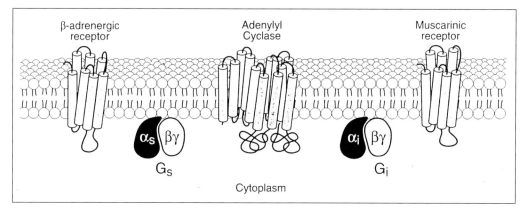

1st Messenger	→	Receptor	→	G protein	→	Effector	→	2nd Messenger
a	Adrenaline	→	β-Adrenergic	→	G_s	→	↑Adenylyl Cyclase	→	↑cAMP
b	Acetylcholine	→	Muscarinic	→	G_i	→	↓Adenylyl Cyclase	→	↓cAMP
c	Light	→	Rhodopsin	→	G_t	→	↑PDE	→	↓cGMP

Fig. 1 Relay diagram of G-protein signalling. The first line is based on Rodbell's early concept of informational processing, which is restated in biochemical terms in the second line. The lettered subheadings below are specific examples of signalling relays for hormone-sensitive adenylyl cyclase and the visual transduction systems. Horizontal arrows indicate the direction of the flow of information, upward-pointing arrows indicate increases in effector activity or levels of second messenger, and downward-pointing arrows indicate decreases in effector activity or levels of second messenger.

Fig. 2 A schematic view of plasma membrane components that participate in the hormone-sensitive adenylyl cyclase system. G-protein-coupled receptors have a similar structure, which includes seven putative trans-membrane helices and regions of homology in the cytoplasmic loops. Lipid modifications (not shown) promote interactions of G proteins with the inner surface of the plasma membrane. Crystal structures of G proteins have been solved and are described in Chapter 9. Adenylyl cyclases presumably span the membrane 12 times, with the active site oriented on the cytoplasmic side.

enzyme, but in addition, GTP was required (1). Gilman and colleagues purified the first G protein, G_s, by reconstituting hormone-stimulated adenylyl cyclase activity in membranes from a mutant cell line lacking the G protein (2). Bitensky detected a light-activated cGMP PDE in the retina and called attention to parallels between the visual transduction pathway and the hormone-sensitive adenylyl cyclase system (3). These observations led to the purification of transducin (G_t) (4, 5).

1.2   G protein diversity

Every discovery of a new G protein has underscored common characteristics of this class of signalling molecule. When G proteins were first identified and purified by following activity, names were assigned with subscripts chosen to evoke functional roles such as G_s and G_i (mediators of hormonal stimulation and inhibition, respectively, of adenylyl cyclase). G-protein oligomers were defined by their α subunits. G_s thus refers to a heterotrimer composed of α_s and any combination of β and γ isoforms.

More recent discoveries, both in purification of protein and cloning of cDNAs, led to the identification of G proteins without known functions. Names were chosen at the whim of the discoverer, such as G_o ('o' for other) coined by Sternweis (6) and G_q coined by Simon's group (purportedly chosen in jest to mimic the acronym of a popular American magazine) (7). As more cDNAs encoding G protein α subunits were cloned, Simon began to number them (starting at 11 and now at 16). Sixteen distinct genes encode α subunits in mammals and 20 or more proteins are synthesized through alternative splicing of mRNA. The family of α subunits is commonly divided into four groups based on their amino-acid sequence identity: G_s, G_i, G_q, and G_{12} (Fig. 3). These classifications of primary structure roughly correlate with function.

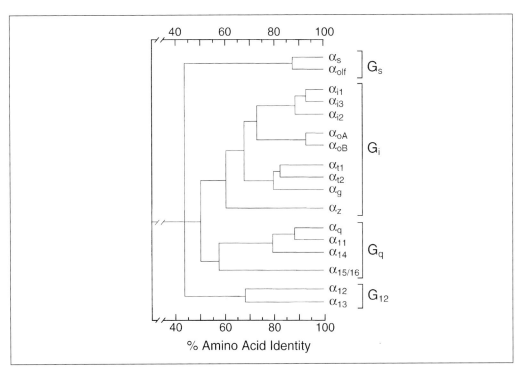

Fig. 3 Sequence relationships between mammalian G-protein α subunits and subfamily groupings. The branch junctions approximate the values calculated for each pair of sequences. α_{15} and α_{16} were originally assumed to be products of separate genes but are now considered to be the same product but from different species (mouse and human, respectively) (126). This figure is adapted from reference (127).

Table 1 Properties of mammalian G_s and G_i subfamilies of α subunits

Subfamily/ subunit	Molecular mass (kDa)	Distribution in tissue	Representative receptors[a]	Effector/role[b]
G_s				
$\alpha_{s(S)}$ [c]	44.2	Ubiquitous	βAR,	↑ Adenylyl cyclase
$\alpha_{s(L)}$ [c]	45.7	Ubiquitous	glucagon, TSH	
$\alpha_{s(XL)}$	94	Brain, adrenal, pituitary		Secretion?
α_{olf}	44.7	Olfactory, neuroepithelium	odorant	↑ Adenylyl cyclase
G_i				
α_{i1}	40.3	Nearly ubiquitous		↓ Adenylyl cyclase
α_{i2}	40.5	Ubiquitous	M_2Cho, α_2AR	↑ K^+ channels (?)
α_{i3}	40.5	Nearly ubiquitous	Others	↑ Phospholipase A_2 (?)
α_{oA}	40.0	Brain, others	Met-Enk	↓ Ca^{2+} channels
α_{oB}	40.1	Brain, others	α_2AR, others	↓ Adenylyl cyclase, others?
α_{t1}	40.0	Retinal rods	Rhodopsin	↑ cGMP-specific
α_{t2}	40.1	Retinal cones	Cone opsins	Phosphodiesterase
α_g	40.5	Taste buds	Taste (?)	?
α_z	40.9	Brain, adrenal, platelets	M_2Cho (?), others (?)	↓ Adenylyl cyclase, others (?)

[a] Receptor abbreviations: βAR, β-adrenergic; M_2Cho, M_2-muscarinic cholinergic; α_2AR, α_2-adrenergic; Met-Enk, methionine-enkephalin; TSH, thyroid-stimulating hormone.
[b] ↑ = activation; ↓ = inhibition.
[c] Splice variations: $\alpha_{s(S)}$ = short forms of α_s; $\alpha_{s(L)}$ = long forms of α_s.

Alternative splicing produces carboxyl terminal variants of α_o (8, 9) and long and short versions of α_s (10,11). The splice variants of α_o are activated by different receptors in intact cells (12) whereas only modest differences in the activities of the α_s variants have been observed. A single report of alternatively spliced α_{i2} has appeared, with evidence for differential subcellular localization of the variants in transfected cells (13). The focus of this chapter is on the G_s and G_i subfamilies (Table 1, except for G_z, which is covered in the next chapter).

Five genes encoding β subunits have been identified (Table 2). Alternative transcriptional start sites in the β_5 gene give rise to short and long forms of the β_5 protein (14). Among the three subunits, G_γ exhibits the greatest diversity in amino-acid sequence. Eleven different γ subunits are known (there is no γ_6 by current definition, Table 2) and if all of these could combine with the six β isoforms, then there would be 66 potential combinations. It appears that most pairs can form but there are exceptions (15). Most combinations of G protein α and $\beta\gamma$ pairs can be formed (recombinant proteins) but it is not known if all of these actually do form at concentrations of subunits present endogenously in mammalian cells. Although individual cells may not express all isoforms, most express multiple isoforms of α, β, and γ subunits.

1.3 Mechanistic model

Figure 4 shows a current model for the mechanism of G-protein-mediated signal transduction in which nucleotide-driven conformational changes in the α subunit

Table 2 Properties of β and γ subunits

Subunit[a]	Molecular mass (kDa)[b]	Amino-acid identity (%)[c]	Tissue distribution
β			
β_1	37	100	Ubiquitous
β_2	37	90	Nearly ubiquitous
β_3	37	83	Nearly ubiquitous
β_4	37	89	Nearly ubiquitous
β_5	39	52	Brain
β_{5L}	44	52	Retinal rods
γ			
γ_1	8.4	42	Retinal rods
γ_2	7.9	100	Brain, adrenal
γ_3	8.5	80	Brain, testis
γ_4	8.3	77	Nearly ubiquitous
γ_5	7.3	48	Ubiquitous
γ_7	7.5	70	Ubiquitous
γ_8	7.0	70	Nasal epithelium, vomeronasal organ
γ_{cone}	7.7		Retinal cones
γ_{10}	6.0	53	Ubiquitous
γ_{11}	8.0	33	Nearly ubiquitous
γ_{12}	8.0	63	Ubiquitous

[a] The trend has been to name γ subunits in the order in which their cDNAs were cloned. Accordingly, the γ_6 subunit (reported initially as a protein) has been renamed γ_2. Two sequences were nearly simultaneously reported and designated γ_8. Here the first clone to be published is designated γ_8 (128) and the second as γ_{cone} (129).
[b] The mass of γ subunits is approximate because most papers do not specify whether the reported mass is of the primary or processed protein product. The mass of final product will be influenced by removal of the initiator methionine and the final three amino acids and by appendage of isoprenyl and methyl groups (covered in Chapter 2).
[c] Comparisons are made between β_1 or γ_2 (set at 100%) and other like subunits. Amino-acid identity is approximate because, in some cases, comparisons are made between subunits of different species, which may introduce greater variance than if data were available for like species.

drive interactions with upstream receptor or downstream effector. Two cycles are superimposed in this scheme: GTP binding and hydrolysis, and subunit dissociation and reassociation. In the basal state (Fig. 4a) the ligand-binding site of the receptor is unoccupied, the three G-protein subunits form an oligomer, and GDP is bound to the α subunit. Activation is triggered when agonist binds to the receptor, which acts as a guanine nucleotide exchange factor (Figure 4b). Liganded receptor binds to its cognate G protein and promotes release of GDP from G_α (the rate-limiting step), thus allowing this subunit to bind the more abundant guanine nucleotide in the cell, GTP. A conformational change accompanies binding of GTP and leads to the dissociation of G_α from receptor and the high-affinity complex of β and γ subunits (Fig. 4c). These liberated subunits are competent to modulate the activity of effectors. In the example shown in Fig. 4c, G_α-GTP interacts with the effector (although $G_{\beta\gamma}$ also modulates some effectors). The intrinsic GTPase activity of G_α acts as a molecular clock to terminate the signal by returning this subunit to the inactive GDP-bound state with increased affinity for $G_{\beta\gamma}$. Reassociation of G_α-GDP with $G_{\beta\gamma}$ restores the system to the basal state to await a new cycle of activation.

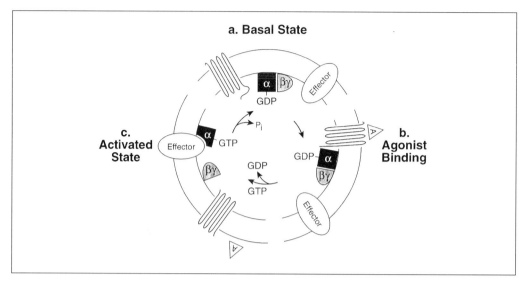

Fig. 4 Simple mechanistic model of G-protein-mediated signal transduction. The circular parallel lines represent the plasma membrane. The receptor is represented by the serpentine lines spanning the membrane seven times and agonist is shown as the letter A surrounded by a triangle. The two superimposed cycles of GTP binding and hydrolysis and G-protein subunit dissociation are described in the text (Section 1.3). This figure is adapted, with permission, from reference (32).

2. Covalent modifications of α subunits

Covalent modifications, both permanent and reversible, influence the function and localization of G proteins. The best-characterized modifications of α subunits are ADP-ribosylation, catalysed pathologically by toxins, and lipid modifications that occur physiologically (Table 3, described below). Phosphorylation is restricted mostly to G_z and the G_{12} subfamily of α subunits, covered in Chapter 2. Covalent modifications of β subunits are not well established but γ subunits are known to be prenylated and carboxymethylated, also described in the next chapter.

2.1 ADP-ribosylation

Knowledge of the G-protein isoform specificity and functional consequences of the ADP-ribosylation reactions has greatly aided identification of the particular G proteins that participate in signalling pathways. G protein α subunits are often classified by whether they are ADP-ribosylation targets for cholera toxin, pertussis toxin, or neither. In fact, division of the G protein topics between this chapter and the next are along these lines: G proteins detailed here are sensitive to cholera or pertussis toxin, whereas those in Chapter 2 are insensitive.

Table 3   Covalent modifications of G protein α subunits

G Protein	Toxin-mediated ADP-ribosylation	Myristoylation[a]	Palmitoylation[b]
α_s	Cholera	—	C3
α_i	Pertussis	G2	C3
α_o	Pertussis	G2	C3
α_t	Pertussis, cholera	G2	—
α_z	—	G2	+
α_q	—	—	C9, C10
α_{12}	—	—	+
α_{13}	—	—	+

[a] G2 refers to site of myristoylation, in one-letter amino-acid code and number for position in sequence.
[b] The site of palmitoylation, where known, is indicated by a one-letter code for amino acid followed by a number indicating position.

2.1.1   Pertussis toxin

The S1 subunit of pertussis toxin is an ADP-ribosyl transferase that catalyses the transfer of the ADP-ribose moiety of NAD^+ to relevant α subunits. A cysteine residue, four amino acids removed from the carboxyl terminus of α_i, α_o, and α_t, is the site modified by pertussis toxin (this cysteine residue is not present in other α subunits) (16). ADP-ribosylation of the cysteine site blocks interactions between the G protein and receptor. Pertussis-toxin treatment of cells is therefore useful for identifying whether a signalling pathway involves G_i, G_o, or G_t.

2.1.2   Cholera toxin

In contrast to pertussis toxin, cholera toxin selectively modifies the multiple forms of α_s and α_t (17) on a conserved arginine residue. This modification of arginine 201 in α_s (numbering of long isoform) irreversibly inhibits GTPase activity, thus locking the protein in an active form. Toxin treatment of cells is useful for determining whether G_s is involved in a signalling pathway because the pathway would be stimulated by the toxin in the absence of agonist.

Since an analogous arginine residue is present in other α subunits, it is not obvious why they are poor substrates for cholera toxin. Obvious possibilities include differences in sequence surrounding the site of modification or differences in their ability to interact with the necessary cofactor, ADP-ribosylating factor (ARF). *In vitro* modification of purified G_s requires the toxin, ARF, lipid, NAD, and GTP (18, 19). In the absence of guanine nucleotides, G_i or G_o can become *in vitro* substrates for the toxin, especially when a cognate receptor is stimulated. This observation has proven useful (when radioactive NAD is used as ADP-ribosyl donor) to identify which α subfamily member is coupled to a particular receptor in membrane preparations (20).

2.2 Acylation

All G-protein α subunits are acylated at or near their amino termini by fatty acids, myristate and/or palmitate. These lipid modifications facilitate protein–protein and protein–membrane interactions of G_α. A key difference between the two acylations is that myristoylation is a stable modification that lasts the lifetime of the protein whereas palmitoylation is reversible and subject to regulation.

2.2.1 Myristoylation

The 14-carbon saturated fatty acid, myristic acid (C14:0), is appended to the amino terminus of members of the $G_{\alpha i}$ subfamily. Following cleavage of the initiator methionine, myristate is cotranslationally attached to the resulting amino-terminal glycine (G2) via a stable amide bond (21). The reaction is catalysed by myristoyl-CoA:protein N-myristoyltransferase, expressed endogenously in yeast to humans (but not in bacteria) (22). The enzyme has a narrow acyl-CoA specificity; it prefers not to transfer palmitate, which is expressed abundantly in eukaryotic cells. There is no absolute consensus sequence specifying N-myristoylation of proteins, but glycine at position two is required and a hydroxyamino acid at position 6 is preferred. Myristoylation is thought to facilitate membrane association of G_i subfamily members, because mutation of G2 prevents myristoylation and directs the protein to the cytoplasmic fraction of transfected mammalian cells (23). G_i subfamily members can be expressed alone or coexpressed with yeast N-myristoyl transferase and purified from E. coli (24). Direct comparisons of myristoylated and non-myristoylated α subunits from the bacteria indicate that myristate promotes interaction of G_α with $G_{\beta\gamma}$ (24), effector (25, 26), and GTPase-activating proteins (GAPs) (27, 28). $G_{\alpha t}$ is atypical because it is heterogeneously acylated at the amino-terminal glycine by C12:0, C14:2, and C14:1, as well as C14:0 (29, 30). This heterogeneous acylation was found to be tissue (retina) and species specific (31).

2.2.2 Palmitoylation

Palmitoylation is of particular interest among lipid modifications of proteins because it is readily reversible and has the potential to be regulated (32, 33). Palmitoylation is limited to a small subset of cellular proteins, most of which are found to be associated with the plasma membrane. Signalling proteins are common targets of this modification and include G-protein-coupled receptors, G_α, adenylyl cyclase, Ras and non-receptor tyrosine kinases. Reversible palmitoylation of proteins is achieved by the relatively unstable esterification of cysteine thiol groups (this is termed thioacylation) by C16:0, palmitate. Although often referred to as palmitoylation, thioacylation is often found to be heterogeneous. No primary sequence motif has been identified to predict most sites of palmitoylation in proteins. Frequently, sites of palmitoylation are found close to another lipid modification, such as myristate, or near hydrophobic stretches of amino acids that are putative transmembrane domains (such as those found in G-protein-coupled receptors). A sequence motif that serves as a predictor for a small subset of palmitoylated proteins is Met–Gly–Cys at the amino terminus of a protein. This motif is found in the G_i subfamily and in some non-receptor tyrosine kinases.

Palmitoylation of G_α is considered biologically significant because removal of palmitate is a receptor-regulated phenomenon. For example, activation of G_s by the β-adrenergic receptor is accompanied by an increase in the turnover of palmitate on the α subunit of only this particular G protein (34–36). A simple mechanistic hypothesis for the increase in turnover is that the activated protein is a better substrate for a constitutive palmitoylthioesterase. Duncan and Gilman have purified and cloned the cDNA encoding a cytoplasmic protein with acyl protein thioesterase activity (37). Although it is not yet established whether the enzyme is the one that is primarily responsible for the turnover of palmitate on G proteins *in vivo*, it does efficiently utilize palmitoylated G_α as substrate and is distinct from the previously described lysosomal enzyme (38).

Enzymes catalysing protein palmitoylation *in vitro* have defied purification. When combined with the lack of specificity (for fatty acyl-CoAs and sites in proteins) and reports of protein autoacylation (in presence of palmitoyl-CoA but no enzyme source) some investigators have been prompted to seriously question whether the transfer reaction is catalysed enzymatically in cells (39). Regardless of the mechanism *in vivo*, thioacylation of G_α is regulated by receptor and functions for the modification are emerging. Palmitoylation of G_α promotes interaction of the protein with $G_{\beta\gamma}$ (40) and with membranes (23) and, conversely, this modification represses interaction with regulators of G-protein signalling (RGS proteins, described in Section 4.2) (28).

3. Targets

Various reagents have proven useful for identifying the involvement of G proteins when a signalling pathway is to be dissected. It will be useful to list these before consideration of proven or potential functions that are regulated by G proteins (Table 4).

Early studies gave rise to the notion that regulation of effector activity was the exclusive province of α subunits. Purified α_s or α_t stimulated the activity of adenylyl cyclase or cGMP PDE, respectively, while βγ did not (5, 41). Further experiments indicated that $G_{\beta\gamma}$ stabilized the inactive GDP-bound form of G_α by reducing the rate of dissociation of GDP (42). Thus $G_{\beta\gamma}$ was relegated to a mundane task of dampening the activity of lively α subunits. The 'α only' dogma was cracked in the late 1980s when Logothetis *et al.* discovered that $G_{\beta\gamma}$ could regulate the activity of a K^+ channel in cardiac myocytes (43). Initially this observation was met with hot debate (44) but modulation of effector activity by $G_{\beta\gamma}$ is now well accepted. As the debate transpired, genetic analysis of the pheromone signalling pathway in budding yeast yielded evidence that $G_{\beta\gamma}$, rather than G_α, was the primary regulator of downstream events in this system (45, 46). With the cDNA cloning and expression of individual isoforms of mammalian adenylyl cyclases, regulation of a subset of these enzymes by βγ has provided additional examples of important roles for βγ in effector modulation.

Table 4 Reagents affecting G-protein function

Reagent	Mechanism of action
GTPγS	Hydrolysis resistant analog of GTP. Stably activates many GTPases independently of receptor. Does not enter intact cells.
GDPβS	Analog of GDP that cannot be converted to GTP. Prevents activation of GTPases.
AlF$_4^-$	Activates GDP-associated Gα by occupying the position of the γ phosphate of GTP. Occasionally has proven useful with intact cell preparations. Can alter activity of phosphatases, kinases and other enzymes. Does not activate small monomeric GTPases (130) in the absence of a GTPase-activating protein (131)
Mastoparan	Peptide (from wasp venom) mimic of stimulated receptor that promotes nucleotide exchange on Gα with preference for activation of α_0 and α_i. Effect is blocked by pertussis toxin. Useful on intact cells or purified components. High concentrations can permeabilize and kill cells.
Pertussis Toxin	Catalyzes ADP ribosylation of α_0 and α_i thereby preventing their activation by receptor. Can be utilized with intact cells or purified components. ADP-ribosylated α_0 and α_i can be activated by GTPγS.
Cholera Toxin	Catalyzes ADP ribosylation of α_s thus blocking GTPase activity and permanently activating this G proteiin. Can be utilized with intact cells or purified components.
Heterologous expressioin of α_i	α_i, which is not endogenously expressed in cultured cells, sequesters βγ, thus preventing βγ-mediated functional effects.
Expression of βARK C-terminus	PH domain of βARK (β-adrenergic receptor kinase) binds and sequesters βγ, thereby preventing βγ-mediated functional effects.
Overexpression of βγ	Sequesters Gα or directly moduates effector(s).

Binding of GTP is accompanied by a conformational change in G_α (covered in Chapter 9), thus promoting the ability of the protein to bind and modulate the activity of an effector. Because no difference in the structure of βγ in the free (47) versus heterotrimeric state (48, 49) was observed, it was hypothesized (and subsequently demonstrated) that binding sites for G_α and effectors are overlapping on the $G_{\beta\gamma}$ complex (50, 51). In the heterotrimeric GDP-bound state, G_α presumably occludes sites on $G_{\beta\gamma}$ for binding of effector.

3.1 Adenylyl cyclases

Adenylyl cyclases catalyse the conversion of ATP to cAMP, the second messenger that mediates diverse cellular responses, primarily by activating cAMP-dependent protein kinases. Adenylyl cyclase activity in yeast is regulated by the monomeric GTPase, Ras, while this function is performed by heterotrimeric G proteins in mammalian cells. cDNAs have been isolated for nine isoforms of mammalian adenylyl cyclases; they are designated types I to IX (or ACI to ACIX) in chronological order of cDNA cloning. Additional members of the family will probably be identified as different tissues are examined or less stringent screening conditions are employed. Mammalian adenylyl cyclases are expressed in all tissues, but at very low levels (0.01–0.001% of membrane protein, except in olfactory neuroepithelium where it represents 0.1%). Low levels of expression and unavailability of satisfactory anti-bodies have limited most studies of tissue distribution to detection of mRNA (52). The ability to prepare recombinant proteins has permitted study of enzyme structure and function and has revealed unexpected complexity of multimodal regulation of cAMP synthesis.

The amino-acid sequences of adenylyl cyclase family members contain 12 stretches of hydrophobic residues, each presumed to represent a transmembrane span, reminiscent of a variety of transporters and channels (52). Each isozyme is anticipated to have a short cytosolic amino terminus, followed by two repeats of a module composed of six transmembrane helices, followed by an approximately 40 kDa cytosolic domain (as depicted in Fig. 2, N terminus on left, C terminus on right). The large cytoplasmic domains together form a catalytic unit.

Activation by α_s seems to represent the only natural regulatory feature that is common to all members of the adenylyl cyclase family. Activation by the plant diterpene, forskolin, and inhibition by certain adenosine analogues (P-site inhibitors) are also shared by all isoforms (except IX) of the enzyme. Forskolin is a useful reagent for direct activation of adenylyl cyclase, independent of hormone or G proteins. It is fortunate that the biochemically intractable transmembrane domains of adenylyl cyclase can be discarded for production of soluble, enzymatically active, catalytic domains in E. coli (53). Importantly, the two soluble domains purified from bacteria are regulated by $G\alpha_s$, forskolin, and P-site inhibitors. In the crystal structure of the complex, α_s is nestled in a cleft of one of the two soluble cyclase domains, distant from the catalytic site formed by both domains (54). It is hypothesized that α_i would bind to the analogous position on the other soluble cyclase domain (26, 54).

Most natural regulators of adenylyl cyclase activity are isoform specific, as summarized in Table 5 (52). For instance, α_i inhibits ACI, V, and VI but not ACII. $G_{\beta\gamma}$ inhibits ACI but not ACV or VI. The concentration of $G_{\beta\gamma}$ required to inhibit ACI is higher than the concentration of α_s necessary for stimulation. The concentration of $G_{\beta\gamma}$ released from activated G_s (a low-abundance G protein) would be insufficient to

Table 5 Type-specific regulation of mammalian adenylyl cyclases

Regulator	Effect	AC subtype	Comments
Forskolin	Stimulation	All except IX	Synergism with α_s (ACII, IV, V, VI); not ACI
α_s	Stimulation	All	Synergism with forskolin (ACII, IV, V, VI); not ACI
			Synergism with Ca^{2+}/CaM (ACI, III, VIII)
α_i	Inhibition	ACI, V, VI; not ACII	Effect on ACI is modest with Ca^{2+}/CaM; poor with $G\alpha_s$
α_z	Inhibition	ACI, V	
α_o	Inhibition	ACI; not ACII, V, VI	See α_i
$\beta\gamma$	Inhibition	ACI; not ACV, VI	Dependent on α_s
	Stimulation	ACII, IV, VII; not ACV, VI	
Ca^{2+}/CaM	Stimulation	ACI, III, VIII	Synergism with α_s; ACIII requires α_s or forskolin
Ca^{2+}	Inhibition	ACV, VI	
P-site analogues	Inhibition	All	
PKC	Stimulation	ACII, V, VII	
PKA	Inhibition	ACV, VI	

inhibit the type I enzyme but the amount released from co-activation of one of the more abundant G proteins (such as G_o or G_i, in brain where ACI is expressed) could be sufficient. Conversely, and surprisingly, $G_{\beta\gamma}$ was found to stimulate conditionally types II, IV, and VII. Calcium and protein kinases are also type specific in their regulation of adenylyl cyclase activities. These properties permit an adenylyl cylase to respond specifically when two convergent pathways are activated simultaneously.

3.2   Light-activated cGMP phosphodiesterase

The components of phototransduction are highly concentrated in a stack of discs in the outer segment of retinal rod cells (Fig. 5). These outer segments are readily detached for biochemical study and contain approximately 2000 discs per cell. The receptor, rhodopsin, accounts for a phenomenal 70% of outer-segment membrane protein (55). When a photon strikes rhodopsin, the covalently bound chromophore, 11-*cis* retinal, isomerizes to the all-*trans* form, thus driving a conformational change in rhodopsin. A large number of G_t molecules can interact in rapid succession with a single photoexcited rhodopsin. Activated G_t, in turn, switches on the potent PDE (composed of α, β, and γ subunits) that rapidly hydrolyses cGMP. GTP-bound α_t activates the enzyme by binding and sequestering the inhibitory PDE γ subunit, thereby relieving constraint on the activity of PDE$\alpha\beta$. The resulting reduction in concentration of cytosolic cGMP induces closure of ion channels of the plasma membrane, to generate a nerve signal. Channel closure also induces a drop in the cytosolic calcium level, leading to the activation of guanylyl cyclase and the reopening of channels.

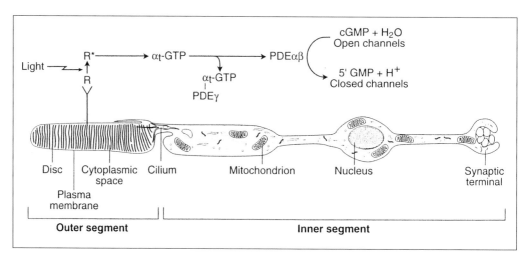

Fig. 5 Diagram of a vertebrate rod cell and flow of information in visual transduction. Approximately 10^8 rod cells are tightly packed in a retina and are aligned with the path of incoming light. The outer segment is filled with several hundred stacked disc membranes, which are the sites of photo signal transduction. Photoisomerization of rhodopsin triggers a cascade leading to cGMP hydrolysis and closure of membrane channels, which generates a nerve signal (see text for details).

3.3   Ion channels

Acetylcholine slows the heart rate by stimulation of m2 muscarinic receptors in cardiac pacemaker cells. These and other G_i subfamily-linked receptors (such as somatostatin, μ-opioid, and $α_2$-adrenergic receptors) activate a K^+ channel (I_{KACh}) in cardiac myocytes. Work from several laboratories has established that $G_{βγ}$ is the physiological activator of I_{KACh} (15). This channel is dramatically gated by $G_{βγ}$ with channel openings increasing up to 1000-fold upon application of the purified complex to a patch of membrane. Direct binding of $G_{βγ}$ to channel subunits is the mechanism by which G proteins regulate the flow of K^+ ions of cardiac myocytes. Based on electrophysiological recordings and Western and/or Northern blot analysis of tissues, the protein-gated K^+ channel has also been shown to be present in the brain and pancreas (15).

One group of investigators has reported that purified preparations of hetero-trimeric G_s or $α_s$ activate calcium channels in cardiac and skeletal muscle T-tubules (without the need for cytosolic factors) (56). Several G-protein-linked receptors inhibit N- and P/Z-type calcium channels in neurons. In many cases no cytoplasmic second messenger was responsible for modulation of channel activity and experiments point to $G_{βγ}$ as the mediator of inhibition (57, 58). G_o appears to be the source of $G_{βγ}$ in some experiments, but G_i cannot be ruled out in all cases (59). Vasoactive intestinal peptide, which slows gating of voltage dependent N-type calcium channels in rat sympathetic neurons, seems to involve G_s (60).

3.4   Other targets of $G_{βγ}$

G-protein βγ subunits have been implicated in a number of signalling reactions other than those cited above. G_i is considered to be the primary potential source of free $G_{βγ}$ dimers because of its high relative abundance in most cell types.

3.4.1   Phospholipases

PLC

Numerous G-protein-linked receptors stimulate phosphoinositide-specific phospho-lipase C-β (PLC-β), an effector enzyme that catalyses hydrolysis of the minor lipid phosphatidylinositol 4,5-bisphosphate to form two second messengers, inositol 1,4,5-triphosphate and diacylglycerol. Members of the G_i and G_q subfamilies directly regu-late PLC-β. Activated G_q subfamily members stimulate PLC-β via α subunits whereas G_i-mediated stimulation of the enzyme is conferred by βγ. Details of the regulation of PLC-β are covered in Chapter 2.

PLA$_2$

GTPγS stimulates phospholipase A_2 (PLA$_2$) activity in permeabilized rat basophilic leukaemia cells. Consistent with regulation by a G protein, antigen-mediated crosslinking of receptors for IgE could synergize with low concentrations of GTPγS (61). Since IgE receptors are not typical (seven membrane-spanning) G-protein-coupled receptors, it is unclear whether a small or large GTPase is involved in regulation of the phospholipase. There is, however, a single report that purified retinal $\beta\gamma$, reconstituted in G-protein-depleted retinal membranes, activates PLA_2 (62). It is not known whether the effect of $\beta\gamma$ on enzyme activity is direct.

3.4.2   Kinases

GRKs

Phosphorylation of G-protein-coupled receptors is one mechanism by which signalling is regulated (63). A family of G-protein-coupled receptor kinases (GRKs 1–6) specifically phosphorylate serine and threonine residues on the conformationally active receptor. After phosphorylation, arrestin proteins bind to receptor and sterically prevent further G-protein activation. There is general agreement that the activity of two members of the family (GRK2 and GRK3, otherwise known as βARK1 and βARK2) is enhanced by $G_{\beta\gamma}$. Varied mechanisms of enhancement have been proposed including decreased K_m, increased V_{max}, and translocation of the kinase (15). Several GRKs, that apparently reside in the cytoplasm in unstimulated cells, are translocated to the membrane fraction following agonist binding to receptor (63). It is thought that the GRKs form a complex with membrane-bound $G_{\beta\gamma}$ upon activation of cognate G proteins. The binding site for $G_{\beta\gamma}$ is in a domain of approximately 125 amino acids at the carboxyl terminus of GRKs 2 and 3. This domain of the GRKs overlaps with a sequence motif, termed the pleckstrin homology (PH) domain, a consensus motif first identified in the eponymous cytoskeletal protein, pleckstrin (64). Although PH domains are relatively variable in sequence, the three-dimensional structures determined for three such domains are quite similar. The structure of PH domains from pleckstrin, spectrin, and dynamin are composed of β-sheets and a carboxy-terminal amphophilic α-helix. $G_{\beta\gamma}$ binds (*in vitro*) to PH domains from many proteins (but not to all candidates) (65). GRK1 (rhodopsin kinase), in contrast to isoforms 2 and 3, does not contain a PH domain nor is its activity influenced by $G_{\beta\gamma}$. This kinase isoform is instead membrane bound, at least in part, by virtue of its prenyl moiety appended to the carboxyl terminus (66).

Btk

Bruton's tyrosine kinase (Btk) is a member of the PH domain-containing family of non-receptor protein kinases that lack a myristoylation signal (which characterizes Src-related non-receptor kinases). The signalling function of Btk is not defined but defects in this kinase are responsible for human and murine B-cell deficiencies. An *in vitro* binding assay and an *in vivo* competition assay, indicates that $G_{\beta\gamma}$ interacts with the PH domain of Btk (67). Btk is stimulated by cell transfection with $G_{\beta\gamma}$ and in reconstitutions with purified $G_{\beta\gamma}$ (and an additional unidentified membrane component[s]) (68).

MAPK

Mitogen-activated protein kinase (MAPK) cascades play key roles in the regulation of cell proliferation and differentiation. A yeast genetics approach first revealed a

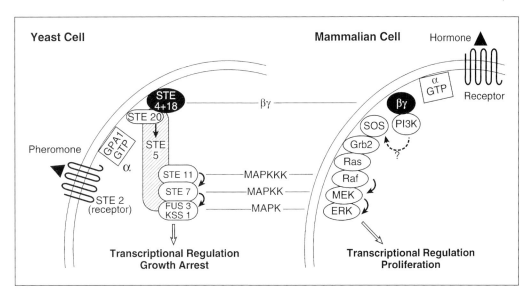

Fig. 6 Models for $G_{\beta\gamma}$-mediated activation of MAPK pathways in yeast and mammalian cells. The parallel semicircular lines represent the plasma membrane. Homologues between species are pointed out by horizontal lines and identified by their common names in the middle (between the diagrams of each cell). The solid black arrows indicate phosphorylation steps. The connection between PI3K and SOS is uncertain (indicated by dashed arrow and question mark).

function for $G_{\beta\gamma}$ in the cell cycle. *Saccharomyces cerevisiae* can exist in any of three distinct cell types: the haploid mating type MATa or MATα or as the diploid MATa/MATα. Prior to mating, MATa and MATα cells both secrete specific peptide pheromones that bind to G-protein-coupled receptors on cells of the opposite mating type. Binding of peptide ligand triggers a programme of developmental processes required for cell fusion to form diploids, including morphological and transcriptional changes and growth arrest in the G_1 phase of the cell cycle. Null mutations in the yeast homologues of β and γ (*STE4* and *STE18* genes, respectively) disrupt the mating and pheromone response (46). Disruption of the yeast gene for the G-protein α subunit (*GPA*, also known as *SCG1*) leads to a constitutive mating response, suggesting that the yeast α homologue was a negative regulator of $G_{\beta\gamma}$ and that $G_{\beta\gamma}$ serves as the direct signalling element (45, 69). Genetic approaches identified the kinase Ste20p as acting between the G protein and the MAPK module (Fig. 6) (70). Co-immunoprecipitation and *in vitro* binding experiments have identified a physical association between G_β and Ste20p. A binding site for G_β, in a non-catalytic region of Ste20p, has been identified and is conserved in the mammalian Ste20p homologues, the p21-activated protein kinases (PAKs) (71). Mutations in G_β or Ste20p that prevent this association block activation of the MAPK cascade. The function of the pheromone-induced interaction between G_β and Ste20p may be to bring this kinase into the vicinity of its downstream target Ste11 which (like $G_{\beta\gamma}$) interacts with the putative scaffolding protein Ste5p (Fig. 6).

Mitogenic signalling initiated by G-protein-coupled receptors in mammalian cells can potentially occur by multiple mechanisms (that are likely cell-type specific). Activation of Ras has been shown to constitute one major pathway which is employed by G-protein-coupled receptors (Fig. 6) but the mechanistic details are not as well understood as the pathway initiated by receptor tyrosine kinases (described in Chapter 3). Both pertussis-toxin-sensitive and -insensitive G proteins may participate in stimulation of the MAPK cascade (in particular, the extracellular-signal-regulated kinase (ERK) cascade) (70). There is evidence that G$_\alpha$ or G$_{\beta\gamma}$ subunits may relay the signal directly or indirectly (via second messengers). G$_i$-linked receptors, such as the lysophosphatidic acid and thrombin receptors, the α_2-adrenoceptor and the m2 muscarinic receptor, activate ERKs in a pertussis-toxin-sensitive manner. The preponderance of evidence indicates that G$_{\beta\gamma}$, rather than G$_\alpha$, links the G$_i$-coupled receptors to the MAPK cascade. Potential intermediaries between $\beta\gamma$ and the ERK cascade include a variety of tyrosine kinases and phosphoinositide 3-kinase (PI3K).

PI3K

This lipid kinase phosphorylates phosphatidylinositol 4,5-bisphosphate to form the putative membrane phospholipid signalling molecule, phosphatidylinositol 3,4,5-triphosphate. PI3K is activated by mitogens that stimulate either receptor tyrosine kinases or G-protein-coupled receptors. Activation of the ERK cascade by G$_i$-coupled receptors or overexpressed $\beta\gamma$ dimers is attenuated by inhibitors of PI3K such as wortmannin (70). Several species of PI3K have been cloned and characterized. Heterodimeric PI3Kα and PI3Kβ, consisting of p110 catalytic subunits and different p85 adaptor subunits, are regulated by receptors with intrinsic or associated tyrosine kinase activity (72). PI3Kγ is activated *in vitro* by G protein α or $\beta\gamma$ subunits but does not interact with p85 (73). This PI3K homologue harbours a putative PH domain (not found in the other isoforms) which may mediate interaction with G$_{\beta\gamma}$.

Overexpression of PI3Kγ in mammalian cells activates MAPK in a $\beta\gamma$-dependent fashion, and expression of a catalytically inactive mutant of PI3Kγ abolishes the stimulation of ERK by G$_{\beta\gamma}$ or in response to stimulation of m2 muscarinic receptors (74). Free $\beta\gamma$ is thought to recruit PI3Kγ to the plasma membrane, enhancing the activity of a tyrosine kinase (designated as a question mark and dashed arrow in Fig. 6), which in turn leads to the activation of the Shc–Grb2–Sos–Ras pathway, resulting in increased ERK activity.

3.4.3 Dynamin

The flow of proteins through vesicular import and export pathways is perturbed by probes of G-protein function such as aluminium fluoride, pertussis toxin, cholera toxin, mastoparan, and sequestration of $\beta\gamma$ (75–77). It is unclear how G proteins function in these complex systems because neither the relevant nucleotide exchange factors nor effectors have been identified. However, G$_{\beta\gamma}$ has been shown to interact functionally with dynamin, a 100 kDa GTPase thought to play a role in the fission of vesicles during endocytosis. G$_{\beta\gamma}$ negatively regulates dynamin GTPase activity stimulated by phosphatidylinositol 4,5-bisphosphate (78). It is presumed (but not yet established) that $\beta\gamma$ exerts its effects by binding to the PH domain of dynamin. This example provides one potential mechanism for the direct regulation of endocytosis by G proteins.

3.4.4 Phosducin

Phosducin was originally discovered as a major retinal phosphoprotein that could be copurified with $G_{\beta\gamma}$ from retina. Expression was initially considered to be restricted to the retina and the developmentally related pineal gland, but more recently phosducin has been shown to be expressed ubiquitously. In the retina, phosducin inhibits activation of G_t. It is thought that binding (or 'trapping') of $G_{\beta\gamma}$ by phosducin prevents reassociation of the heterotrimer, a step necessary for reactivation by rhodopsin. Phosducin, at higher concentrations, can also bind and regulate the activity of G_α (79). The crystal structure of a complex of phosducin and $G_{\beta\gamma}$ has been solved (80). The amino-terminal portion of phosducin binds to the same surface of G_β as G_α, which would explain why phosducin prevents reassociation of G_α with $G_{\beta\gamma}$. Phosducin also inhibits the hormone-sensitive adenylyl cyclase system, presumably by a mechanism similar to that of the visual system. Interestingly, phosphorylation of phosducin by cAMP-dependent protein kinase reduces the ability of phosducin to bind $G_{\beta\gamma}$. This negative regulation of phosducin activity could allow freed $G_{\beta\gamma}$ to regulate other pathways, such as PI3K, thus implicating phosducin as a molecule with the capacity to integrate signals from different pathways (81).

4. Regulation of signalling by GAPs

The appropriate response to an extracellular stimulus is dependent on the intensity and duration of the signal. Regulation of signalling occurs at multiple levels, including receptor, G protein, and effector. At the level of the G protein, GTPase activating proteins (GAPs) provide a mechanism for enhancing temporal resolution of signalling by increasing selectivity between receptor and G protein, increasing rates of activation and deactivation of effectors, and attenuation of signalling (82).

Physiological rates of termination of most G-protein signalling in cells are much faster than the measured rate of hydrolysis of GTP by the relevant isolated G protein. For example, the lack of after image upon closing your eyes indicates the rapidity required for termination of G_t activity. The intrinsic GTPase activity of α_t hydrolyses GTP and returns the protein to the inactive state on a time scale of tens of seconds and is thus far too slow to account for how quickly you lose an image after closing your eyes (83). This discrepancy in timing between *in vivo* termination of signalling and *in vitro* inactivation of α_t, strongly suggests a need for GAP activity.

GAPs were first recognized as regulators of protein synthesis factors and the monomeric GTPases such as Ras. The more recently discovered GAPs for heterotrimeric G proteins are divided into two groups. One group is comprised of certain effectors that can accelerate the GTPase activity of their cognate G_α proteins and the second is a large, newly discovered family of proteins known as RGS (regulators of G-protein signalling) proteins. The physiological significance of all the 20 different mammalian RGS proteins that have been identified so far is not yet known.

4.1 Effectors as GAPs

Only select effectors appear to function as GAPs. For example PLC exhibits GAP activity for α_q (covered in Chapter 2) while adenylyl cyclase does not appear to have this activity for α_s or α_i. The effector, cGMP PDE has been reported to regulate the intrinsic GTPase activity of α_t, but the mechanism is not yet well established. Part of the difficulty in the analysis is due to the resistance of α_t to exchange nucleotide (that is, to load the protein with radioactive GTP for GTPase assays), except in the presence of rhodopsin-containing membranes. Arshavsky and Bownds reconstituted photoreceptor membranes that retained G_t but were depleted of effector. Single-turnover GTP hydrolysis was accelerated by the addition of cGMP PDE γ subunits, suggesting that the effector acted as a GAP for G_t. Further experiments revealed, however, that the PDE γ subunit could form a tight complex with α_t without accelerating GTP hydrolysis. More recently PDE γ has been proposed to co-operate with another piece of the puzzle (an unidentified protein) present in the photoreceptor membranes (83). Wensel and colleagues have proposed that the unidentified puzzle piece is RGS9 (84).

4.2 RGS proteins as GAPs

These proteins were functionally discovered by genetic approaches as negative regulators of G-protein signalling in yeast (SST2p in *S. cerevisiae*) and worms (EGL10 in *Caenorhabditis elegans*). This story is a particularly good example of how information from very different approaches have brought new awareness of common mechanism in signal transduction (85). A yeast two-hybrid approach identified a mammalian G_α interacting protein, termed GAIP, that exhibited homology to Sst2p, in what is now considered the RGS core domain (86). Homology to this core domain is now considered a hallmark for identifying new members of the RGS family.

Recombinant core domain and intact RGS proteins have been purified and demonstrated to stimulate GTP hydrolysis of select G protein α subunits (87, 88). The core domain is absolutely required for promoting GTPase activity of G_α. The crystal structure confirms that the RGS core domain binds to regions of G_α that participate in hydrolysis of GTP (89). The highly variable amino- and carboxyl-terminal regions, that flank the core domain of RGS proteins, are presumed to contribute to the regulation, selectivity, and localization of these proteins. In general, conventional assays have not revealed a high degree of specificity of RGS proteins for particular G_α isoforms. For instance, most of the tested RGS proteins perform as GAPs for G_i and G_q but not for G_s or G_{12} subfamily members. Similarly, when overexpressed in cultured cells, the same RGS proteins attenuate signalling through G_i or G_q, but not G_s-mediated pathways (90).

Northern blot analysis indicates that most RGS proteins are expressed ubiquitously but retina-specific expression of RET-RGS1, RGS-r (also known as RGS16), and RGS9 has been reported. Truncated forms of each of these three proteins have been shown to accelerate GTP hydrolysis in urea-washed (GAP-depleted) photoreceptor membranes (84, 91, 92), but only the activity of RGS9 was enhanced by addition of cGMP PDE γ (84). These results suggest a unique role for RGS9 in phototransduction.

The physiological role played by the multitude of other RGS proteins (in retina and other tissues) remains to be elucidated. Presumably the plethora of isoforms furnishes greater specificity of action *in vivo*, than that revealed by *in vitro* assays of purified proteins or experiments involving heterologous expression. Perhaps as different experimental approaches are taken more specificity of action of each RGS protein will be revealed. One such example is provided by Wilkie and colleagues. They found that the potency of RGS4 injected into cells, to attenuate signalling (via a single G protein, G_q), is dependent on which receptor activates the G protein (93). Their results suggest that signalling via a particular receptor may be more tightly regulated *in vivo* when a certain RGS protein is expressed and localized appropriately. This area of G-protein research is particularly active and should uncover mechanisms that underlie fidelity and temporal resolution of signalling.

5. Organization

A long-standing issue is whether receptors, G proteins, and effectors exist as organized complexes or are randomly distributed and freely diffusible in the plasma membrane. In highly specialized cells, such as the retinal photoreceptors, selectivity of signal transduction is achieved by compartmentalization of the receptor subtypes and G_t isoforms in different cells. Rhodopsin (which mediates vision in dim light) is found with α_{t1} in rod photoreceptors (Fig. 5). The three opsins (receptors) that mediate human colour vision reside in those photoreceptor cells with cone-shaped outer segments, together with α_{t2} (94). Less-specialized cells carry out diverse functions, and several G-protein-mediated signalling pathways coexist. Individual cells can express simultaneously a large array of proteins that participate in G-protein-mediated signalling: a large number of receptors; a substantial variety of G protein α, β, and γ subunits; distinct effectors that exist in multiple isoforms; and regulators such as members of the large families of receptor kinases, arrestins, and RGS proteins. Although the critical outcomes that result from the interactions between these proteins can and, at times, must be studied using purified proteins reconstituted *in vitro*, other aspects of G-protein-mediated signalling are not observable in such well-defined systems. In general, the specificity of receptor–G protein interactions in reconstituted systems is less stringent than that observed in membranes or cells (95). For example, Kleuss *et al.* used antisense oligonucleotides to demonstrate that somatostatin and muscarinic receptors are coupled to different combinations of alternatively spliced α_o and different β and γ subunits to regulate calcium channels in GH3 cells (12, 96, 97). Such remarkable specificity for receptor and G-protein subunit interactions has not been achieved in reconstituted systems. In neuronal cells several G_i-coupled receptors appear to interact with distinct pools of the same G protein (Fig. 7) (95). Anomalously, the magnitude of stimulation of adenylyl cyclase activity by hormones is much greater in intact cells than in isolated membranes.

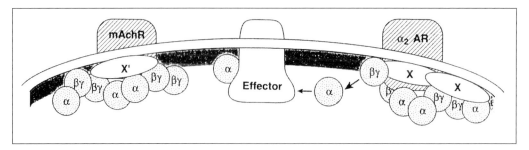

Fig. 7 Possible membrane organization of receptors and G proteins. The interactions of receptors and G proteins do not appear to be governed entirely by a random collision of proteins in the plasma membrane. Rather, sorting of the distinct components of the G-protein-mediated signal transduction system into organized regions or supramolecular complexes may be important for the specificity of receptor-G protein interactions in cells. Different receptors (e.g. α_2 adrenergic and m4 muscarinic) that are capable of coupling to the same G protein *in vitro* do not seem to share the same pool of G protein in intact cells. It will be important to identify any additional cellular machinery (X or X') that may be responsible for the organization of these systems in cells. The interaction of one receptor with multiple G proteins suggested in this model is consistent with the significant excess of G proteins over receptors in most tissues. It is also possible that some effector molecules could be incorporated into the complexes of receptors and G proteins. This figure and legend are adapted with permission from reference (95).

These and other phenomena, such as the apparent compartmentalization of second messengers (98), suggest the need for non-random, functional organization of signalling proteins in the plasma membrane to achieve the requisite magnitude of response, functional specificity, and regulation that characterize these systems (Fig. 7) (95).

A non-random distribution of receptors, G proteins, and effector activity in subdomains of the plasma membrane have been reported but a requirement for this organization to generate functional specificity has not yet been established. Immuno-fluorescence microscopy has revealed a punctate pattern of G protein α_i and β subunits consistent with the concentration of these proteins at certain sites on the plasma membrane (99). The pattern was similar, but not identical, to that for caveolin, a protein marker for specializations of the plasma membrane termed caveolae. These structures are morphologically identified in cross-section (by electron microscopy) as flask-shaped invaginations of the plasma membrane that can be open or closed to the extracellular milieu. Functionally, they were first implicated in cellular transport processes, such as transcytosis in endothelial cells and potocytosis in epithelial cells. An impressive and growing body of evidence indicates that a variety of molecules, known to function directly or indirectly in signal transduction, are enriched in caveolae (100). Much of this evidence rests on fractionation experiments; caveolar membrane is characterized by low buoyant density and resistance to detergent solubilization.

It has been proposed that caveolin binds directly to and regulates the activity of G proteins and an extraordinary number of other signalling molecules, including receptors, Ras, protein kinases, adenylyl cyclase, and endothelial nitric oxide synthase (101). In this model, caveolin interaction would provide a mechanism for the focal localization and functional organization of signalling proteins (caveolin might function as protein X or X' in Fig. 7). However, the contribution of this paradigm to regulation of the adenylyl cyclase system has been questioned. Localization (to low density subdomains of the plasma membrane) and hormone regulation of the system was found to be normal in S49 lymphoma cells that do not express caveolin (99). In addition, the effects of caveolin-derived peptides on the activity of G proteins, reported by the one group (102), could not be replicated by another (99). Fidelity of G-protein signalling most likely requires some organization and compartmentation of the protein players, but the molecular mechanism for achieving functional specificity at the plasma membrane remains to be elucidated.

6. Genetic analysis

Yeast (*S. cerevisiae*) and worms (*C. elegans*) are two model systems that have been used successfully to dissect signalling processes by genetic means. These organisms can be screened for behavioural changes, such as mating in yeast and egg laying in worms, to identify mutations in signalling proteins. As noted in Section 4.2, this strategy was vital to the discovery of RGS proteins, thereby providing new insights relevant to signalling mechanisms in higher organisms. Transgenic and knockout mice provide another genetic approach to understanding physiological functions of signalling proteins. Genetic deficiencies of α_s and α_i subfamily members in knockout mice and heritable diseases in humans are described below and in Section 7.2, respectively.

6.1 α_s

Homozygous knockout of the α_s gene appears to be embryonically lethal in mice, as might be predicted for a ubiquitously expressed gene, the product of which is the sole hormone-responsive activator of adenylyl cyclase. Curiously, heterozygous disruption of the α_s gene is associated with early postnatal lethality. Yu *et al.* found distinct phenotypes, dependent on whether mice inherited the maternal or paternal knockout allele, indicating that the α_s gene was imprinted (103). The process of imprinting is an epigenetic phenomenon by which the alleles (paternal and maternal) are differentially expressed. The paternal α_s allele was not expressed in renal cortex and adipose tissue that inherited the maternal knockout gene. Thus, heterozygous mice that inherit the knockout allele from their mother express significantly less α_s messenger in these tissues compared to siblings that inherit the wild-type allele from their mother. Decreased α_s protein expression was correlated with a reduction in hormone-stimulated cAMP production in renal cortex (hormone resistance). By contrast, α_s in renal medulla did not appear to be imprinted because the maximal physiological response to vasopressin was normal in heterozygous mice in either the maternal or paternal allele and the level of expression of α_s was indistinguishable.

Tissue-specific imprinting of the α_s gene in humans may explain why the same genetic lesion presents with different disease symptoms, depending on maternal or paternal inheritance of the mutant allele (described in Section 7.2).

6.2 α_{i2}

Gene disruption and tissue-specific expression of antisense RNA have been utilized to ablate expression of α_{i2} in mice. The Birnbaumer group disrupted the α_{i2} gene and found that agonist-mediated inhibition of adenylyl cyclase was partially, but not completely, diminished in tissue homogenates (104). While the inhibition of cardiac adenylyl cyclase was blunted by about 65% in all $\alpha_{i2}-/-$ animals tested, only 50% of these animals showed impaired inhibition of adipocyte adenylyl cyclase (to three different agonists tested). No difference in the pertussis-toxin-sensitive inhibition of adenylyl cyclase by lysophosphatidic acid was observed in fibroblasts derived from wild-type and knockout mice. Transformed embryonic fibroblast lines were obtained from both wild-type and $\alpha_{i2}-/-$ mice by transfecting the complete SV40 genome into primary embryonic fibroblasts. Western blotting analysis of cloned cells demonstrated not only a reduction of α_{i2}, but also of β subunits. In contrast, the level of α_{i3} was increased 30–50%. This increase in α_{i3} and/or the presence of α_{i1} may contribute to the maintenance of G-protein-mediated inhibition of adenylyl cyclase in the $\alpha_{i2}-/-$ mice.

Knockout mice lacking α_{i2} display growth retardation and develop a lethal diffuse colitis with features closely resembling ulcerative colitis in humans, including the development of adenocarcinoma of the colon (105). The disease is associated with a local increase in memory $CD4^+$ T cells and certain proinflammatory cytokines (106). There are a variety of mechanisms whereby inactivation of the α_{i2} gene might cause a defect in the regulation of immune responses and play a role in the development of inflammation or transformation. For example, G_i proteins regulate certain events in T-cell activation and thymocyte differentiation (107) and participate in regulation of the MAPK pathway (Section 3.4.2) through which many cytokines signal.

Malbon and co-workers generated mice expressing α_{i2} antisense RNA under the control of the cAMP-sensitive promoter for phosphoenolpyruvate carboxykinase (PEPCK). Expression of this transgene began at birth (when the PEPCK promoter is activated) and suppression of α_{i2} was limited to target tissues, including liver and fat (108). The basal level of cAMP in adipocytes isolated from the transgenic mice was threefold higher than that of control animals (109). Similar to the knockout mice described above, considerable but incomplete blunting of agonist-mediated inhibition of adenylyl cyclase activity occurred in adipocytes from antisense RNA-expressing mice (109). Transgenic and knockout mice also shared a reduction in body mass relative to wild-type animals (105, 108). The lower body weight of transgenic mice could not be explained by reductions in the mass of targeted organs, nor by differences in food consumption, suggesting that a reduction in α_{i2} expression induces a metabolic alteration adversely affecting neonatal growth. The Malbon group concluded that α_{i2} acts as a positive regulator of insulin action because the deficiency of

α_{i2} in target tissues of the transgenic mice produced hyperinsulinaemia, impaired glucose tolerance, and resistance to insulin *in vivo* (issues not addressed in the reports on the knockout mice) (110). In addition, insulin-stimulated phosphorylation of insulin receptor substrate 1 (a proximal element of insulin signalling) in transgenic mice was attenuated *in vivo*. This attenuation was consistent with increased protein–tyrosine phosphatase activity in adipose, liver, and skeletal muscle from transgenic mice. These mice may serve as a model for the study of integration between tyrosine kinase and G-protein-mediated pathways and for non-insulin-dependent diabetes mellitus in humans.

6.3 α_{o}

As the most abundant G protein in neurons, α_{o} constitutes about 1% of brain membrane protein and 10% of growth-cone membrane in mammals. Although the lack of both splice variants of α_{o} lowered the survival rate of mice in two studies, it is surprising that the mice were born with no gross histological abnormalities in the nervous system (111, 112). In the study reported by Fishman and colleagues, only mild abnormalities in neurological function were noted, in the form of tremors and occasional seizures (111). In the second report, describing independently derived $\alpha_{o}-/-$ mice, Birnbaumer and co-workers noted tremors, hyperalgesia, impairment of motor control (falling from rotating rods or one-inch-wide beams), unsuccessful mating, hyperactivity, and an unusual turning behaviour in $\alpha_{o}-/-$ mice (an internet site for viewing short records of the turning behaviour is listed in the references) (112). These investigators also found a decrease in the inhibition of non-L-type calcium channel currents by an opioid receptor agonist in dorsal root ganglion cells from α_{o}-deficient mice. It is not known if the hyperalgesia is related to the altered calcium channel currents or other changes in neuronal sensitivity.

In addition to neurons, α_{o} is expressed only in endocrine cells and heart, at levels comparable to those of other G proteins. The Fishman group found that α_{o} is essential for normal muscarinic regulation of L-type calcium channels in ventricular myocytes (111). These channels are activated by the α-adrenergic receptor (via cAMP) and this response is diminished by muscarinic agonists. In myocytes from $\alpha_{o}-/-$ mice the stimulatory effect of isoproterenol on L-type calcium channels was intact but the inhibitory effect of the muscarinic agonist, carbamylcholine (carbachol), was nearly abolished. These results raised the possibility that in wild-type cells, G_{o} may function to mediate muscarinic inhibition of adenylyl cyclase activity, thereby reducing cAMP levels and the activity of L-type calcium channels. This possibility was dismissed by the Birnbaumer group, who found that carbachol-inhibited adenylyl cyclase activity was indistinguishable in ventricular muscle homogenates from their wild-type and $\alpha_{o}-/-$ mice (112). Further investigation and comparison of the two independent lines of $\alpha_{o}-/-$ mice may reveal the mechanism by which G_{o} mediates muscarinic inhibition of L-type calcium channel activity and resolve discrepancies between phenotypes.

7. Disease in humans caused by defects in G proteins

Either activating or inactivating defects in G-protein α subunits can give rise to diseases in humans. Examples of G-protein-associated pathology include the effects of the bacterial toxins (introduced in Section 2.1) and genetic mutations.

7.1 Bacterial toxins

Elucidation of the molecular basis of intoxication by the pathogens, *Vibrio cholerae* and *Bordetella pertussis*, was a landmark achievement. In addition to being clinically relevant, the chemical and functional modification of G proteins by the toxins produced by these organisms have proven to be of great value for the identification of players in signalling pathways (Table 3).

7.1.1 Cholera

The activity of α_s is increased in intestinal epithelial cells of patients with secretory diarrhoea caused by infection with *Vibrio cholerae*. The bacteria secrete an oligomeric protein exotoxin composed of an A subunit and five B subunits. The B oligomer is responsible for toxin binding to the cell surface ganglioside G_{M1}. Following entry into the cell and processing by proteolysis and reduction, the A component catalyses the ADP ribosylation of arginine 201 in α_s (19). This reduces GTPase activity drastically, thereby creating constitutively activated α_s. The persistent production of cAMP that ensues leads ultimately to the devastating fluid and electrolyte loss that causes the diarrhoea.

7.1.2 Pertussis

Pertussis, or whooping cough, is a communicable, acute infection of the respiratory tract caused by *Bordetella pertussis*. The microbe possesses several virulence factors that may contribute to pathogenicity, either by permitting colonization of the respiratory tract or through direct toxic effects (113). Two of the virulence factors, pertussis toxin and adenylate cyclase toxin, affect levels of cAMP in host cells. Pertussis toxin (also known as islet-activating protein) is an A–B-type protein exotoxin (similar to cholera toxin). The B component binds to the surface of the host cell and enables the A component to enter. Inside the cell, the components dissociate and a disulphide bond within the A component is reduced, thus revealing its cryptic ADP–ribosyl transferase activity. Among the cellular targets for ADP ribosylation by the active A component are members of the G_i subfamily of α subunits. Pertussis-toxin-mediated ADP ribosylation of α_i subfamily members prevents their interaction with receptor, thus attenuating the ability of inhibitory hormones to depress adenylyl cyclase activity (and often potentiating the effects of stimulatory hormones on the enzyme).

   Adenylate cyclase toxin is a novel toxin found on the external side of the cell membrane of *Bordetella* species (113). This toxin enters mammalian host cells where it is activated by calmodulin and catalyses the uncontrolled production of cAMP. Adenylate cyclase toxin has been shown to suppress neutrophil and monocyte activities, which may contribute to the persistence of infection by inhibiting clearance of the organism. The toxin may also contribute to pathogenesis by causing accumulation of cAMP in the mucosa, leading to increased fluid secretion and enhancement of the entry of pertussis toxin.

7.2 Genetic mutations

Determinants of the clinical manifestations of mutations in G protein α subunits include the range of expression of the involved gene, and whether the mutation is somatic or germline (114). Such mutations in G_s and G_i subfamilies of α subunits cause diseases ranging from hormone resistance to night blindness.

7.2.1 Hormone resistance

Pseudohypoparathyroidism is a dramatic example of decreased G_s function. For one subtype of the disease, PHP Ia, affected subjects show resistance not only to parathyroid hormone but also to several other hormones, such as thyroid-stimulating hormone and gonadotrophins, whose actions are mediated by receptors coupled to G_s. Affected individuals also show phenotypic features, including obesity, short stature, skeletal abnormalities, which are collectively called Albright hereditary osteodystrophy (AHO). However, within an AHO kindred certain individuals may show these phenotypic features without hormone resistance. A variety of loss-of-function mutations in the α_s gene have been identified in kindreds with AHO. Most mutations impair mRNA and/or protein formation; a minority appear to be missense mutations (114). These heterozygous mutations cause an approximately 50% reduction in α_s function (usually measured in easily obtained erythrocytes from patients) which is thought to impair the cAMP response to hormone stimulation.

Based on multigenerational studies of families with AHO, only maternal inheritance appears to result in the complete syndrome including hormone resistance (115). These observations led to the suggestion that the α_s gene may be imprinted (a topic introduced in Section 6.1). Hayward *et al.* demonstrated that while expression of α_s occurs from both alleles, the human gene is indeed imprinted in a promoter-specific fashion (116). They identified a region approximately 35 kb upstream of the originally described exon 1 that was methylated exclusively on a maternal allele. Transcripts from this region are alternatively spliced onto exon 2, yielding mRNA species that are derived exclusively from the paternal allele, including one that encodes a protein homologous to the large α_s-related protein, $XL_{\alpha s}$, first identified in rat (117). The differential imprinting of the α_s gene may contribute to the anomalous inheritance of PHP Ia.

A rare combination of seemingly disparate clinical syndromes has been reported for two unrelated male patients: PHP Ia and precocious masculinization (testotoxicosis, due to gonadotrophin-independent hypersecretion of testosterone by testicular Leydig cells). In both patients, a mutation in the α_s gene caused a replacement of an alanine at position 366 with serine and a 50% decrease in erythrocyte G_s activity (the remaining 50% was due to the normal G_s allele). Bourne and colleagues established that the mutant α$_s$ constitutively activated adenylyl cyclase *in vitro*, causing hormone-independent cAMP accumulation when expressed in cultured cells, and accounting for the testotoxicosis phenotype (as cAMP stimulates testosterone secretion) (118). Although the mutant α$_s$ was quite stable at testis temperature (32 °C), it was rapidly degraded at body temperature of 37 °C, explaining the PHP Ia phenotype caused by loss of G$_s$ activity. *In vitro* experiments indicated that accelerated release of GDP causes both the constitutive activity and the thermolability of α$_s$ (118). The point mutation has been likened to a quarterback with an injured finger, causing the protein to lose its grip on the nucleotide (119). Thus the normal requirement for agonist-stimulated receptor to promote release of GDP is circumvented by this injured form of α$_s$, thereby facilitating GTP binding.

7.2.2 McCune–Albright syndrome

The McCune–Albright syndrome is characterized by hyperfunction of one or more endocrine glands (pituitary somatotrophs, adrenal cortex, thyroid, and gonads) coupled with café-au-lait skin hyperpigmentation and bone deformity (polyostotic fibrous dysplasia). The cause of such pleiotropic manifestations in a sporadic disorder was obscure until it was recognized that, while a germline-activating mutation in the α$_s$ gene might be lethal, a somatic mutation occurring early in embryogenesis, could lead to constitutively increased cAMP formation in many tissues (114). Missense mutation of arginine 201 (the site of ADP ribosylation by cholera toxin) is one of the constitutively activating mutations identified in patients with McCune–Albright syndrome. Patients express mutations in a mosaic pattern that correlates with the cellular abnormalities observed. The spectrum of affected tissues may be determined by the time in development that the mutations occur. Timing may range from early in embryogenesis (leading to a wide distribution with pleiotropic, potentially severe manifestations) to those occurring as later monofocal events (leading to localized manifestations such as growth hormone-secreting pituitary tumours) (114).

7.2.3 Tumours

A subset of pituitary adenomas are characterized by elevated cAMP levels and growth hormone (GH) production (120). Cells from these tumours are relieved of the requirement for GH-releasing hormone to stimulate adenylyl cyclase, growth hormone secretion, and proliferation of normal pituitary somatotrophs. Prompted by the idea that α$_s$ activated by mutation could relieve the requirement for GH-releasing hormone and thereby contribute to the growth of these tumours, Bourne, Vallar, and colleagues sought and found mutations in α$_s$ genes. In about 40% of the GH-releasing hormone independent tumours, α$_s$ was constitutively activated by a point mutation in either of two residues: arginine 201 (the ADP ribosylation site) or glutamine 227, which is equivalent to glutamine 61 of Ras, a frequent site of GTPase-inhibiting mutations in the Ras proteins (121). Some autonomously functioning thyroid adenomas have similar mutations (122). These results are reminiscent of the events that convert *Ras* genes to oncogenes, covered in Chapter 3.

7.2.4 Night blindness

Patients with congenital stationary night blindness have normal daytime vision, mediated by cone photoreceptors, but are blind when ambient light is so dim that a normal subject would utilize only rod photoreceptors to see without colour discrimination. Although the disease is heterogeneous, one form of dominantly inherited congenital night blindness (the Nougaret form) is caused by a missense mutation in rod α_t (α_{t1}) (123). The mutation is in glycine 38 which is conserved in most G proteins and in Ras (glycine 12). It is presumed that this mutation reduces GTPase activity of α_t, as has been demonstrated for the analogous mutation in α_s (124). Mutant α_{t1} has become the third component of the rod phototransduction cascade (joining constitutively active rhodopsin and the α subunit of rod cGMP PDE) where a defect is implicated as a cause of dominant stationary night blindness (125).

Acknowledgements

The author wishes to acknowledge helpful suggestions made by many colleagues in the Department of Pharmacology at University of Texas Southwestern Medical Center at Dallas. Particular thanks go to Marc Mumby, Suchetana Mukhopadhyay, and Thomas Wilkie who made special efforts. Work in the author's laboratory is supported by a grant from the National Institute of General Medical Sciences.

References

1. Rodbell, M., Birnbaumer, L., and Pohl, S. L. (1969) *Hormones, receptors, and adenyl cyclase activity in mammalian cells.* Fogarty International Center Proceedings, Vol. 4, p. 162. National Institutes of Health, Bethesda, MD.
2. Northup, J. K., Sternweis, P. C., Smigel, M. D., Schleifer, L. S., Ross, E. M., and Gilman, A. G. (1980) Purification of the regulatory component of adenylate cyclase. *Proc. Natl Acad. Sci. USA,* **77**, 6516.
3. Wheeler, G. L. and Bitensky, M. W. (1977) A light-activated GTPase in vertebrate photoreceptors: regulation of light-activated cyclic GMP phosphodiesterase. *Proc. Natl Acad. Sci. USA,* **74**, 4238.
4. Kuhn, H. (1980) Light- and GTP-regulated interaction of GTPase and other proteins with bovine photoreceptor membranes. *Nature,* **283**, 587.
5. Fung, B. K.-K., Hurley, J. B., and Stryer, L. (1981) Flow of information in the light-triggered cyclic nucleotide cascade of vision. *Proc. Natl Acad. Sci. USA,* **78**, 152.
6. Sternweis, P. C. and Robishaw, J. D. (1984) Isolation of two proteins with high affinity for guanine nucleotides from membranes of bovine brain. *J. Biol. Chem.,* **259**, 13806.
7. Strathmann, M. and Simon, M. I. (1990) G protein diversity: a distinct class of α subunits is present in vertebrates and invertebrates. *Proc. Natl Acad. Sci. USA,* **87**, 9113.
8. Hsu, W. H., Rudolph, U., Sanford, J., Bertrand, P., Olate, J., Nelson, C., Moss, L. G., Boyd, A. E., III, Codina, J., and Birnbaumer, L. (1990) Molecular cloning of a novel splice variant of the α subunit of the mammalian G_o protein. *J. Biol. Chem.,* **265**, 11220.

9. Strathmann, M., Wilkie, T. M., and Simon, M. I. (1990) Alternative splicing produces transcripts encoding two forms of the α subunit of GTP-binding protein G_o. *Proc. Natl Acad. Sci. USA,* **87**, 6477.

10. Bray, P., Carter, A., Simons, C., Guo, V., Puckett, C., Kamholz, J., Spiegel, A., and Nirenberg, M. (1986) Human cDNA clones for four species of $G_{\alpha s}$ signal transduction protein. *Proc. Natl Acad. Sci. USA,* **83**, 8893.

11. Robishaw, J. D., Smigel, M. D., and Gilman, A. G. (1986) Molecular basis for two forms of the G protein that stimulates adenylate cyclase. *J. Biol. Chem.,* **261**, 9587.

12. Kleuss, C., Hescheler, J., Ewel, C., Rosenthal, W., Schultz, G., and Wittig, B. (1991) Assignment of G-protein subtypes to specific receptors inducing inhibition of calcium currents. *Nature,* **353**, 43.

13. Montmayeur, J. P. and Borrelli, E. (1994) Targeting of $G_{\alpha i2}$ to the golgi by alternative spliced carboxyl-terminal region. *Science,* **263**, 95.

14. Watson, A. J., Aragay, A. M., Slepak, V. Z., and Simon, M. I. (1996) A novel form of the G protein β subunit $G\beta_5$ is specifically expressed in the vertebrate retina. *J. Biol. Chem.,* **271**, 28154.

15. Clapham, D. E. and Neer, E. J. (1997) G protein beta/gamma subunits. *Annu. Rev. Pharm. Toxicol.,* **37**, 167.

16. West, R. E. Jr, Moss, J., Vaughan, M., Liu, T., and Liu, T.-Y. (1985) Pertussis toxin-catalyzed ADP-ribosylation of transducin. *J. Biol. Chem.,* **260**, 14428.

17. Van Dop, C., Tsubokawa, M., Bourne, H. R., and Ramachandran, J. (1984) Amino acid sequence of retinal transducin at the site ADP-ribosylated by cholera toxin. *J. Biol. Chem.,* **259**, 696.

18. Kahn, R. A. and Gilman, A. G. (1984) Purification of a protein cofactor required for ADP-ribosylation of the stimulatory regulatory component of adenylate cyclase by cholera toxin. *J. Biol. Chem.,* **259**, 6228.

19. Moss, J., Haun, R. S., Tsai, S.-C., Welsh, C. F., Lee, F. S., Price, S. R., and Vaughan, M. (1994) Activation of cholera toxin by ADP-ribosylation factors: 20-kDa guanine nucleotide-binding proteins. *Meth. Enzymol.,* **237**, 44.

20. Milligan, G. (1988) Techniques used in the identification and analysis of function of pertussis toxin-sensitive guanine nucleotide binding proteins. *Biochem. J.,* **255**, 1.

21. Buss, J. E., Mumby, S. M., Casey, P. J., Gilman, A. G., and Sefton, B. M. (1987) Myristoylated α subunits of guanine nucleotide-binding regulatory proteins. *Proc. Nat. Acad. Sci. USA,* **84**, 7493.

22. Johnson, D. R., Bhatnager, R. S., Knoll, L. J., and Gordon, J. I. (1994) Genetic and biochemical studies of protein N-myristoylation. *Annu. Rev. Biochem.,* **63**, 869.

23. Wedegaertner, P. B., Wilson, P. T., and Bourne, H. R. (1995) Lipid modifications of trimeric G proteins. *J. Biol. Chem.,* **270**, 503.

24. Linder, M. E., Pang, I.-H., Duronio, R. J., Gordon, J. I., Sternweis, P. C., and Gilman, A. G. (1991) Lipid modifications of G protein subunits. Myristoylation of G_{oa} increases its affinity for βγ. *J. Biol. Chem.,* **266**, 4654.

25. Taussig, R., Iñiguez-Lluhi, J., and Gilman, A. G. (1993) Inhibition of adenylyl cyclase by $G_{i\alpha}$. *Science,* **261**, 218.

26. Dessauer, C. W., Tesmer, J. J. G., Sprang, S. R., and Gilman, A. G. (1998) Identification of a $G_{i\alpha}$ binding site on type V adenylyl cyclase. *J. Biol. Chem.,* **273**, 25831.

27. Wang, J., Tu, Y. P., Woodson, J., Song, X. L., and Ross, E. M. (1997) A GTPase-activating protein for the G protein $G_{\alpha z}$-identification. *J. Biol. Chem.,* **272**, 5732.

28. Tu, Y., Wang, J., and Ross, E. M. (1997) Inhibition of brain G_z GAP and other RGS proteins by palmitoylation of G protein α subunits. *Science,* **278**, 1132.

29. Neubert, T. A., Johnson, R. S., Hurley, J. B., and Walsh, K. A. (1992) The rod transducin α subunit amino terminus is heterogeneously fatty acylated. *J. Biol. Chem.* **267**, 18274.

30. Kokame, K., Fukada, Y., Yoshizawa, T., Takao, T., and Shimonishi, Y. (1992) Lipid modification at the N-terminus of photoreceptor G-protein α subunit. *Nature,* **359**, 749.

31. Johnson, R. S., Ohguro, H., Palczewski, K., Hurley, J. B., Walsh, K. A., and Neubert, T. A. (1994) Heterogeneous *N*-acylation is a tissue- and species-specific posttranslational modification. *J. Biol. Chem.,* **269**, 21067.

32. Mumby, S. M. (1997) Reversible palmitoylation of signaling proteins. *Curr. Opinion Cell Biol.,* **9**, 148.

33. Milligan, G., Parenti, M., and Magee, A. I. (1995) The dynamic role of palmitoylation in signal transduction. *Trends Biochem. Sci.* 20, 181.

34. Degtyarev, M. Y., Spiegel, A. M., and Jones, T. L. Z. (1993) Increased palmitoylation of the G_s protein α subunit after activation by the β-adrenergic receptor or cholera toxin. *J. Biol. Chem.,* **268**, 23769.

35. Mumby, S. M., Kleuss, C., and Gilman, A. G. (1994) Receptor regulation of G protein palmitoylation. *Proc. Natl Acad. Sci. USA,* **91**, 2800.

36. Wedegaertner, P. B. and Bourne, H. R. (1994) Activation and depalmitoylation of $G_{s\alpha}$. *Cell,* **77**, 1063.

37. Duncan, J. A. and Gilman, A. G. (1998) A cytoplasmic acylprotein thioesterase that removes palmitate from G protein α subunits and p21[ras]. *J. Biol. Chem.,* **273**, 15830.

38. Verkruyse, L. A. and Hofmann, S. L. (1996) Lysosomal targeting of palmitoyl-protein thioesterase. *J. Biol. Chem.,* **271**, 15831.

39. Jackson, C. S., Zlatkine, P., Bano, C., Kabouridis, P., Mehul, B., Parenti, M., Milligan, G., Ley, S. C., and Magee, A. I. (1995) Dynamic protein acylation and the regulation of localization and function of signal-transducing proteins. *Biochem. Soc. Trans.,* **23**, 568.

40. Iiri, T., Backlund, P. S., Jones, T. L. Z., Wedegaertner, P. B., and Bourne, H. R. (1996) Reciprocal regulation of $G_{s\alpha}$ by palmitate and the beta/gamma subunit. *Proc. Nat. Acad. Sci. USA,* **93**, 14592.

41. Sternweis, P. C., Northup, J. K., Smigel, M. D., and Gilman, A. G. (1981) The regulatory component of adenylate cyclase: purification and properties. *J. Biol. Chem.,* **256**, 11517.

42. Higashijima, T., Ferguson, K. M., Sternweis, P. C., Smigel, M. D., and Gilman, A. G. (1987) Effects of Mg^{2+} and the βγ-subunit complex on the interactions of guanine nucleotides with G proteins. *J. Biol. Chem.,* **262**, 762.

43. Logothetis, D. E., Kurachi, Y., Galper, J., Neer, E. J., and Clapham, D. E. (1987) The βγ subunits of GTP-binding proteins activate the muscarinic K^+ channel in heart. *Nature,* **325**, 321.

44. Birnbaumer, L. (1987) Viewpoint: Which G protein subunits are the active mediators in signal transduction? *Trends Pharmacol. Sci.,* **8**, 209.

45. Dietzel, C. and Kurjan, J. (1987) The yeast *SCG1* gene: a G_α-like protein implicated in the a- and α-factor response pathway. *Cell,* **50**, 1001.

46. Whiteway, M., Hougan, L., Dignard, D., Thomas, D. Y., Bell, L., Saari, G. C., Grant, F. J., O'Hara, P., and MacKay, V. L. (1989) The *STE4* and *STE18* genes of yeast encode potential β and γ subunits of the mating factor receptor-coupled G protein. *Cell,* **56**, 467.

47. Sondek, J., Bohm, A., Lambright, D. G., Hamm, H. E., and Sigler, P. B. (1996) Crystal structure of a G_α protein beta/gamma dimer at 2.1 Å resolution. *Nature,* **379**, 369.

48. Lambright, D. G., Sondek, J., Bohm, A., Skiba, N. P., Hamm, H. E., and Sigler, P. B. (1996) The 2.0 A crystal structure of a heterotrimeric G protein. *Nature,* **379**, 311.

49. Wall, M. A., Coleman, D. E., Lee, E., Iniguez-Lluhi, J. A., Posner, B. A., Gilman, A. G., and Sprang, S. R. (1995) The structure of G protein heterotrimer $G_{i\alpha1}\beta_1\gamma_2$. *Cell,* **83**, 1047.

50. Ford, C. E., Skiba, N. P., Bae, H., Daaka, Y., Reuveny, E., Shekter, L. R., Rosal, R., Weng, G., Yang, C.-S., Iyengar, R., Miller, R. J., Jan, L. Y., Lefkowitz, R. J., and Hamm, H. E. (1998) Molecular basis for interactions of G protein beta/gamma subunits with effectors. *Science*, **280**, 1271.

51. Li, Y., Sternweis, P. M., Charnecki, S., Smith, T. F., Gilman, A. G., Neer, E. J., and Kozasa, T. (1998) Sites for $G\alpha$ binding on the G protein β subunit overlap with sites for regulation of phospholipase $C\beta$ and adenylyl cyclase. *J. Biol. Chem.*, **273**, 16265.

52. Sunahara, R. K., Dessauer, C. W., and Gilman, A. G. (1996) Complexity and diversity of mammalian adenylyl cyclases. *Annu. Rev. Pharm. Toxicol.*, **36**, 461.

53. Tang, W.-J. and Gilman, A. G. (1995) Construction of a soluble adenylyl cyclase activated by $G_{s\alpha}$ and forskolin. *Science*, **268**, 1769.

54. Tesmer, J. J. G., Sunahara, R. K., Gilman, A. G., and Sprang, S. R. (1997) Crystal structure of the catalytic domains of adenylyl cyclase in a complex with $G_{s\alpha}$.GTPgammaS. *Science*, **278**, 1907.

55. Godchaux, W., III and Zimmerman, W. F. (1979) Soluble proteins of intact bovine rod cell outer segments. *Exp. Eye Res.*, **28**, 483.

56. Mattera, R., Graziano, M. P., Yatani, A., Zhou, Z., Graf, R., Codina, J., Birnbaumer, L., Gilman, A. G., and Brown, A. M. (1989) Splice variants of the α subunit of the G protein G_s activate both adenylyl cyclase and calcium channels. *Science*, **243**, 804.

57. Ikeda, S. R. (1996) Voltage-dependent modulation of N-type calcium channels by G protein beta/gamma subunits. *Nature*, **380**, 255.

58. Herlitze, S., Garcia, D. E., Mackie, K., Hille, B., Scheuer, T., and Catterall, W. A. (1996) Modulation of Ca^{2+} channels by G-protein beta/gamma subunits. *Nature*, **380**, 258.

59. Hille, B. (1998) Modulation of ion-channel function by G-protein-coupled receptors. *TINS*, **17**, 531.

60. Zhu, Y. and Ikeda, S. R. (1994) VIP inhibits N-type Ca^{2+} channels of sympathetic neurons via a pertussis toxin-insensitive but cholera toxin-sensitive pathway. *Neuron*, **13**, 657.

61. Narasimhan, V., Holowka, D., and Baird, B. (1990) A guanine nucleotide-binding protein participates in IgE receptor-mediated activation of endogenous and reconstituted phospholipase A2 in a permeabilized cell system. *J. Biol. Chem.*, **265**, 1459.

62. Jelsema, C. L. and Axelrod, J. (1987) Stimulation of phospholipase A_2 activity in bovine rod outer segments by the $\beta\gamma$ subunits of transducin and its inhibition by the α subunit. *Proc. Natl Acad. Sci. USA*, **84**, 3623.

63. Pitcher, J. A., Freedman, N. J., and Lefkowitz, R. J. (1998) G protein-coupled receptor kinases. *Annu. Rev. Biochem.*, **67**, 653.

64. Inglese, J., Koch, W. J., Touhara, K., and Lefkowitz, R. J. (1995) G-beta-gamma interactions with PH domains and Ras-MAPK signaling pathways. *Trends Biochem. Sci.*, **20**, 151.

65. Touhara, K., Inglese, J., Pitcher, J. A., Shaw, G., and Lefkowitz, R. J. (1994) Binding of G protein beta/gamma subunits to pleckstrin homology domains. *J. Biol. Chem.*, **269**, 10217.

66. Inglese, J., Koch, W. J., Caron, M. G., and Lefkowitz, R. J. (1992) Isoprenylation in regulation of signal transduction by G-protein coupled receptor kinases. *Nature*, **359**, 147.

67. Tsukada, S., Simon, M. I., Witte, O. N., and Katz, A. (1994) Binding of beta/gamma subunits of heterotrimeric G proteins to the PH domain of Bruton tyrosine kinase. *Proc. Natl Acad. Sci. USA*, **91**, 11256.

68. Langhans-Rajasekaran, S. A., Wan, Y., and Huang, X.-Y. (1995) Activation of Tsk and Btk tyrosine kinases by G protein beta/gamma subunits. *Proc. Natl Acad. Sci. USA*, **92**, 8601.

69. Miyajima, I., Nakafuku, M., Nakayama, N., Brenner, C., Miyajima, A., Kaibuchi, K., Arai, K.-I., Kaziro, Y., and Matsumoto, K. (1987) *GPA1*, a haploid-specific essential gene,

encodes a yeast homolog of mammalian G protein which may be involved in mating factor signal transduction. *Cell,* **50**, 1011.

70. Sugden, P. H. and Clerk, A. (1997) Regulation of the ERK subgroup of MAP kinase cascades through G protein-coupled receptors. *Cell. Signal.,* **9**, 337.

71. Leeuw, T., Wu, C., Schrag, J. D., Whiteway, M., Thomas, D. Y., and Leberer, E. (1998) Interaction of a G-protein β-subunit with a conserved sequence in Ste20/PAK family protein kinases. *Nature,* **391**, 191.

72. Kapeller, R. and Cantley, L. C. (1994) Phosphatidylinositol 3-kinase. *Bioessays,* **16**, 565.

73. Stoyanov, B., Volinia, S., Hanck, T., Rubio, I., Loubtchenkov, M., Malek, D., Stoyanova, S., Vanhaesebroeck, B., Dhand, R., Nurnberg, B., Gierschik, P., Seedorf, K., Hsuan, J. J., Waterfield, M. D., and Wetzker, R. (1995) Cloning and characterization of a G protein-activated human phosphoinositide-3 kinase. *Science,* **269**, 690.

74. Lopez-Ilasaca, M., Crespo, P., Pellici, P. G., Gutkind, J. S., and Wetzker, R. (1997) Linkage of G protein-coupled receptors to the MAPK signaling pathway through PI 3-kinase gamma. *Science,* **275**, 394.

75. Lin, H. C., Duncan, J. A., Kozasa, T., and Gilman, A. G. (1998) Sequestration of the G protein beta/gamma subunit complex inhibits receptor-mediated endocytosis. *Proc. Natl Acad. Sci. USA,* **95**, 5057.

76. Bomsel, M. and Mostov, K. (1992) Role of heterotrimeric G proteins in membrane traffic. *Mol. Biol. Cell,* **3**, 1317.

77. Nuoffer, C. and Balch, W. E. (1994) GTPases: Multifunctional molecular switches regulating vesicular traffic. *Annu. Rev. Biochem.,* **63**, 949.

78. Lin, H. C. and Gilman, A. G. (1996) Regulation of dynamin I GTPase activity by G protein beta-gamma subunits and phosphatidylinositol 4,5-bisphosphate. *J. Biol. Chem.,* **271**, 27979.

79. Bauer, P. H., Bluml, K., Schroder, S., Hegler, J., Dees, C., and Lohse, M. J. (1998) Interactions of phosducin with the subunits of G-proteins. *J. Biol. Chem.,* **273**, 9465.

80. Gaudet, R., Bohm, A., and Sigler, P. B. (1996) Crystal structure at 2.4 angstrom resolution of the complex of transducin beta-gamma and its regulator, phosducin. *Cell,* **87**, 577.

81. Hawes, B. E., Touhara, K., Kurose, H., Lefkowitz, R. J., and Inglese, J. (1994) Determination of the $G_{beta/gamma}$-binding domain of phosducin. *J. Biol. Chem.,* **269**, 29825.

82. Ross, E. M., Wang, J., Tu, Y., and Biddlecome, G. H. (1998) Guanosine triphosphatase-activating proteins for heterotrimeric G proteins. *Adv. Pharm.* **42**, 458.

83. Angleson, J. K. and Wensel, T. G. (1993) A GTPase-accelerating factor for transducin, distinct from its effector cGMP phosphodiesterase, in rod outer segment membranes. *Neuron,* **11**, 939.

84. He, W., Cowan, C. W., and Wensel, T. G. (1998) RGS9, a GTPase accelerator for phototransduction. *Neuron,* **20**, 95.

85. Berman, D. M. and Gilman, A. G. (1998) Mammalian RGS proteins: Barbarians at the gate. *J. Biol. Chem.,* **273**, 1269.

86. De Vries, L., Mousli, M., Wurmser, A., and Farquhar, M. G. (1995) GAIP, a protein that specifically interacts with the trimeric G protein $G\alpha_{i3}$, is a member of a protein family with a highly conserved core domain. *Proc. Natl Acad. Sci. USA,* **92**, 11916.

87. Berman, D. M., Wilkie, T. M., and Gilman, A. G. (1996) GAIP and RGS4 are GTPase-activating proteins for the G_i subfamily of G protein α subunits. *Cell,* **86**, 445.

88. Popov, S., Yu, K., Kozasa, T., and Wilkie, T. M. (1997) The RGS domains of RGS4, RGS10 and GAIP retain GAP activity *in vitro. Proc. Natl Acad. Sci. USA,* **94**, 7216.

89. Tesmer, J. J. G., Berman, D. M., Gilman, A. G., and Sprang, S. R. (1997) Structure of RGS4

bound to AlF$_4^-$ activated G$_{i\alpha1}$: Stabilization of the transition state for GTP hydrolysis. *Cell*, **89**, 251.

90. Huang, C., Hepler, J. R., Gilman, A. G., and Mumby, S. M. (1997) Attenuation of G$_i$ and G$_q$-mediated signaling by expression of RGS4 or GAIP in mammalian cells. *Proc. Natl Acad. Sci. USA*, **94**, 6159.

91. Wieland, T., Chen, C.-K., and Simon, M. I. (1997) The retinal specific protein RGS-r competes with the gamma subunit of cGMP phosphodiesterase for the α subunit of transducin and facilitates signal termination. *J. Biol. Chem.*, **272**, 8853.

92. Faurobert, E. and Hurley, J. B. (1997) The core domain of a new retina specific RGS protein stimulates the GTPase activity of transducin *in vitro*. *Proc. Natl Acad. Sci. USA*, **94**, 2945.

93. Zeng, W., Xu, Popov, S., Mukhopadhyay, S., Chidiac, P., Swistok, J. *et al.* (1998) The N-terminal domain of RGS4 confers receptor selective inhibition of G protein signaling. *J. Biol. Chem.*, **273**, 34687.

94. Lerea, C. L., Somers, D. E., Hurley, J. B., Klock, I. B., and Bunt-Milam, A. H. (1986) Identification of specific transducin α subunits in retinal rod and cone photoreceptors. *Science*, **234**, 77.

95. Neubig, R. R. (1994) Membrane organization in G-protein mechanisms. *FASEB J.*, **8**, 939.

96. Kleuss, C., Scherubl, H., Hescheler, J., Schultz, G., and Wittig, B. (1992) Different β subunits determine G protein interaction with transmembrane receptors. *Nature*, **358**, 424.

97. Kleuss, C., Scherubl, H., Hescheler, J., Schultz, G., and Wittig, B. (1993) Selectivity in signal transduction determined by γ subunits of heterotrimeric G proteins. *Science*, **259**, 832.

98. Neer, E. J. (1995) Heterotrimeric G proteins—organizers of transmembrane signals. *Cell*, **80**, 249.

99. Huang, C., Hepler, J. R., Chen, L. T., Gilman, A. G., Anderson, R. G. W., and Mumby, S. M. (1997) Organization of G proteins and adenylyl cyclase at the plasma membrane. *Mol. Biol. Cell*, **8**, 2365.

100. Anderson, R. G. W. (1998) The caveolae membrane system. *Annu. Rev. Biochem.*, **67**, 199.

101. Okamoto, T., Schlegel, A., Scherer, P. E., and Lisanti, M. P. (1998) Caveolins, a family of scaffolding proteins for organizing 'preassembled signaling complexes' at the plasma membrane. *J. Biol. Chem.*, **273**, 5419.

102. Li, S.W., Okamoto, T., Chun, M. Y., Sargiacomo, M., Casanova, J. E., Hansen, S. H., Nishimoto, I., and Lisanti, M. P. (1995) Evidence for a regulated interaction between heterotrimeric G proteins and caveolin. *J. Biol. Chem.*, **270**, 15693.

103. Yu, S., Yu, D., Lee, E., Eckhaus, M., Lee, R., Corria, Z., Accili, D., Westphal, H., and Weinstein, L. S. (1998) Variable and tissue-specific hormone resistance in heterotrimeric G$_s$ protein α-subunit (G$_s$α) knockout mice is due to tissue-specific imprinting of the G$_s$α gene. *Proc. Natl Acad. Sci. USA*, **95**, 8715.

104. Rudolph, U., Spicher, K., and Birnbaumer, L. (1996) Adenylyl cyclase inhibition and altered G protein subunit expression and ADP-ribosylation patterns in tissues and cells from G$_{i\alpha2}$-/- mice. *Proc. Nat. Acad. Sci. USA*, **93**, 3209.

105. Rudolph, U., Finegold, M. J., Rich, S. S., Harriman, G. R., Srinivasan, Y., Brabet, P., Boulay, G., Bradley, A., and Birnbaumer, L. (1995) Ulcerative colitis and adenocarcinoma of the colon in G$_{\alpha i2}$-deficient mice. *Nature Genet.*, **10**, 143.

106. Rudolph, U., Finegold, M. J., Rich, S. S., Harriman, G. R., Srinivasan, Y., Brabet, P., Bradley, A., and Birnbaumer, L. (1995) Gi2α protein deficiency: a model of inflammatory bowel disease. *J. Clin. Immunol.*, **15**, 101S.

107. Chaffin, K. E. and Perlmutter, R. M. (1998) A pertussis toxin-sensitive process controls thymocyte emigration. *Eur. J. Immunol.,* **21**, 2565.

108. Moxham, C. M., Hod, Y., and Malbon, C. C. (1993) Induction of $G_{\alpha i2}$-specific antisense RNA *in vivo* inhibits neonatal growth. *Science,* **260**, 991.

109. Moxham, C. M., Hod, Y., and Malbon, C. C. (1993) $G_{i\alpha 2}$ mediates the inhibitory regulation of adenylylcyclase *in vivo*. Analysis in transgenic mice with $G_{i\alpha 2}$ suppressed by inducible antisense RNA. *Dev. Genet.,* **14**, 266.

110. Moxham, C. M. and Malbon, C. C. (1996) Insulin action impaired by deficiency of the G-protein subunit $G_{i\alpha 2}$. *Nature,* **379**, 840.

111. Valenzuela, D., Han, X., Mende, U., Fankhauser, C., Mashimo, H., Huang, P., Pfeffer, J., Neer, E. J., and Fishman, M. C. (1997) $G_{\alpha o}$ is necessary for muscarinic regulation of Ca^{2+} channels in mouse heart. *Proc. Natl Acad. Sci. USA,* **94**, 1727.

112. Jiang, M., Gold, M. S., Boulay, G., Spicher, K., Peyton, M., Brabet, P., Srinivasan, Y., Rudolph, U., Elison, G., and Birnbaumer, L. (1998) Multiple neurological abnormalities in mice deficient in the G protein G_o. *Proc. Natl Acad. Sci. USA,* **95**, 3269.

113. Vogel, F. R. (1994) Bordetella pertussis. In Pulmonary infections and immunity, (ed. H. Chmel, M. Bendinelli, and H. Friedman), p. 149. Plenum Press, New York.

114. Spiegel, A. M. (1997) The molecular basis of disorders caused by defects in G proteins. *Horm. Res.,* **47**, 89.

115. Davies, S. J. and Hughes, H. E. (1993) Imprinting in Albright's hereditary osteodystrophy. *J. Med. Genet.,* **30**, 101.

116. Hayward, B. E., Kamiya, M., Strain, L., Moran, V., Campbell, R., Hayashizaki, Y., and Bonthron, D. T. (1998) The human *GNAS1* gene is imprinted and encodes distinct paternally and biallelically expressed G proteins. *Proc. Natl Acad. Sci. USA,* **95**, 10038.

117. Kehlenbach, R. H., Matthey, J., and Huttner, W. B. (1994) $XL_{\alpha s}$ is a new type of G protein. *Nature,* **372**, 804.

118. Iiri, T., Herzmark, P., Nakamoto, J. M., Van Dop, C., and Bourne, H. R. (1994) Rapid GDP release from $G_{s\alpha}$ in patients with gain and loss of endocrine function. *Nature,* **371**, 164.

119. Clapham, D. E. (1994) Why testicles are cool. *Nature,* **371**, 109.

120. Vallar, L., Spada, A., and Giannattasio, G. (1987) Altered G_s and adenylate cyclase activity in human GH-secreting pituitary adenomas. *Nature,* **330**, 566.

121. Landis, C. A., Masters, S. B., Spada, A., Pace, A. M., Bourne, H. R., and Vallar, L. (1989) GTPase inhibiting mutations activate the α chain of G_s and stimulate adenylyl cyclase in human pituitary tumors. *Nature,* **340**, 692.

122. Russo, D., Arturi, F., Wicker, R., Chazenbalk, G. D., Schlumberger, M., DuVillard, J. A., Caillou, B., Monier, R., Rapoport, B., and Filetti, S. (1995) Genetic alterations in thyroid hyperfunctioning adenomas. *J. Clin. Endocrinol. Metab.,* **80**, 1347.

123. Dryja, T. P., Hahn, L. B., Reboul, T., and Arnaud, B. (1996) Missense mutation in the gene encoding the alpha subunit of rod transducin in the Nougaret form of congenital stationary night blindness. *Nature Genet.,* **13**, 358.

124. Graziano, M. P. and Gilman, A. G. (1989) Synthesis in *Escherichia coli* of GTPase-deficient mutants of $G_{s\alpha}$. *J. Biol. Chem.,* **264**, 15475.

125. Gregory-Evans, K. and Bhattacharya, S. S. (1998) Genetic blindness. Current concepts in the pathogenesis of human outer retinal dystrophies. *Trends Genet.,* **14**, 103.

126. Wilkie, T. M., Gilbert, D. J., Olsen, A. S., Chen, X. N., Amatruda, T. T., Korenberg, J. R., Trask, B. J., Dejong, P., Reed, R. R., Simon, M. I., Jenkins, N. A., and Copeland, N. G. (1992) Evolution of the mammalian G protein α subunit multigene family. *Nature Genet.,* **1**, 85.

127. Hepler, J. R. and Gilman, A. G. (1992) G Proteins. *Trends Biochem. Sci.,* **17**, 383.

128. Ryba, N. J. P. and Tirindelli, R. (1995) A novel GTP-binding protein gamma-subunit, G-gamma-8, is expressed during neurogenesis in the olfactory and vomeronasal neuroepithelia. *J. Biol. Chem.,* **270**, 6757.

129. Ong, O. C., Yamane, H. K., Phan, K. B., Fong, H. K. W., Bok, D., Lee, R. H., and Fung, B. K. K. (1995) Molecular cloning and characterization of the G protein gamma subunit of cone photoreceptors. *J. Biol. Chem.,* **270**, 8495.

130. Kahn, R. A. (1991) Fluoride is not an activator of the smaller (20–25 kDa) GTP-binding proteins. *J. Biol. Chem.,* **266**, 15595.

131. Mittal, R., Zhmadian, M. R., Goody, R. S., and Wittinghofer, A. (1996) Formation of a transition-state analog of the ras GTPase reaction by ras-GDP, tetrafluoroaluminate, and GTPase-activating proteins. *Science,* **273**, 115.

2 | G proteins II: G_q, G_{12}, and G_z

JENNIFER L. GLICK, THOMAS E. MEIGS, AND PATRICK J. CASEY

1. Introduction

The ability of the cell to sense and communicate with its environment is critical for its survival, and therefore for the survival of the organism. Simpler life forms such as *Dictyostelium* sense toxic or unfavourable environments and rely on the process of chemotaxis to steer away from these areas and toward areas where conditions are more suitable. During the development of multicellular organisms, progenitor cells must migrate appropriately and coordinate their differentiation with one another in order to produce a viable embryo. Adult organisms rely on sensory cells such as photoreceptors, the olfactory epithelium, and taste buds, as well as a central nervous system, to guide them out of danger and track down food or potential mating partners. Further still, in multicellular organisms, cells must be able to respond to complex signals that originate from many regions of the body. A classic example of this is the well-known 'fight or flight' response, wherein an animal senses imminent danger and the body immediately prepares itself for the necessary action (for instance, with a faster heartbeat to increase blood flow to its musculature). Many of the processes involved in this type of response (vision, memory, smell, taste, reflex) are in large part regulated by heterotrimeric G proteins (G proteins).

G proteins are directly responsible for the transmission of many extracellular signals to the intracellular environment (1–3). This is accomplished through the binding of molecules that elicit these signals (e.g. neurotransmitters, peptide hormones, chemicals such as cAMP, or odorants) to a class of cell-surface receptors, the G protein-coupled receptors. These receptors span the plasma membrane and, in response to ligand binding, activate their associated G protein that resides on the inner leaflet of the plasma membrane, thereby passing the signal from the extra-cellular milieu to the cytoplasm of the cell (3, 4). The receptor-activated G protein then interacts with downstream 'effector' proteins that generally are capable of modulating the cellular level of specific signalling compounds termed the second messengers; these second messengers trigger a cascade of events that ultimately enables the cell to respond appropriately to the incoming signal (Fig. 1).

This chapter is a continuation of the discussion of G proteins begun in Chapter 1.

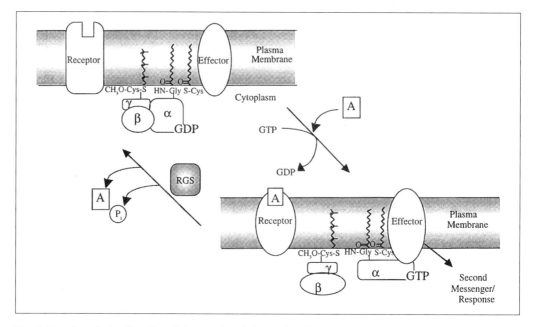

Fig. 1 Overview of signalling through heterotrimeric G proteins. Binding of a specific agonist (A) to its receptor results in activation of the associated G protein by binding of GTP. G protein subunit dissociation is a consequence of this activation, with the result that both the α-GTP and $\beta\gamma$ complexes are now available for regulation of appropriate effector molecules ($\beta\gamma$ regulation is not shown). The effector in general alters the level of a second messenger molecule, leading to the cellular response to the agonist. The signal is terminated by hydrolysis of GTP by the α subunit; in some cases this hydrolysis can be accelerated by association of the α subunit with a member of the regulators of G protein signalling (RGS) family of proteins. Following GTP hydrolysis, the α-GDP and $\beta\gamma$ complexes reunite and the system returns to the basal state to await another incoming signal (see text for details).

Since the preceding chapter provides a thorough introduction to G proteins and the mechanisms by which they transduce signals, the reader is referred there for more general information on G proteins. The previous chapter also describes the biochemical and genetic characterization of the 'classical' (i.e. G_i and G_s) subfamilies of G proteins. These G proteins are classical in the sense that they were the first to be discovered and they are the most widely characterized. However, the advent of modern cloning techniques has revealed a second 'generation' of G proteins, namely the G_q subfamily (consisting of $G_{q\alpha}$, $G_{11\alpha}$, $G_{14\alpha}$, $G_{15\alpha}$, and $G_{16\alpha}$), the G_{12} subfamily (consisting of $G_{12\alpha}$ and $G_{13\alpha}$) and $G_{z\alpha}$, which is classified as a G_i subfamily member (Table 1) (5–8). Interestingly, unlike the classical G proteins, these newer G proteins were generally found to be resistant to ADP-ribosylation, a type of modification catalysed by certain bacterial toxins, namely pertussis toxin, which modifies and inactivates the G_i proteins, and cholera toxin, which modifies and activates $G_{s\alpha}$ (see Chapter 10). The identification of these new α subunits led to the subsequent discovery of a set of novel signalling pathways that are governed by G proteins and are resistant to the two bacterial toxins. Unlike the G protein α subunits that were

Table 1 The G-protein α subunit families: the known G protein α subunits as defined by their sequence identity

Family	Member	Effector	Distribution
G_s	$G_{s\alpha}$	↑ AC	Ubiquitous
	G_{olf}	↑ AC	Olfactory epithelia
G_i	$G_{t\alpha}$	↑ cGMP-PDE	Retina
	G_{gust}	↑ cGMP-PDE	Taste buds
	$G_{i1\alpha}$	↓ AC, others?	Nearly ubiquitous
	$G_{i2\alpha}$	↓ AC, others?	Ubiquitous
	$G_{i3\alpha}$	↓ AC, others?	Nearly ubiquitous
	$G_{o\alpha}$	↓ Ca²⁺ channels?	Brain
	$G_{z\alpha}$	↓ AC?, others?	Brain, retina, platelet, adrenal medulla
G_q	$G_{q\alpha}$	↑ PLC-β, ↑ Btk?	Ubiquitous
	$G_{11\alpha}$	↑ PLC-β	Ubiquitous
	$G_{14\alpha}$	↑ PLC-β	Ubiquitous
	$G_{15\alpha}$	↑ PLC-β	Ubiquitous
	$G_{16\alpha}$	↑ PLC-β	Ubiquitous
G_{12}	$G_{12\alpha}$	(RhoGEF, Btk, RasGAP1^M?)	Ubiquitous
	$G_{13\alpha}$	RhoGEF	Ubiquitous

AC, adenylyl cyclase; cGMP-PDE, guanosine 3',5'-cyclic monophosphate phosphodiesterase; PLC-β, phospholipase C-β; Btk, Bruton's tyrosine kinase; RhoGEF, Rho guanine nucleotide exchange factor; RasGAP1^M, a specific GTPase-activating protein for Ras.

identified via traditional techniques of protein purification and reconstitution, most of the so-called 'second generation' of α subunits were identified by DNA cloning and were originally 'orphans' for which a function was not immediately known. The task of assigning specific receptors and $\beta\gamma$ subunits, as well as specific cellular functions, to novel α subunits has proven to be quite challenging in many cases. Nevertheless, a great deal has been learned about these novel α subunits over the years, and this chapter attempts a summary of the current state of characterization of the G proteins that are not subject to ADP-ribosylation, namely the G_q and G_{12} subfamily members as well as $G_{z\alpha}$. While these α subunits are also generally insensitive to the effects of cholera toxin, for the purposes of this chapter they will be referred to collectively as pertussis toxin (PTX)-resistant G proteins (7).

2. Covalent modifications of PTX-resistant G proteins

2.1 Lipid modifications of G proteins

One type of covalent modification that occurs on G proteins is addition of lipids (Figs 1 and 2, see also Table 2). All α subunits are myristoylated or palmitoylated, or both, at or near their amino termini, and γ subunits are prenylated on their C termini (9, 10). Myristoylation is a cotranslational addition of the 14-carbon myristoyl group onto the amino-terminal glycine of all members of the G_i subfamily via an amide linkage that persists for the lifetime of the protein (11). Myristoylation is critical for

Lipid	Structure	Subunits	Linkage	Position of Modification
N-Myristoyl	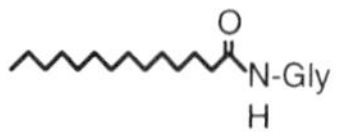 N-Gly	G_{i2}, G_{o2}, $G_{z\alpha}$, $G_{t\alpha}$	Amide	Amino-terminal Gly
S-Palmitoyl (S-acyl)	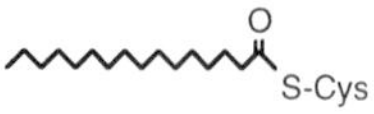 S-Cys	All except $G_{t\alpha}$	Thioester	Internal, no defined consensus
S-Prenyl	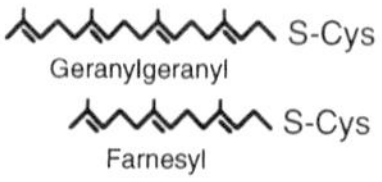 S-Cys (Geranylgeranyl) S-Cys (Farnesyl)	G_γ	Thioether	Carboxy-terminal Cys of CaaX motif

Fig. 2 Lipid modifications found on G protein subunits. The structures of the types of lipids found on G protein subunits are shown, along with the chemical type of linkage to the protein and (where known) the position of the protein that is modified (see text for details).

membrane association of these α subunits, as well as for association with the βγ subunit (12–14), and may play a direct role in regulating effectors (15). Palmitoylation (or, more correctly, S-acylation) occurs on all α subunits with the exception of transducin ($G_{t\alpha}$), and involves the post-translational attachment of the 16-carbon palmitoyl or similar acyl chain through a thioester linkage to cysteine residue(s) near

Table 2 Modifications of G protein α subunits: the known modifications of α subunits that play a role in regulating their function

α Subunit	N-terminal sequence	Lipidation	C-terminal sequence	Toxin sensitivity	Kinase substrate
G_s	MG**C**LGNSKTE	Palm-	RMHLPQYELL	CTX	EGFR; Src?
G_t	MGAGASAEEK	Myr- and variants	KENLKD**C**GLF	CTX, PTX	PKC?
G_{i1}	MG**C**TLSAEDK	Myr-, Palm-	KNNLKD**C**GLF	CTX, PTX	None
G_{i2}	MG**C**TLSAEDK	Myr-, Palm-	KNNLKD**C**GLF	CTX, PTX	PKC? Src?
G_{i3}	MG**C**TLSAEDK	Myr-, Palm-	KNNLKE**C**GLF	CTX, PTX	None
G_o	MG**C**TLSAEER	Myr-, Palm-	ANNLRG**C**GLY	PTX	None
G_z	MG**C**RQSSEEK	Myr-, Palm-	QNNLKYIGLC	None	PKC
G_q	MTLDSIMA**CC**	Palm-	QLNLKEYNLV	None	Src
G_{11}	MTLESIMA**CC**	Palm-	QLNLKEYNLV	None	None
$G_{15/16}$	MARSLTWG**CC**	Palm-	ARYLDREINLL	None	None
G_{12}	MSGVVRTLSR**C**LL	Palm-	QENLKDIMLQ	None	PKC
G_{13}	MADFLPSRSV**C**VL	Palm-	HDNLKQLMLQ	None	None

Under 'N-terminal sequence', glycine (G) residues modified by myristoylation and cysteine (C) residues subject to palmitoylation are in bold.
Under 'C-terminal sequence', cysteine (C) residues modified by pertussis toxin-catalysed ADP-ribosylation are in bold.
Myr-, myristoylated; Palm-, palmitoylated; CTX, cholera toxin substrate; PTX, pertussis toxin substrate; PKC, protein kinase C; EGFR, epidermal growth factor receptor.

the N terminus (16–18). Palmitoylation is a dynamic modification, which suggests that it may be a mode of regulation. However, efforts to understand the mechanism of palmitoylation of α subunits have been hampered because there does not appear to be a consensus sequence for the addition of a palmitate, and, while protein palmitoyltransferases have been identified, the enzyme remains to be cloned and fully characterized (19, 20). Like myristoylation, the isoprenoid modification of γ subunits is a stable modification and is responsible for the membrane association of the βγ complex (9, 10). Further information on the role of acylation events, specifically in G_i subfamily signalling, is provided in Chapter 1; the following sections summarize the currently available data on acylation of PTX-resistant G_α proteins and on prenylation of the βγ complex.

2.1.1 Lipid modifications of G_q proteins

While all the members of the G_q subfamily are subject to palmitoylation, the role of this lipidation in membrane localization and signalling for this subfamily has been investigated almost exclusively using $G_{q\alpha}$. These studies involved mutating the two cysteine residues (Cys9 and Cys10, see Table 2) in the amino-terminus of $G_{q\alpha}$ that can be palmitoylated, and investigating the localization and function of the mutated proteins. In all cases, mutation of both Cys9 and Cys10 to serine or alanine resulted in complete loss of palmitoylation of the protein (21–23). However, the data regarding the effects of these mutations on G_q-mediated signalling and membrane localization have been surprisingly contradictory. In one study, expression of the C9S/C10S double-mutant form of $G_{q\alpha}$ in HEK293 cells resulted in a protein that was completely cytosolic and unable to mediate α_2-adrenergic stimulation of PLC-β (21). In another study, expression of the double mutant C9A/C10A in COS-7 cells resulted in a protein that was primarily particulate (23), although this double mutant was similarly unable to mediate receptor-stimulated activation of PLC-β in those cells (23). A third study also demonstrated that the C9A/C10A mutant of $G_{q\alpha}$, when expressed in HEK293 or Sf9 cells, was primarily particulate (22). However, unlike the other two studies, this and other mutant forms of $G_{q\alpha}$ were purified and reconstituted into vesicle systems to investigate receptor-mediated stimulation of PLC-β. Although the double cysteine mutant was a poor regulator of PLC-β, a truncated form of $G_{q\alpha}$ yielded the surprising finding that, while removal of the first six amino acids abolished palmitoylation, this non-palmitoylated form of $G_{q\alpha}$ was able to stimulate PLC-β normally. These results indicate that the palmitoylation status of $G_{q\alpha}$ may not be important for signalling to PLC-β *per se*, rather it is the cysteine residues themselves that are important (22).

2.1.2 Lipid modifications of G_{12} proteins

G_{12} subfamily proteins are palmitoylated (17), but, like $G_{q\alpha}$, lack the N-terminal myristoylation site found in the G_i subfamily. Palmitoylation of G_{12} proteins appears to be important for certain G_{12}-mediated responses. Mutation of the putative site of palmitoylation, Cys11, inhibited a constitutively active variant of $G_{12\alpha}$ from eliciting its characteristic transformation of NIH3T3 cells (24). Interestingly, if the palmitoylation-deficient $G_{12\alpha}$ was further altered by creating an artificial myristoylation site near the N terminus, the transforming phenotype was completely restored, indicating that myristoylation could functionally replace palmitoylation in $G_{12\alpha}$. In another study, $G_{13\alpha}$ was found to be further modified by the attachment of arachidonate in a thioester linkage (25). This result reinforces the notion that many types of fatty acyl chains can modify proteins on Cys residues, although the functional significance of this heterogeneity in acylation is unknown.

2.1.3 Lipid modifications of $G_{z\alpha}$

$G_{z\alpha}$, like the other members of the G_i subfamily, is subject to both myristoylation and palmitoylation (12, 18). Both the myristoylation and palmitoylation sites of $G_{z\alpha}$ have been mutated and the effects of those mutations on membrane association and signalling abilities of $G_{z\alpha}$ have been investigated. Mutation of the palmitoylation site (Cys3) to an alanine had no effect on the ability of $G_{z\alpha}$ to be myristoylated; however, mutation of the myristoylation site (Gly2) to an alanine abolished the attachment of palmitate to the protein (26). Both mutations resulted in mistargetting of $G_{z\alpha}$; the G2A mutation caused the protein to partially accumulate in intracellular membranes, and the C3A mutant enzyme was found to be cytosolic (26). While the partially mistargetted C3A mutant was able to signal, the G2A variant was not. Coexpression of βγ, however, restored the ability of the G2A (i.e. non-myristoylated) form of $G_{z\alpha}$ to be both correctly targetted to the plasma membrane and palmitoylated; this co-expression of βγ also restored the signalling capacity of the protein. Together, these observations indicate that both myristoylation and βγ play an important role in proper membrane localization, palmitoylation, and signalling of $G_{z\alpha}$ (26).

2.1.4 Lipid modifications of $G_{\beta\gamma}$

Lipid modification of βγ subunits also plays an important role in the localization and function of this subunit complex. G-protein γ subunits possess a C-terminal CaaX motif—where 'C' is a cysteine residue, 'a' is generally an aliphatic residue, and 'X' is a methionine, leucine, or serine—that directs the covalent addition of a 15-carbon farnesyl (when X = Met, Ser) or 20-carbon geranylgeranyl group (when X = Leu) to the cysteine residue through a thioether linkage (27). Following prenylation, the proteins are further processed by the clipping of the -aaX residues and methylation of the C-terminal carboxylate (28, 29). The retinal γ subunits are farnesylated, while the non-retinal γ subunits (except γ_{11}, which is farnesylated) are geranylgeranylated (30–32). The influences of this prenyl modification on $G_{\beta\gamma}$ are twofold. First, prenylation of γ is required for proper membrane association, as mutation of the cysteine in the CaaX motif results in a $G_{\beta\gamma}$ complex that is found in the cytosol (33, 34). In addition, non-prenylated $G_{\beta\gamma}$ complexes are unable to regulate effector molecules (35–37) and show markedly reduced ability to associate with G_α (35, 38). Interestingly, the presence of a prenyl group is not required for assembly of the βγ complex itself (33, 34, 38).

The role of proteolysis and methylation steps in γ subunit processing in $G_{\beta\gamma}$ signalling is less clear; however, the final step of carboxymethylation is the only one

of the three processing steps that is reversible. The reversible nature of this modification provides a potential regulatory step for $G_{\beta\gamma}$ function. Initial studies have
indicated that carboxymethylation of $\beta\gamma$ does influence the properties of the subunit
complex, but the effects reported have been quite minimal. The isoprenylated, non-
methylated form of $G_{t\beta\gamma}$ was stably associated with membranes and less effective
than the fully processed form in supporting receptor-mediated GTP exchange on $G_{t\alpha}$,
indicating that carboxymethylation of $\beta\gamma$ is important for membrane association and
interaction with α (39, 40). In a similar type of study, the non-methylated $G_{\beta\gamma}$ was
much less active toward its downstream effectors PLC-β and PI3 kinase, again
indicating that carboxymethylation is important for the ability of $\beta\gamma$ to interact with
its signalling partners (41). Whether carboxymethylation is regulated in the cell,
however, is still under debate.

2.2 Phosphorylation of G proteins

The most ubiquitous cellular mechanism for regulation of a protein by covalent
modification is through the phosphorylation of the protein by one or more of the
many protein kinases that exist in cells. G protein signalling pathways are no
exception, and phosphorylation of all of the major components (i.e. receptor, G
protein, effector) have been documented. In terms of the G proteins themselves, the
cellular kinases implicated include protein kinase C (PKC) and the non-receptor
tyrosine kinase, Src (Table 2). While the biological significance of many of these
phosphorylation events is not yet clear, phosphorylation of the various G protein
subunits appears to allow for an additional level of regulation and a potential
mechanism for crosstalk between G protein-coupled pathways and other pathways
mediated by protein kinases.

2.2.1 Phosphorylation by protein kinase C

Several α subunits serve as PKC substrates, the most prominent of which are $G_{z\alpha}$ and
$G_{12\alpha}$. $G_{z\alpha}$ is rapidly and stoichiometrically modified by PKC *in vitro*, as well as in
platelets that have been treated with PKC-activating agents such as thrombin
(42–44). The effect of phosphorylation of $G_{z\alpha}$ on its signalling properties has been
studied extensively. While phosphorylation has no effect on the intrinsic rates of
nucleotide exchange and GTP hydrolysis properties, or on the ability of $G_{z\alpha}$ to inhibit
adenylyl cyclase, the phosphorylated form of $G_{z\alpha}$ does have a markedly reduced
affinity for the $\beta\gamma$ subunit (44, 45). Additionally, $G_{z\alpha}$ that has been phosphorylated
by PKC is resistant to the GTPase-stimulating activity of RGS proteins (see Section
6.2) (46, 47). While these observations, so far limited to *in vitro* systems, suggest a role
for phosphorylation in some aspect of signalling through $G_{z\alpha}$, the cellular effects of
PKC-mediated phosphorylation on $G_{z\alpha}$ signalling remain unclear. As noted above,
PKC-mediated phosphorylation has also been observed for $G_{12\alpha}$ *in vitro* and in
phorbol ester-treated NIH3T3 cells in which $G_{12\alpha}$ was stably expressed. Like $G_{z\alpha}$,
phosphorylation of $G_{12\alpha}$ reduced the affinity of this α subunit for $\beta\gamma$ (45). In addition,

activation of platelets by thrombin or thromboxane A_2 resulted in phosphorylation of $G_{12\alpha}$ and $G_{13\alpha}$ (48).

The retinal G protein, $G_{t\alpha}$, is also phosphorylated *in vitro* by PKC (49). However, the stoichiometry of labelling was not reported, and only the βγ-free, GDP-bound form of $G_{t\alpha}$ (which presumably exists only for extremely short periods of time during nucleotide exchange) served as a substrate, making the physiological relevance of the phosphorylation difficult to interpret. Another G protein that has been reported to be a PKC substrate is $G_{i2\alpha}$ (50). This α subunit was found to be somewhat phosphorylated under basal conditions in both hepatocytes and neuroblastoma cells, and the level of phosphorylation increased upon treatment of these cells with agents that activate PKC (51–53). However, recombinant $G_{i2\alpha}$ is a poor PKC substrate, and phosphorylation of $G_{i2\alpha}$ appears to be primarily a hepatocyte-specific event (54, 55). Recently, $G_{q\alpha}$ and $G_{11\alpha}$ phosphorylation in response to the metabotrophic agonist carbachol was demonstrated. This phosphorylation event, presumably mediated through PKC, was shown to influence the activation of the G protein by the receptor, but not to affect the regulation of its effector, phospholipase Cβ (56).

PKC-mediated phosphorylation may also be important in the activities of specific βγ complexes. The γ subunit $γ_{12}$ can be phosphorylated by PKC *in vitro* and in cells that natively express $γ_{12}$ (57, 58); other γ subunits expressed in the cells were not substrates. The finding that a $α_{i1}β_1γ_{12}$ heterotrimer containing a phosphorylated γ subunit more efficiently coupled to the A1 adenosine receptor, and a free phosphorylated $β_1γ_{12}$ complex more effectively activated adenylyl cyclase (59), provided compelling evidence for a functional consequence of this γ subunit phosphorylation. Interestingly, phosphorylation of $γ_{12}$ had no effect on the ability of the $β_1γ_{12}$ subunit to activate phospholipase Cβ, suggesting that phosphorylation may shift the preference of the βγ subunit from one effector to another (59).

2.2.2 Phosphorylation by tyrosine kinases

G proteins have been shown to be substrates for other protein kinases, most notably tyrosine kinases. The EGF (epidermal growth factor) receptor can phosphorylate $G_{s\alpha}$ *in vitro* (60, 61); this phosphorylation event activated $G_{s\alpha}$, and has been proposed to be the mechanism by which activation of the EGF receptor increases cellular cAMP levels (62). $G_{q\alpha}$ is phosphorylated in Rat-1 cells transformed with the v-*src* oncogene (63), and the phosphorylation results in $G_{q\alpha}$ being more active toward PLC-β in *in vitro* assays.

2.2.3 Other phosphorylation events

In addition to α and γ, G protein β subunits have been demonstrated to be phosphorylated. Incubation of membranes from a variety of cell types with γ-[^{32}P]GTP or [^{35}S]GTPγS resulted in a phosphorylation or thiophosphorylation, respectively, of a histidine residue on the β subunit (64, 65). Dephosphorylation of β subsequently occurred with the transfer of the phosphate to a GDP molecule that was membrane associated, suggesting that the β subunit may participate in a phosphate transfer reaction. One obvious possibility for the potential phosphate recipient is the GDP-bound form of the α subunit, which would provide a novel mechanism for α subunit activation that is alternative to GTP/GDP exchange. However, recent studies have shown that the GDP bound to G_s and G_i proteins does not serve as the phosphate acceptor, thus leaving the identity of the acceptor a mystery at this time (66).

3. Signalling through G_q proteins

3.1 Regulation of phosphoinositide-specific phospholipase C activity

The effects of a wide variety of hormones and neurotransmitters, such as vasopressin, acetylcholine, and thromboxane A_2, are mediated through the phosphoinositide-specific phospholipase C (PI-PLC) class of enzymes (Fig. 3) (8). These enzymes catalyse the hydrolysis of phosphatidylinositol 4,5-bisphosphate (PIP_2) to produce inositol 1,4,5-triphosphate (IP_3) and diacylglycerol (DAG). Liberated IP_3 is able to bind to receptors at the endoplasmic/sarcoplasmic reticulum; these receptors form an IP_3-gated calcium channel. The binding of IP_3 results in channel opening, causing an

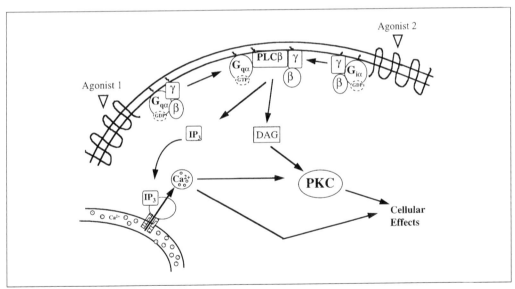

Fig. 3 Overview of G protein signalling through the phospholipase Cβ/IP_3/Ca^{2+} pathway. Shown is an overview of two distinct mechanisms for G protein control of PLC-β activity. In general, agonists that regulate via a PTX-insensitive process (designated Agonist 1; examples include those for α_1-adrenergic, m1 muscarinic, and H-1 histamine receptors) lead to activation of G_q proteins, and it is the GTP-liganded α subunit that directly stimulates PLC-β activity. Agonists that regulate via a PTX-sensitive process (designated Agonist 2; examples include those for α_2-adrenergic, m2 muscarinic, and A_1-adenosine receptors) lead to activation of G_i proteins, and in this case it is the released $\beta\gamma$ complex that stimulates PLC-β. In both cases, PLC-β action on a phosphoinositide substrate has the potential to produce two distinct second messengers, the inositol phosphate (IP_3) which triggers release of Ca^{2+} from internal stores, and diacylglycerol (DAG) which participates in the regulation of protein kinase C (PKC). See text for details.

increase in intracellular calcium levels. Calcium is a potent second messenger in cells; increases in Ca^{2+} lead to a variety of effects, including secretion, contraction, proliferation, and neuronal excitation (67). The second product of the PI-PLC reaction, diacylglycerol, potently activates protein kinase C, which is then able to phosphorylate a large number of cellular substrates and thereby modulate a wide variety of signalling pathways (68).

Members of the PLC-β subfamily of PI-PLCs (there are four, termed $\beta1$–$\beta4$) are regulated by α subunits of the G_q subfamily (69, 70). Receptors coupled to G_q are thus able to raise IP_3 levels in cells in a PTX-resistant fashion (Fig. 3). There is some controversy as to whether the individual members of the G_q subfamily differ in their ability to regulate the individual PLC-β isozymes. Some studies have suggested that PLC-β isozymes are activated equally well by the individual G_q subfamily members (71–73), while others suggest a difference in PLC-β susceptibility (70, 74). In addition to regulation of PLC-β, there is also a report that $G_{q\alpha}$ can activate a specific type of non-receptor tyrosine kinase termed Bruton's tyrosine kinase, or Btk (75). However, more recent data suggest that $G_{12\alpha}$ may be the more relevant activator of this molecule (see below).

One possible mechanism for selectivity in signalling through distinct members of the G_q subfamily α subunits would be an ability of specific receptors to discriminate between the various isoforms; indeed, there is evidence from transfection studies that certain receptors can do this (76–78). G_q subfamily members possess different expression patterns; $G_{q\alpha}$ and $G_{11\alpha}$ are ubiquitously expressed, while expression of $G_{14\alpha}$ is more restricted and $G_{15\alpha}$ and $G_{16\alpha}$ are expressed exclusively in haemato-poietic cells (8). Hence, it is possible that selectivity in the $G_{q\alpha}$/PLC-β-mediated pathways results from a combination of receptor selectivity and tissue-specific expression. It has also been reported recently that RGS proteins (see below) can play a role in receptor selectivity for G_q proteins (79).

Some agonists elicit PLC-β/Ca^{2+} responses that are PTX-sensitive, indicating that the signal is through a G_i-type protein rather than a G_q family member (Fig. 3). These responses are a result of the activation of PLC-β by G protein $\beta\gamma$ subunits liberated upon activation of G_i proteins (80, 81). $G_{\beta\gamma}$ subunits can activate all four PLC-β isoforms, with PLC-$\beta3$ and PLC-$\beta2$ being the most sensitive to $G_{\beta\gamma}$ stimulation (82, 83). However, $G_{\beta\gamma}$ subunits appear to be less effective at activating PLC-β than are the G_q subfamily α subunits (83).

3.2   Genetic studies on the G_q subfamily

Of all the PTX-resistant G proteins, the G_q subfamily has been the best characterized genetically. A role for the G_q proteins in cardiac function was initially suggested from studies with agonists such as angiotensin II and endothelin that bind G_q-coupled receptors and stimulate hypertrophy in cultured cardiac myocytes. A critical role for $G_{q\alpha}$ proteins in cardiac development and hypertrophy was confirmed via gene disruption and overexpression studies. Moderate overexpression of $G_{q\alpha}$ in mice resulted in cardiac hypertrophy, characterized by enlarged heart size and decreased contractility (84). Higher levels of $G_{q\alpha}$ expression resulted in massive cardiac dilation and congestive heart failure (84, 85). In addition to playing a role in cardiac performance in adults, $G_{q\alpha}$ and $G_{11\alpha}$ are also important in cardiac development. Mice lacking $G_{q\alpha}$, $G_{11\alpha}$, $G_{14\alpha}$, and $G_{15\alpha}$, as well as double knockout combinations, have been generated. Surprisingly, while mice lacking $G_{q\alpha}$ have high mortality rates and suffer from platelet aggregation defects, internal bleeding, ataxia, and other motor impairments (86, 87), heart tissue from these animals appears to be normal. Similarly, heart tissue from $G_{11\alpha}$ (the other G_q subfamily member that is expressed in the heart) knockouts appear normal. However, mice lacking both $G_{q\alpha}$ and $G_{11\alpha}$ die *in utero* and the embryos exhibit severe heart abnormalities (88). Mice with a minimum of one intact $G_{q\alpha}$ or $G_{11\alpha}$ allele survive to birth but die shortly afterward; by contrast, mice carrying two intact alleles of either gene survive to adulthood. These studies indicate that there is indeed redundancy between $G_{q\alpha}$ and $G_{11\alpha}$ in the developing heart (88). However, this redundancy is not likely to hold for all tissues, since mice lacking $G_{q\alpha}$ exhibit severe health defects while those lacking $G_{11\alpha}$ do not.

The gene disruption studies have also revealed an additional level of redundancy between the G_q subfamily members, that being in their ability to couple to receptors. When pancreatic and submandibular gland cells from $G_{11}^{-/-}$, $G_{11}^{-/-}$ $G_{14}^{-/-}$, $G_{14}^{-/-}$ $G_{15}^{-/-}$, and $G_q^{-/-}$ $G_{15}^{-/-}$ knockout mice were analysed for their ability to regulate Ca^{2+} in response to various stimuli, there was no difference between the wild-type mice and any of the knockout strains (89); Ca^{2+} influx in cells from wild-type and each of the above knockout strains was identical when the cells were stimulated with carbachol, epinephrine, bombesin, or cholecystokinin. These experiments indicate that the various G_q subfamily members are able to couple to any of these three receptors as needed (89).

4. Signalling through G_{12} proteins

4.1 Biological consequences of activating G_{12} proteins

Of the four classes of heterotrimeric G proteins, perhaps the least understood with regard to cell signalling is the G_{12} subfamily, which consists of two proteins, $G_{12\alpha}$ and $G_{13\alpha}$ (6). Genes encoding $G_{12\alpha}$ and $G_{13\alpha}$ were first identified by PCR-based approaches (90) and later the full mouse cDNAs for these proteins were isolated by conventional hybridization cloning (91). The expression of both subtypes is quite ubiquitous; both are observed in a wide variety of tissues (6). Recombinant $G_{12\alpha}$ and $G_{13\alpha}$ have been produced in Sf9 cells, and detailed biochemical studies of the recombinant proteins revealed that both have quite slow rates of guanine nucleotide exchange and GTP hydrolysis compared to other G proteins, with the exception of $G_{z\alpha}$ (see below) (92, 93).

Most of the information on signalling pathways mediated by G_{12} proteins has come from rather indirect experimental approaches, the most common strategy being transient expression of mutationally active variants of $G_{12\alpha}$ and $G_{13\alpha}$ (the Q229L mutant of $G_{12\alpha}$ and Q226L mutant of $G_{13\alpha}$). The general strategy is transfection of these 'QL' variants into different cell types followed by monitoring of a variety of phenotypic responses in the cells. Several such studies have implicated G_{12} subfamily proteins in signalling pathways controlled by a specific member of the MAP kinase family termed c-*Jun* N-terminal kinase (JNK, see Fig. 4). JNK is believed to play a role in cellular stress responses, differentiation, and apoptosis (94). JNK is rapidly and constitutively activated in NIH3T3 cells transfected with mutationally activated forms of either $G_{12\alpha}$ or $G_{13\alpha}$ (95). Additional studies have suggested that this activation of JNK is dependent upon functional Ras (95). JNK activation has also been observed in P19 mouse embryonal carcinoma cells transfected with activated $G_{12\alpha}$ or $G_{13\alpha}$; in addition, the expression of the activated G proteins induced differentiation of the cells (96). Furthermore, expression of antisense RNA to either $G_{12\alpha}$ or $G_{13\alpha}$ in P19 cells blocked stimulation of JNK activity by retinoic acid, a treatment that induces expression of G_{12} proteins (96). In terms of apoptotic responses, transfection of activated $G_{13\alpha}$ into CHO or COS-7 cells was found to significantly increase the number of cells undergoing apoptosis. Coexpression of a dominant-negative form of the monomeric GTPase Rho significantly decreased the number of apoptotic cells, suggesting that activated $G_{13\alpha}$ signalled through Rho proteins (97). A potential mechanism through which $G_{12\alpha}$ and $G_{13\alpha}$ signal to the apoptotic machinery was revealed by recent studies showing that the activity of a specific protein kinase, termed apoptosis signal-regulating kinase 1 (ASK1), was elevated in COS-7 cells expressing mutationally activated forms of the G proteins; in

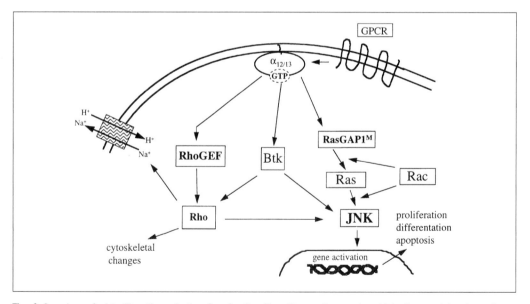

Fig. 4 Overview of signalling through the G_{12} family. Signalling pathways in which G_{12} proteins have been implicated are shown. However, it must be emphasized that many of the arrows are tentative and this picture is likely to change substantially in the coming years. GPCR, G-protein-coupled receptor; Btk, Bruton's tyrosine kinase; RhoGEF, a guanine nucleotide exchange factor for Rho proteins; RasGAP1^M, a specific form of GTPase-activating protein for Ras proteins; JNK, c-*Jun* N-terminal kinase. See text for details.

addition, expression of the proteins induced apoptosis in these cells through ASK1-dependent and MEKK1 (a MAP kinase knase) dependent pathways (98).

Another cellular response apparently mediated through G$_{12}$ proteins is Na$^+$/H$^+$ exchange across the plasma membrane (Fig. 4). Transfection of mutationally activated G$_{13\alpha}$ into HEK293 cells resulted in stimulation of the activity of the ubiquitous Na$^+$/H$^+$ exchanger NHE-1 (99). This effect was further investigated by utilizing phorbol ester treatment to deplete COS-1 cells of PKC activity. Under these conditions activated G$_{13\alpha}$, but not activated G$_{12\alpha}$, was able to stimulate NHE-1 activity when transfected into these cells, suggesting that G$_{12\alpha}$ and G$_{13\alpha}$ conduct signals to NHE-1 through distinct PKC-dependent and independent pathways, respectively (100). Further studies revealed that activated G$_{13\alpha}$ itself utilized two distinct pathways for stimulation of NHE-1: a Cdc42-dependent, MEKK1-dependent pathway and a Rho-dependent, MEKK1-independent pathway (101). Additionally, stimulation of NHE-1 by activated G$_{12\alpha}$ was shown to require both functional Ras and a phosphatidylcholine-specific phospholipase (102).

Other signalling pathways potentially mediated by G$_{12}$ proteins have been identified. Expression of activated G$_{12\alpha}$ or G$_{13\alpha}$ in HEK293 cells stimulated tyrosine phosphorylation of several focal adhesion assembly proteins, including p125 focal adhesion kinase, paxillin, and p130 Crk-associated substrate. These phosphorylation events were found to be dependent on functional Rho, and could also be blocked by treatment of the cells with cytochalasin D, an agent that specifically disrupts the actin cytoskeleton (103). Other cellular responses in which G$_{12}$ proteins have been implicated include stimulation of phospholipase A$_2$ (104) and signalling from angiotensin II type 1 receptors to phospholipase C-β (105).

From a number of studies, including several described above, it has become clear that G$_{12}$ subfamily signalling is linked somehow to the members of the Rho family of monomeric GTPases. This family includes Rho, Rac, and Cdc42, and all these proteins are known to participate in cellular processes that control cell morphology and shape changes (see Chapter 4) (106). Expression of mutationally activated forms of G$_{12\alpha}$ or G$_{13\alpha}$ in Swiss 3T3 cells resulted in formation of stress fibres and focal adhesion assemblies (107), processes known to be dependent on Rho activity (106). Furthermore, the response to expression of the activated G$_{12}$ proteins was inhibited by microinjection of the *Clostridium botulinum* C3 exoenzyme, which specifically inactivates Rho by catalysing the ADP-ribosylation of a key asparagine residue on the protein (see Chapter 10). In other studies, JNK activation elicited by expression of mutationally activated G$_{12\alpha}$ was blocked by coexpression of dominant negative Rac (108). Furthermore, coexpression of activated G$_{13\alpha}$ and phospholipase D in COS-7 cells led to activation of the phospholipase in a fashion dependent on Rho and Rac, but not Cdc42 (109). In another study that also provides evidence for distinct roles for the two G$_{12}$ family members, the formation of stress fibres and focal adhesion assemblies in Swiss 3T3 cells elicited by treatment with lysophosphatidic acid (LPA) was blocked in cells by microinjection of an antibody specific to G$_{13\alpha}$, while a G$_{12\alpha}$-specific antibody did not inhibit the LPA-induced responses (110).

A potential link between G$_{12}$ proteins and oncogenesis was first revealed by the isolation of a wild-type (i.e. unactivated) form of $G_{12\alpha}$ in an expression cloning strategy designed to identify potential oncogenes in a soft-tissue sarcoma cell line (111). However, significant occurrence of mutated or overexpressed G_{12} proteins in human tumours has not yet been shown. Mutationally activated $G_{12\alpha}$ and $G_{13\alpha}$ have been found to be potent transforming agents in many cells and animal tumorigenesis models. Expression of mutationally activated $G_{12\alpha}$ or $G_{13\alpha}$ in NIH3T3 or Rat-1 cells resulted in focus formation and growth in soft agar, two experimental hallmarks of a transformed phenotype in cultured cells (104, 112, 113). Additionally, injection of NIH3T3 cells expressing activated $G_{12\alpha}$ into nude mice led to the animals developing large tumours within 2 weeks (104, 112). In another report, wild-type $G_{12\alpha}$ and constitutively active Rac were found to co-operate in transformation of NIH3T3 cells (114), suggesting that monomeric GTPases in the Rho family are also important in this activity of G_{12} proteins.

Quite recently, several studies have appeared describing a direct functional interaction between G_{12} proteins and specific molecules that may function as downstream effector proteins (Fig. 4). One of these putative effectors is a previously identified guanine nucleotide exchange factor specific for Rho, termed p115RhoGEF (115). In the most recent studies, p115RhoGEF, through an RGS domain (see section 6.2), was found to form a complex with $G_{13\alpha}$; the interaction could only be detected if $G_{13\alpha}$ was in its active conformation (116). When p115RhoGEF, Rho, and activated $G_{13\alpha}$ were combined in an *in vitro* system, $G_{13\alpha}$ was found to increase the ability of p115RhoGEF to activate nucleotide exchange on Rho. Other G_α proteins, including $G_{12\alpha}$, were unable to produce the effect, although increasing levels of $G_{12\alpha}$ did inhibit the $G_{13\alpha}$-induced stimulation of p115RhoGEF activity (116). Furthermore, the RGS domain of p115GEF was found to function as a GAP for both $G_{13\alpha}$ and $G_{12\alpha}$ *in vitro*, although this effect was much more pronounced for $G_{13\alpha}$ (117). These studies have not only revealed an apparent direct molecular target of activated $G_{13\alpha}$, but also provide the first example of a GTPase-activating protein for the G_{12} subfamily. Together, these studies provide a molecular mechanism that may account for the ability of G_{12} proteins to signal to Rho-dependent cellular pathways.

Two additional molecules have recently been identified that exhibit the properties expected of true effectors of $G_{12\alpha}$ (Fig. 4). This discovery came from studies examining the specificity of the aforementioned activation of the Bruton's tyrosine kinase (Btk) by $G_{q\alpha}$. Examination of other G protein α subunits for their ability to activate Btk led to the surprising finding that activated $G_{12\alpha}$ was also quite efficient in this regard (118). Delineation of the region of Btk required for regulation by $G_{12\alpha}$ revealed a domain composed of a pleckstrin-homology domain and an adjacent region termed a Btk motif. Furthermore, this region was found to be conserved in a specific form of a Ras GAP termed GAP1^m and binding of $G_{12\alpha}$ to GAP1^m was found to stimulate the GAP activity of the molecule toward Ras (118). In addition to identifying a specific kinase whose activity is responsive to $G_{12\alpha}$, these studies revealed an additional direct link between heterotrimeric and monomeric GTPases, in this case through a GAP for a specific monomeric GTPase rather than through an exchange factor as in the case of the Rho GEF.

4.2 Genetic studies on the G$_{12}$ subfamily

The generation of mice lacking G$_{13\alpha}$ has provided several interesting insights into the role of this G protein (119). Crossing of mice heterozygous for the disrupted G$_{13\alpha}$ allele did not yield any offspring that were homozygous for disrupted G$_{13\alpha}$, indicating that the knockout has a recessive lethal phenotype. Embryos homozygous for G$_{13\alpha}$ disruption die at around embryonic day 10, and examination of the embryos just prior to this point in development revealed that they lacked a properly developed and functional vascular system. While these embryos showed normal differentiation of progenitor cells to form endothelial cells, the endothelial cells were apparently unable to undergo the growth, sprouting, and migration that are characteristic of angiogenesis. Studies using fibroblasts cultured from the G$_{13\alpha}$-deficient embryos revealed that the cells had lost the ability to migrate in response to thrombin. Given that thrombin receptor stimulation has been demonstrated to activate G$_{13\alpha}$ and G$_{12\alpha}$ in platelet membranes (120), this observation in G$_{13\alpha}$-deficient cells provides a possible physiological function for the coupling of the thrombin receptor and G$_{13\alpha}$.

Additional clues to potential functions of G$_{12}$ proteins come from studies of a G$_{12}$ homologue in *Drosophila*. This protein, termed Concertina, shares *c*. 55% identity in sequence with G$_{12\alpha}$ and G$_{13\alpha}$. *Drosophila* lacking functional Concertina are unable to proceed through gastrulation in the embryonic stage of development, apparently because of an inability to coordinate cell shape changes and form a ventral furrow (121). Further genetic studies have indicated that a specific guanine nucleotide exchange factor for the *Drosophila* Rho1 protein, termed DRhoGEF, is also required for gastrulation, and that the morphological defects seen in organisms lacking DRhoGEF strongly resemble those observed in embryos lacking Concertina (122). In addition, the gastrulation defects in embryos lacking DRhoGEF were similar to those previously observed in embryos lacking the ligand Fog (123), a protein that shares a degree of topological similarity with thrombin (124). Taken together, these findings suggest that a Fog/Concertina/DRhoGEF/Rho1 pathway in *Drosophila* parallels G$_{12}$-mediated pathways in mammalian cells.

5. Signalling through G$_{z\alpha}$

The molecular targets of G$_{z\alpha}$, like those of G$_{12\alpha}$ and G$_{13\alpha}$, are not clear. G$_{z\alpha}$ is a member of the G$_i$ subfamily of G proteins, and as such it shares high sequence homology with the other G$_i$ proteins. However, G$_{z\alpha}$ possesses several properties that distinguish it from the other G$_i$ subfamily members. These properties (see below), and the rather unique tissue distribution of G$_{z\alpha}$, suggest that this G protein plays a specialized role in cell signalling. While members of the G$_i$ subfamily are rather ubiquitously expressed, Northern blot and immunoblot analysis have revealed that G$_{z\alpha}$ is expressed almost exclusively in brain, retina, platelets, and adrenal medulla, although expression has been detected in other tissues and in several cell lines (125–130) (Table 3).

Table 3 Tissues and cell lines in which $G_{z\alpha}$ expression has been detected

Tissue/cell line	Method of detection	Reference
Brain	Immunoblot, Northern	125–128
Retina	Immunoblot, Northern	126, 128
Platelet	Immunoblot, Northern	129
Adrenal medulla	Immunoblot, Northern	126, 128
Spermatazoa	Immunoblot	174
Pituitary	Northern	175
Megakaryocytes	Northern	129
RBL-2H3	Northern	176
HEL	Northern	129
Dami	Northern	129
MEG-01	Immunoblot	172
SH-SY5Y	Immunoblot	173
GH3	Northern	177
GH4C1	Immunoblot	138

RBL-2H3, a basophilic leukaemia cell line; HEL, a human erthyrocyte leukaemia cell line; SH-SY5Y, neuroblastoma; GH3 and GH4, pituitary adenoma

Characterization of purified recombinant $G_{z\alpha}$ revealed several biochemical properties of the protein that were quite unusual for G protein α subunits. The intrinsic rate of nucleotide exchange for $G_{z\alpha}$ is slow (0.02 min^{-1}) and, perhaps importantly, this rate of exchange falls to nearly zero in the presence of high concentrations (>50 μM) of Mg^{2+} (127). This Mg^{2+} effect is not seen with other trimeric G proteins, although the Ras proteins show a similar Mg^{2+} sensitivity (131). In addition to being a slow exchanger, $G_{z\alpha}$ is also a slow GTPase compared to other α subunits. The unusually slow kinetics of hydrolysis by $G_{z\alpha}$ has prompted speculation that:

- this protein might participate in more long-lived cellular responses; and
- that there could exist a GTPase activating protein (GAP) for $G_{z\alpha}$, similar to the GAPs that exist for the Ras family of GTPases; such a GAP has indeed now been identified (see below).

Another characteristic unique to $G_{z\alpha}$ over the other G_i subfamily members is its insensitivity to modification by PTX. As described in Section 2, $G_{z\alpha}$ lacks the otherwise conserved cysteine near the C-terminus that serves as the site of PTX-catalysed ADP-ribosylation of the $G_{i\alpha}$ subunits. Also, as described in Section 2 , $G_{z\alpha}$ is an excellent substrate for PKC, another characteristic that sets it apart from the other G_i family members. A final characteristic of $G_{z\alpha}$ that sets it apart from other α subunits is its sensitivity to unsaturated, acidic lipids (132). $G_{z\alpha}$ is selectively inactivated by these fatty acids, with arachidonate being particularly potent. While the physiological relevance of this inactivation is not understood, the phenomenon may be quite relevant considering the high level of arachidonate that is produced upon stimulation of the cell types in which $G_{z\alpha}$ is expressed.

Table 4 Receptors linked to G_z in heterologous tissues: those listed have been shown to couple to $G_{z\alpha}$

Receptor	Technique	Cell line/tissue	Effect	Reference
μ-Opioid	Antisense oligo injection	Rat brain	↓ Analgesia	170
	Antibody injection	Rat brain	↓ Analgesia	170
	Transfection	HEK293	↓ cAMP; ↑ cAMP (βγ)	136
δ-Opioid	Transfection	HEK293	↓ cAMP; ↑ cAMP (βγ)	134
	Transfection	Ltk- fibroblasts	↑ IP formation (βγ)	134
κ-Opioid	Transfection	HEK293	↓ cAMP	135
5-HT	Co-infection	Sf9	n.d.	142
	Immunoprecipitation	Sf9	n.d.	141
Melatonin	Transfection	HEK293	↓ cAMP; ↑ cAMP (βγ)	137
fMLP	Transfection	HEK293	↑ cAMP (βγ)	139
α_2-Adrenergic	Transfection	HEK293	↓ cAMP	133
A_1-Adenosine	Transfection	HEK293	↓ cAMP	133
NKR-P1	Immunoprecipitation	NK cells	n.d.	171
D_2-Dopamine	Transfection	HEK293	↓ cAMP	133
D_5-Dopamine	Immunoprecipitation	GH4C1	n.d.	138

5-HT, 5-Hydroxytryptamine; fMLP, N-formylmethionyl-leucyl-phenylalanine, a chemotactic peptide; HEK, human embryonic kidney; NK, natural killer; Sf9, *Sodoptera frugiperda* (insect cell line); IP, inositol phosphate; cAMP, adenosine 3', 5'-cyclic monophosphate; ↓, decreased; ↑, increased; βγ, effect presumably mediated through release of βγ; n.d., not determined.

There are a few hints in the current literature as to the potential cellular role(s) of $G_{z\alpha}$. The available data come primarily from overexpression studies in heterologous cell lines that have been transfected with $G_{z\alpha}$, receptors, and adenylyl cyclase. It is difficult to assess the physiological relevance of these signalling events when many or all of the components involved have been overexpressed, especially in cell lines that do not natively express $G_{z\alpha}$. However, these data are at the very least a starting point from which we may derive some clues as to G_z signalling function. By far the most well-characterized system in which an activity for $G_{z\alpha}$ has been observed is the inhibition of adenylyl cyclase. Transfection of HEK293 cells with $G_{z\alpha}$ and a variety of receptors, including α_2-adrenergic, D_2-dopamine, A_1-adenosine, δ-opioid, μ-opioid, κ-opioid, and melatonin, resulted in an agonist-dependent, PTX-resistant inhibition of the endogenous adenylyl cyclase in the cells (133–137), indicating that $G_{z\alpha}$ can couple to receptors that normally signal through the G_i family of G proteins (in cells not expressing $G_{z\alpha}$, the responses are still seen but are PTX-sensitive) to regulate adenylyl cyclase (Table 4). Another study demonstrated coupling of $G_{z\alpha}$ to the human D_5-dopamine receptor when the receptor was expressed in a pituitary cell line, GH4C1 (138). Coupling was assessed by co-immunoprecipitation of the receptor with $G_{z\alpha}$ using an antibody against the C-terminal region of $G_{z\alpha}$. This study is of particular interest because the $G_{z\alpha}$ that immunoprecipitated with the D_5 receptor was the endogenous protein present in the cells, and also because the receptor did not co-immunoprecipitate with $G_{i1\alpha}$, $G_{i2\alpha}$, or $G_{i3\alpha}$ (138). Other studies have implicated $G_{z\alpha}$ in additional signalling pathways. For example, ectopically expressed

$G_{z\alpha}$ can couple to type II adenylyl cyclase, presumably through the release of $\beta\gamma$, in response to activation of fMLP and μ-opioid receptors (136, 139).

$G_{z\alpha}$ has also been reported to possess oncogenic capability (140). Transfection of a constitutively activated form of $G_{z\alpha}$ into Swiss 3T3 cells induced focus formation, growth in soft agar, and faster growth rates compared with control cells. A similar, but much less dramatic, effect was observed in NIH3T3 cells. Regulation of adenylyl cyclase was ruled out as a possible explanation in these studies, leaving the pathway that $G_{z\alpha}$ was affecting an open question (140).

Reconstitution of receptor/G protein interactions in Sf9 cells has also been a vehicle used to study $G_{z\alpha}$. In one such study, the 5-HT1$_A$ receptor and various G protein α subunits were expressed in Sf9 cells and agonist-dependent activation of the α subunits was then assessed by treating the cell membranes with agonist in the presence of [^{35}S]-GTPγS, followed by immunoprecipitation of the [^{35}S]-GTPγS-bound α subunit. Using this method, the 5-HT (5-hydroxytryptamine) receptor was shown to couple to G_z in an agonist-dependent fashion. This coupling was fairly selective, as the thrombin and neurokinin-1 (NK1) receptors did not couple to G_z nearly as efficiently, and the β-adrenergic receptor did not couple at all (141). Expression of $G_{z\alpha}$, but not $G_{q\alpha}$ or $G_{s\alpha}$, also increased agonist-binding affinity of the 5-HT receptor, confirming the ability of the receptor to couple to G_z (141, 142).

One major problem with using receptor-based studies as a starting point for elucidating G_z function is that many, if not all, of the receptors described in the preceding sections appear to couple to both PTX-sensitive and -resistant G proteins, particularly if the receptors are overexpressed. Nevertheless, the observation of PTX-resistant coupling of receptors to various responses, including Ca^{2+} release, PLC activation, inositol phosphate production, and K^+ channel activation have been observed regularly and may be indicative of G_z involvement in these processes; the most interesting of these will be summarized briefly below (see also Table 4). Perhaps importantly, many of the pathways described are influenced by PKC in so far as agents known to activate PKC alter the PTX-resistant responses observed. This is intriguing since phosphorylation of $G_{z\alpha}$ by PKC has the capacity to alter its activity *in vitro* (see Section 2.2.1), while there is no such evidence available for $G_{q\alpha}$, the other α subunit most often implicated in PTX-resistant phenomena.

One example of a PTX-resistant response that is modulated by PKC activation is that to purinergic agonists. In superior colliculus (SCC) neurons, a PTX-resistant increase in intracellular Ca^{2+} was observed in response to adenosine (143). The same group has also reported a PTX-resistant activation of K^+ currents by ATP in cerebral neurons (144). Importantly, both responses appeared to be PKC-dependent in that they were abolished by a PKC inhibitor. Another example of a PTX-resistant response that is potentially controlled by G_z, rather than by G_q, is the regulation of a K^+ current in SCC neurons in response to adenosine, where the signal was found to be PTX-resistant and PLC-independent (143).

A specific cell type in which it is widely believed that G_z controls important signalling events is the platelet, the only non-neuronal 'tissue' that expresses significant levels of $G_{z\alpha}$ (145). Platelet activating factor, vasopressin, thrombin, and thromboxane A_2 are all powerful activators of platelets. Two of these agonists, thrombin and thromboxane A_2, are of particular interest because treatment of platelets with these agonists results in the phosphorylation of $G_{z\alpha}$ by PKC (42). Like many of the neuronal responses, stimulation of platelets with these agonists results in the activation of PLC-β. However, treatment with thrombin and thromboxane A_2 activates many pathways, some of which are PTX-sensitive and some of which are PTX-resistant. It is now becoming clear that potentially these receptors can couple to many G proteins (e.g. G_i, G_q, G_{12}, as well as G_z), but the identity of the specific G protein controlling any particular response is still in question.

6. Regulation of signal termination

6.1 Effectors of G proteins as GAPs

In addition to the aforementioned GAP activity of p115 RhoGEF towards $G_{13\alpha}$ (see Section 4.1), there exist at least two other G protein-coupled systems in which the effector participates directly in the termination of signalling by acting as a GAP for the activated α subunit; these are cGMP-phosphodiesterase and $G_{t\alpha}$ (see Chapter 1), and PLC-β and $G_{q\alpha}$ (see Section 3.1). In both cases, the existence of a mechanism to accelerate deactivation was initially suggested by the kinetics of the system. G_q subfamily members possess rather slow rates of intrinsic GTP hydrolysis, too slow in fact to account for the rapidity with which receptor-mediated activation of phospholipase C through $G_{q\alpha}$ is terminated. The resolution to this apparent contradiction came from reconstitution studies in which GTPγS- and GTP-liganded forms of $G_{q\alpha}$ were directly compared in their abilities to activate PLC-β. While GTPγS-liganded $G_{q\alpha}$ strongly and persistently activated PLC-β, activation by the GTP-liganded subunit was transient and less efficacious, indicating that PLC-β itself was involved in the termination of $G_{q\alpha}$ signalling (146). Further studies provided evidence that the GTPase activity of $G_{q\alpha}$ is accelerated up to 300-fold in the presence of PLC-β (147), and that PLC-β is a GAP for the other G_q subfamily members as well (148).

6.2 RGS proteins

The regulators of G-protein signalling (RGS) proteins are the newest group of regulatory factors to be discovered for G protein α subunits. The prototype RGS protein is the product of the *SST2* gene in the budding yeast *Saccharomyces cerevisiae* (149, 150). This gene had been known to be involved in the down-regulation of pheromone-induced growth arrest. Genetic studies placed the SST2 gene product downstream from the receptor but upstream of the βγ complex, suggesting that it might somehow act at the step of subunit dissociation or directly on the G protein α subunit (150). Similar genes in other organisms were also identified that were involved in down-regulation of G protein-mediated processes, such as egg laying in *C. elegans*, and proliferation in *Aspergillus* (151–153). However, the biochemical action of SST2-like proteins (149) was not clarified until the identification and biochemical characterization of mammalian RGS proteins.

Mammalian RGS proteins were discovered by yeast two-hybrid and expression cloning techniques (154, 155). Production of mammalian RGS proteins in *Escherichia coli* expression systems yielded recombinant proteins that possessed GAP activity toward G protein α subunits, most notably G_i and G_q subfamily members (155–158). Analysis of the primary structures of these proteins has revealed that they all possess a so-called 'RGS core domain', consisting of a stretch of *c.* 120 amino acids that is well conserved and has become the defining sequence for an RGS protein (151). However, regions outside this core domain vary widely in their length and character among the more than 15 identified RGSs. Some RGS proteins consist of little more than the core domain, while others have extensive N- and C-terminal regions (152, 159).

Exactly how RGS proteins are targetted to membranes, where they presumably need to be in order to interact with G protein α subunits, is not yet clear and may vary for individual RGS family members. One RGS, termed RetRGS1, does contain a predicted transmembrane domain near its N terminus (160). Additionally, members of a subset of RGS proteins, i.e. GAIP, RetRGS1, and RGSZ1, contain a cysteine-rich 13-amino-acid domain, termed a cysteine string (46, 160, 161). While the role of the cysteine string in these proteins is not yet clear, proteins containing this motif tend to be heavily palmitoylated and membrane associated, suggesting that the purpose of the cysteine string is to provide a mechanism for membrane anchoring (162, 163). In fact, both palmitoylation and membrane association have been demonstrated for the GAIP protein (161).

6.2.1 RGS action on $G_{q\alpha}$

At least three RGS proteins, RGS2, RGS4, and GAIP, have been implicated in the regulation of G_q-mediated signalling. These RGSs possess GAP activity toward $G_{q\alpha}$ (157, 164), and they also block the activation of PLC-β by GTPγS-liganded $G_{q\alpha}$, indicating that the RGS proteins are able to regulate G_q-mediated signalling both by accelerating that G protein's GTPase rate and by physically blocking the interaction of $G_{q\alpha}$ with PLC-β (157, 165–167). RGS4 is tenfold more effective than GAIP as a G_q-GAP, and RGS2 was tenfold more effective than RGS4 in its ability to block PLC-β activation, indicating that RGS2 is likely to be the most effective of the RGSs at regulating G_q-mediated signalling (157, 167). However, side-by-side comparisons of the abilities of RGS2, RGS4, and GAIP to act as GAPs for G_q or to block PLC-β activation by the G protein have not been performed. A recent study has demonstrated that another level of RGS regulation of G_q signalling may exist. In this study, deletion of the N terminus of RGS4 (a region outside the core domain) resulted in a dramatic loss in its ability to block G_q-mediated signalling, as well as an alteration in receptor specificity (79). While full-length RGS4 preferentially inhibited G_q-mediated responses to the agonist carbachol over those to Cholecystokinin (CCK), the N-terminal truncation mutant of RGS4 was equally effective against both pathways. This study indicated that regions outside the RGS core domain of RGS4 are important in high-affinity, receptor-selective regulation of G_q responses (79). Whether such is the case for other RGS proteins remains to be elucidated.

6.2.2 RGS action on $G_{z\alpha}$

As mentioned in Section 5, $G_{z\alpha}$ possesses a quite slow GTPase rate compared to most other α subunits, initially suggesting that there may be GTPase activating proteins for $G_{z\alpha}$. The discovery of the RGS proteins revealed this to be the case, as several RGS proteins were found to possess GAP activity toward $G_{z\alpha}$. At least three RGS proteins, RGS10, RGS4, and RGSZ1, are known to act as $G_{z\alpha}$-GAPs (46, 47, 167, 168). However, there are major differences in the selectivities of these three RGSs. While RGS4 and RGS10 are roughly equal in their abilities to act as GAPs toward members of the $G_{i\alpha}$ subfamily, including $G_{z\alpha}$ (157, 169), RGSZ1 shows highly selective activity toward $G_{z\alpha}$ over the other $G_{i\alpha}$ subunits (46, 47). This finding is the first example of selectivity of an RGS toward individual members of a G_{α} subfamily, and it seems likely that such selectivity will be observed in other RGSs as the biochemical properties of additional family members become elucidated. In addition to its preference for $G_{z\alpha}$ as a molecular target, expression of RGSZ1 is limited to tissues of neuronal origin, with a pattern quite similar to that of $G_{z\alpha}$ (46, 47).

It has become apparent recently that the actions of RGS proteins on G_{α} can be influenced by covalent modifications of the α subunits. This was first observed when the influence of α-subunit palmitoylation on susceptibility to RGS action was examined; S-acylation of $G_{i\alpha}$ subfamily members, including $G_{z\alpha}$, was found to attenuate markedly the ability of an RGS to accelerate their rates of GTP hydrolysis (168). This finding was the first example of a functional consequence of this α-subunit modification on the properties of the protein. Quite recently, it was demonstrated that the aforementioned phosphorylation of $G_{z\alpha}$ by PKC also attenuates RGS action, specifically that of RGSZ1, on the protein (46, 47). These studies have highlighted the complex modes of regulation of G_z signalling in particular, and it seems likely that additional insights/surprises will emerge from the study of this intriguing G protein.

7. Concluding remarks

Heterotrimeric G proteins and the pathways that they regulate comprise ubiquitous, complex systems that are critical in many aspects of cell physiology. While a wealth of information has been obtained both on these proteins and the processes they control in the past 20 years, there remain some fundamental unanswered questions. One of the most important of these is that of specificity. While data obtained from the use of reconstituted systems and from overexpression studies show that G protein-coupled receptors communicate with a variety of G proteins, in native tissues the responses to a specific agonist are generally much more selective. Hence, it is likely that other, as yet unidentified, factors exist that play roles in specificity in G protein signalling pathways. Possibilities include other proteins that might associate with G protein subunits to influence interaction with regulators and/or targets, and cytoskeletal or other elements that might facilitate assembly of signalling complexes in specific cell locations.

A related question is whether there are any as yet undiscovered G protein subunits lurking in the genome. In this age of PCR cloning, it would seem that most of the G protein subunits have been uncovered. However, it is certainly possible that there are distant relatives to the current G proteins still unidentified, and that these could participate in novel forms of signalling. This point is further highlighted by data emerging from the near-complete *C. elegans* genome, which contains 20 genes encoding G$_\alpha$ homologues, including 17 that fall loosely within the G$_i$ group (178).

Finally, there remains a question of redundancy. In particular, why are there several subfamilies of G proteins containing multiple members that seem to interact with the same groups of receptors and effectors? While differences in the patterns of tissue expression can account for some of this apparent redundancy, it is not yet clear whether the closely related genes, and in some cases splice variants, in these sub-families play redundant physiological roles. The increasing feasibility of genetic studies such as gene disruptions in mice knockouts are beginning to provide genuine physiological data showing that closely related isoforms do in fact possess distinct biological functions (see Section 3.2 for an example). The expansion of such studies, along with the development of additional technologies to study these pathways in systems that closely approximate the native cellular environment, will undoubtedly reveal new layers of complexity and new roles in physiology for this intriguing family of proteins.

References

1. Gilman, A. G. (1987) G proteins: transducers of receptor-generated signals. *Annu. Rev. Biochem.*, **56**, 615.
2. Neer, E. J. (1995) Heterotrimeric G proteins: Organizers of transmembrane signaling. *Cell*, **80**, 249.
3. Bourne, H. R. (1997) How receptors talk to trimeric G proteins. *Curr. Opinion Cell. Biol.*, **9**, 134.
4. Lefkowitz, R. J. (1996) G protein-coupled receptors and receptor kinases: from molecular biology to potential therapeutic applications. *Nature Biotechnol.*, **14**, 283.
5. Offermanns, S. and Schultz, G. (1994) What are the functions of the pertussis toxin-insensitive G proteins G12, G13, and Gz? *Mol. Cell. Endocrinol.*, **100**, 71.
6. Dhanasekaran, N. and Dermott, J. M. (1996) Signaling by the G12 class of G proteins. *Cell. Signal.*, **8**, 235.
7. Fields, T. A. and Casey, P. J. (1997) Signaling functions and biochemical properties of pertussis toxin-resistant G proteins. *Biochem. J.*, **321**, 561.
8. Exton, J. H. (1996) Regulation of phosphoinositide phospholipases by hormones, neurotransmitters, and other agonists linked to G proteins. *Annu. Rev. Pharmacol. Toxicol.*, **36**, 481.
9. Wedegaertner, P. B., Wilson, P. B., and Bourne, H. R. (1995) Lipid modifications of trimeric G proteins. *J. Biol. Chem.*, **270**, 503.
10. Casey, P. J. (1994) Lipid modifications of G proteins. *Curr. Opinion Cell. Biol.*, **6**, 219.

11. Wilcox, C., Hu, J.-S., and Olson, E. N. (1987) Acylation of proteins with myristic acid occurs cotranslationally. *Science*, **238**, 1275.

12. Mumby, S. M., Heukeroth, R. O., Gordon, J. I., and Gilman, A. G. (1990) G protein alpha subunit expression, myristoylation, and membrane association in COS cells. *Proc. Natl Acad. Sci. USA*, **87**, 728.

13. Jones, T. L. Z., Simonds, W. F., Merendino, J. J., Brann, M. R., and Spiegel, A. M. (1990) Myristoylation of an inhibitory GTP-binding protein alpha subunit is essential for its membrane attachment. *Proc. Natl Acad. Sci. USA*, **87**, 568.

14. Linder, M. E., Pang, I-H., Duronio, R. J., Gordon, J. I., Sternweis, P. C., and Gilman, A. G. (1991) Lipid modifications of G protein subunits: myristoylation of Go-alpha increases its affinity for beta gamma. *J. Biol. Chem.*, **266**, 4654.

15. Gallego, C., Gupta, S. K., Winitz, S., Eisfelder, B. J., and Johnson, G. L. (1992) Myristoylation of the G alpha i2 polypeptide, a G protein alpha subunit, is required for its signaling and transformation functions. *Proc. Natl Acad. Sci. USA*, **89**, 9695.

16. Parenti, M., Vigano, M. A., Newman, C. M. H., Milligan, G., and Magee, A. I. (1993) A novel N-terminal motif for palmitoylation of G-protein alpha subunits. *Biochem J.*, **291**, 349.

17. Veit, M., Nurnberg, B., Spicher, K., Harteneck, C., Ponimaskin, E., Schultz, G., and Schmidt, M. F. G. (1994) The alpha-subunits of G-proteins G12 and G13 are palmitoylated, but not amidically myristoylated. *FEBS Lett.*, **339**, 160.

18. Linder, M. E., Middleton, P., Hepler, J. R., Taussig, R., Gilman, A. G., and Mumby, S. M. (1993) Lipid modifications of G proteins: alpha subunits are palmitoylated. *Proc. Natl Acad. Sci. USA*, **90**, 3675.

19. Liu, L., Dudler, T., and Gelb, M. H. (1996) Purification of a protein palmitoyltransferase that acts on H-Ras protein and on a C-terminal N-Ras peptide. *J. Biol. Chem.*, **271**, 23269.

20. Dunphy, J. T., Greentree, W. K., Manahan, C. L., and Linder, M. E. (1996) G-protein palmitoyltransferase activity is enriched in plasma membranes. *J. Biol. Chem.*, **271**, 7154.

21. Wedegaertner, P. B., Chu, D. H., Wilson, P. T., Levis, M. J., and Bourne, H. R. (1993) Palmitoylation is required for signaling functions and membrane attachment of Gqalpha and Gsalpha. *J. Biol. Chem.*, **268**, 25001.

22. Hepler, J. R., Biddlecome, G. H., Kleuss, C., Camp, L. A., Hofmann, S. L., Ross, E. M., and Gilman, A. G. (1996) Functional importance of the amino terminus of Gq alpha. *J. Biol. Chem.*, **271**, 496.

23. Edgerton, M. D., Chabert, C., Chollet, A., and Arkintall, S. (1994) Palmitoylation but not the extreme amino-terminus of Gq alpha is required for coupling to the NK2 receptor. *FEBS Lett.*, **354**, 195.

24. Jones, T. L. and Gutkind, J. S. (1998) Galpha12 requires acylation for its transforming activity. *Biochemistry*, **37**, 3196.

25. Hallak, H., Muzbek, L., Laposata, M., Belmonte, E., Brass, L. F., and Manning, D. R. (1994) Covalent binding of arachidonate to G protein alpha subunits of human platelets. *J. Biol. Chem.*, **269**, 4713.

26. Morales, J., Fishburn, C. S., Wilson, P. T., and Bourne, H. R. (1998) Plasma membrane localization of Galpha-z requires two signals. *Mol. Biol. Cell*, **9**, 1.

27. Zhang, F. L. and Casey, P. J. (1996) Protein prenylation: Molecular mechanisms and functional consequences. *Annu. Rev. Biochem.*, **65**, 241.

28. Casey, P. J., Thissen, J. A., Higgins, J. B., Zhang, F., and Moomaw, J. F. (1994) Prenylation and G protein signalling. *Recent Progress Hormone Res.*, **49**, 215.

29. Casey, P. J. and Seabra, M. C. (1996) Protein prenyltransferases. *J. Biol. Chem.*, **271**, 5289.

30. Mumby, S. M., Casey, P. J., Gilman, A. G., Gutowski, S., and Sternweis, P. C. (1990) G protein gamma subunits contain a 20-carbon isoprenoid. *Proc. Natl Acad. Sci. USA*, **87**, 5873.

31. Fukada, Y., Takao, T., Ohguro, H., Yoshizawa, T., Akino, T., and Shimonishi, Y. (1990) Farnesylated gamma subunit of photoreceptor G protein indispensable for GTP-binding. *Nature*, **346**, 658.

32. Ray, K., Kunsch, C., Bonner, L. M., and Robishaw, J. D. (1995) Isolation of cDNA clones encoding eight different human G protein gamma subunits, including three novel forms designated the gamma4, gamma10, and gamma11 subunits. *J. Biol. Chem.*, **270**, 21765.

33. Muntz, K H., Sternweis, P. C., Gilman, A. G., and Mumby, S. M. (1992) Influence of gamma subunit prenylation on association of guanine nucleotide-binding regulatory proteins with membranes. *Mol. Biol. Cell*, **3**, 49.

34. Simonds, W. F., Butrynski, J. E., Gautam, N., Unson, C. G., and Spiegel, A. M. (1991) G protein beta-gamma dimers: Membrane targeting requires subunit coexpression and intact gamma CAAX domain. *J. Biol. Chem.*, **266**, 5363.

35. Iniguez-Lluhi, J. A., Simon, M. I., Robishaw, J. D., and Gilman, A. G. (1992) G protein beta gamma subunits synthesized in Sf9 cells: Functional characterization and the significance of prenylation of gamma. *J. Biol. Chem.*, **267**, 23409.

36. Dietrich, A., Brazil, D., Jensen, O. N., Meister, M., Schrader, M., Moomaw, J. F., Mann, M., Illenberger, D., and Gierschik, P. (1996) Isoprenylation of the G protein gamma subunit is necessary and sufficient for beta-gamma dimer-mediated stimulation of phospholipase C. *Biochemistry*, **35**, 15174.

37. Dietrich, A., Meister, M., Brazil, D., Camps, M., and Gierschik, P. (1994) Stimulation of phospholipase C-beta2 by recombinant guanine-nucleotide-binding protein beta-gamma dimers produced in a baculovirus/insect cell expression system. *Eur. J. Biochem.*, **219**, 171.

38. Higgins, J. B. and Casey, P. J. (1994) In vitro processing of G protein gamma subunits: Requirements for assembly of an active beta-gamma complex. *J. Biol. Chem.*, **269**, 9067.

39. Fukada, Y., Matsuda, T., Kokame, K., Takao, T., Shimonishi, Y., Akino, T., and Yoshizawa, T. (1994) Effects of carboxyl methylation of photoreceptor G protein gamma-subunit in visual transduction. *J. Biol. Chem.*, **269**, 5163.

40. Parish, C. A. and Rando, R. R. (1994) Functional significance of G protein carboxy-methylation. *Biochemistry*, **33**, 9986.

41. Parish, C. A., Smrcka, A. V., and Rando, R. R. (1995) Functional significance of beta-gamma subunit carboxymethylation for the activation of phospholipase C and phospho-inositide 3-kinase. *Biochemistry*, **34**, 7722.

42. Carlson, K. E., Brass, L. F., and Manning, D. R. (1989) Thrombin and phorbol esters cause the selective phosphorylation of a G protein other than Gi in human platelets. *J. Biol. Chem.*, **264**, 13298.

43. Lounsbury, K. M., Schlegel, B., Poncz, M., Brass, L. R., and Manning, D. R. (1993) Analysis of Gz-alpha by site-directed mutagenesis: sites and specificity of protein kinase C-dependent phosphorylation. *J. Biol. Chem.*, **268**, 3494.

44. Fields, T. A. and Casey, P. J. (1995) Phosphorylation of Gzalpha by protein kinase C blocks interaction with the beta gamma complex. *J. Biol. Chem.*, **270**, 23119.

45. Kozasa, T. and Gilman, A. G. (1996) Protein kinase C phosphorylates G12-alpha and inhibits its interaction with G-beta-gamma. *J. Biol. Chem.*, **271**, 12562.

46. Glick, J. L., Meigs, T. E., Miron, A., and Casey, P. J. (1998) RGSZ1, a Gz-selective regulator of G protein signaling whose action is sensitive to the phosphorylation state of Gz-alpha. *J. Biol. Chem.*, **273**, 26008.

47. Wang, J., Ducret, A., Tu, Y., Kozasa, T., Aebersold, R., and Ross, E. M. (1998) RGSZ1, a Gz-selective RGS protein in the brain. *J. Biol. Chem.*, **273**, 26014.

48. Offermanns, S., Hu, Y. H., and Simon, M. I. (1996) Galpha12 and Galpha13 are phosphorylated during platelet activation. *J. Biol. Chem.*, **271**, 26044.

49. Zick, Y., Sagi-Eisenberg, R., Pines, M., Gierschik, P., and Spiegel, A. M. (1986) Multisite phosphorylation of the alpha subunit of transducin by the insulin receptor kinase and protein kinase C. *Proc. Natl Acad. Sci. USA*, **83**, 9294.

50. Katada, T., Gilman, A. G., Watanabe, Y., Bauer, S., and Jakobs, K. H. (1985) Protein kinase C phosphorylates the inhibitory guanine-nucleotide-binding regulatory component and apparently suppresses its function in hormonal inhibition of adenylate cyclase. *Eur. J. Biochem.*, **151**, 431.

51. Pyne, N. J., Murphy, G. J., Milligan, G., and Houslay, M. D. (1989) Treatment of intact hepatocytes with either the phorbol ester TPA or glucagon elicits the phosphorylation and functional inactivation of the inhibitory guanine nucleotide regulatory protein Gi. *FEBS Lett.*, **243**, 77.

52. Strassheim, D. and Malbon, C. C. (1994) Phosphorylation of Galpha-i2 attenuates inhibitory adenylyl cyclase in neuroblastoma/glioma hybrid (NG-108–15) cells. *J. Biol. Chem.*, **269**, 14307.

53. Morris, N. J., Bushfield, M., Lavan, B. E., and Houslay, M. D. (1994) Multi-site phosphorylation of the inhibitory guanine nucleotide regulatory protein Gi-2 occurs in intact rat hepatocytes. *Biochem. J.*, **301**, 693.

54. Lounsbury, K. M., Casey, P. J., Brass, L. F., and Manning, D. R. (1991) Phosphorylation of Gz in human platelets: selectivity and site of modification. *J. Biol. Chem.*, **266**, 22051.

55. Houslay, M. D. (1991) Gi-2 is at the centre of an active phosphorylation/dephosphorylation cycle in hepatocytes: The fine-tuning of stimulatory and inhibitory inputs into adenylate cyclase in normal and diabetic states. *Cell. Signal.*, **3**, 1.

56. Umemori, H., Inoue, T., Kume, S., Sekiyama, N., Nagao, M., Itoh, H., Nakanishi, S., Mikoshiba, K., and Yamamoto, T. (1997) Activation of the G protein Gq/11 through tyrosine phosphorylation of the alpha subunit. *Science*, **276**, 1878.

57. Asano, T., Morishita, R., Ueda, H., Asano, M., and Kato, K. (1998) GTP-binding protein gamma12 subunit phosphorylation by protein kinase C-identification of the phosphorylation site and factors involved in cultured cells and rat tissues in vivo. *Eur. J. Biochem.*, **251**, 314.

58. Yasuda, H., Lindorfer, M. A., Myung, C. S., and Garrison, J. C. (1998) Phosphorylation of the G protein gamma12 subunit regulates effector specificity. *J. Biol. Chem.*, **273**, 21958.

59. Morishita, R., Nakayama, H., Isobe, T., Matsuda, T., Hashimoto, Y., Okano, T., Fukada, Y., Mizuno, K., Ohno, S., and Kozawa, O. (1995) Primary structure of a gamma subunit of G protein,gamma12, and its phosphorylation by protein kinase C. *J. Biol. Chem.*, **270**, 29469.

60. Poppleton, H., Sun, H., Fulgham, D., Bertics, P., and Patel, T.B. (1996) Activation of Gs-alpha by the epidermal growth factor receptor involves phosphorylation. *J. Biol. Chem.*, **271**, 6947.

61. Liebmann, C., Grane, A., Boehmer, A., Kovalenko, M., Adomeit, A., Steinmetzer, T., Nurnberg, B., Wetzker, R., and Boehmer, F. D. (1996) Tyrosine phosphorylation of Gs-alpha and inhibition of bradykinin-induced activation of the cyclic AMP pathway in A431 cells by epidermal growth factor receptor. *J. Biol. Chem.*, **271**, 31098.

62. Nair, B. G., Parikh, B., Milligan, G., and Patel, T. B. (1990) Gs-alpha mediates epidermal growth factor-elicited stimulation of rat cardiac adenylate cyclase. *J. Biol. Chem.*, **265**, 21317.

63. Liu, W. W., Mattingly, R. R., and Garrison, J. C. (1996) Transformation of Rat-1 fibroblasts with the v-src oncogene increases the tyrosine phosphorylation state and activity of the alpha subunit of Gq/G11. *Proc. Natl Acad. Sci. USA*, **93**, 8258.

64. Wieland, T., Ulibarri, I., Gierschik, P., and Jakobs, K. H. (1991) Activation of signal-transducing guanine-nucleotide-binding regulatory proteins by guanosine 5'-[gamma-thio]triphosphate. *Eur. J. Biochem.*, **196**, 707.

65. Nurnberg, B., Harhammer, R., Exner, T., Schulze, R. A., and Wieland, T. (1996) Species- and tissue-dependent diversity of G-protein beta subunit phosphorylation: evidence for a cofactor. *Biochem. J.*, **318**, 717.

66. Hohenegger, M., Mitterauer, T., Voss, T., Nanoff, C., and Freissmuth, M. (1996) Thiophosphorylation of the G protein beta subunit in human platelet membranes: evidence against a direct phosphate transfer reaction to G alpha subunits. *Mol. Pharmacol.*, **49**, 73.

67. Berridge, M. J., Bootman, M. D., and Lipp, P. (1998) Calcium—a life and death signal. *Nature*, **395**, 645.

68. Quest, A. F., Ghosh, S., Xie, W. Q., and Bell, R. M. (1997) DAG second messengers: molecular switches and growth control. *Adv. Exp. Med. Biol.*, **400A**, 297.

69. Singer, W. D., Brown, H. A., and Sternweis, P. C. (1997) Regulation of eukaryotic phosphatidylinositol-specific phospholipase C and phospholipase D. *Annu. Rev. Biochem.*, **66**, 475.

70. Lee, S. B. and Rhee, S. G. (1995) Significance of PIP2 hydrolysis and regulation of phospholipase C isozymes. *Curr. Opinion Cell. Biol.*, **7**, 183.

71. Hepler, J. R., Kozasa, T., Smrcka, A. V., Simon, M. I., Rhee, S. G., Sternweis, P. C., and Gilman, A. G. (1993) Purification from Sf9 cells and characterization of recombinant Gqalpha and G11alpha. *J. Biol. Chem.*, **268**, 14367.

72. Smrcka, A. V. and Sternweis, P. C. (1993) Regulation of purified subtypes of phosphatidylinositol-specific phospholipase C beta by G protein alpha and beta gamma subunits. *J. Biol. Chem.*, **268**, 9667.

73. Lee, C. H., Park, D., Wu, D., Rhee, S. G., and Simon, M. I. (1992) Members of the Gq alpha subunit gene family activate phospholipase C beta isozymes. *J. Biol. Chem.*, **267**, 16044.

74. Kozasa, T., Hepler, J. R., Smrcka, A. V., Simon, M. I., Rhee, S. G., Sternweis, P. C., and Gilman, A. G. (1993) Purification and characterization of recombinant G16 alpha from Sf9 cells: activation of purified phospholipase C isozymes by G-protein alpha subunits. *Proc. Natl Acad. Sci. USA*, **90**, 9176.

75. Bence, K., Ma, W., Kozasa, T., and Huang, X. Y. (1997) Direct stimulation of Bruton's tyrosine-kinase by Gq-alpha subunit. *Nature*, **389**, 296.

76. Offermanns, S. and Simon, M. I. (1995) Galpha-15 and Galpha-16 couple a wide variety of receptors to phospholipase C. *J. Biol. Chem.*, **270**, 15175.

77. Amatruda, T. T., Gerard, N. P., Gerard, C., and Simon, M. I. (1993) Specific interactions of chemoattractant factor receptors with G-proteins. *J. Biol. Chem.*, **268**, 10139.

78. Wu, D., Katz, A., Lee, C. H., and Simon, M. I. (1992) Activation of phospholipase C by alpha 1-adrenergic receptors is mediated by the alpha subunits of Gq family. *J. Biol. Chem.*, **267**, 25798.

79. Zeng, W., Xu, X., Popov, S., Mukhopadhyay, S., Chidiac, P., Swistok, J., Danho, W., Yagaloff, K. A., Fisher, S. L., Ross, E. M., Muallem, S., and Wilkie, T. M. (1998) The N-terminal domain of RGS4 confers receptor selective inhibition of G protein signaling. *J. Biol. Chem.*, **273**, 34687.

80. Camps, M., Carozzi, A., Schnabel, P., Scheer, A., Parker, P. J., and Gierschik, P. (1992) Isozyme-selective stimulation of phospholipase C-beta 2 by G protein beta gamma-subunits. *Nature*, **360**, 684.

81. Katz, A., Dianqing, W., and Simon, M. I. (1992) Subunits beta gamma of heterotrimeric G protein activate beta 2 isoform of phospholipase C. *Nature*, **360**, 686.

82. Boyer, J. L., Graber, S. G., Waldo, G. L., Harden, T. K., and Garrison, J. C. (1994) Selective activation of phospholipase C by recombinant G-protein alpha- and beta gamma-subunits. *J. Biol. Chem.*, **269**, 2814.

83. Park, D., Jhon, D. Y., Lee, C. W., Lee, K. H., and Rhee, S. G. (1993) Activation of phospholipase C isozymes by G protein beta gamma subunits. *J. Biol. Chem.*, **268**, 4573.

84. D'Angelo, D. D., Sakata, Y., Lorenz, J. N., Boivin, G. P., Walsh, R. A., Liggett, S. B., and Dorn, G. W. (1997) Transgenic Galpha-q overexpression induces cardiac contractile failure in mice. *Proc. Natl Acad. Sci. USA*, **94**, 8121.

85. Adams, J. W., Sakata, Y., Davis, M. G., Sah, V. P., Wang, Y., Liggett, S. B., Chien, K. R., Brown, J. H., and Dorn, G. W. (1998) Enhanced Galpha-q signaling: a common pathway mediates cardiac hypertrophy and apoptotic heart failure. *Proc. Natl Acad. Sci. USA*, **95**, 10140.

86. Offermanns, S., Toombs, C. F., Hu, Y.-H., and Simon, M. I. (1997) Defective platelet activation in Galpha-q-deficient mice. *Nature*, **389**, 183.

87. Offermanns, S., Hashimoto, K., Watanabe, M., Sun, W., Kurihara, H., Thompson, R. F., Inoue, Y., Kano, M., and Simon, M. I. (1997) Impaired motor coordination and persistent multiple climbing fiber innervation of cerebellar Purkinje cells in mice lacking Galpha-q. *Proc. Natl Acad. Sci. USA*, **94**, 14089.

88. Offermanns, S., Zhao, L. P., Gohla, A., Sarosi, I., Simon, M. I., Wilkie, T. M., Brown, J. H., and Dorn, G. W. (1998) Embryonic cardiomyocyte hypoplasia and craniofacial defects in G alpha q/G alpha 11-mutant mice. *EMBO J.*, **17**, 4304.

89. Xu, X., Croy, J. T., Zeng, W., Zhao, L., Davignon, I., Popov, S., Yu, K., Jiang, H., Offermanns, S., Muallem, S., and Wilkie, T. M. (1998) Promiscuous coupling of receptors to Gq class alpha subunits and effector proteins in pancreatic and submandibular gland cells. *J. Biol. Chem.*, **273**, 27275.

90. Strathmann, M., Wilkie, T. M., and Simon, M. I. (1989) Diversity of the G-protein family: sequences from five additional alpha subunits in the mouse. *Proc. Natl Acad. Sci.*, **86**, 7407.

91. Strathmann, M. P. and Simon, M. I. (1991) Ga12 and Ga13 subunits define a fourth class of G protein alpha subunits. *Proc. Natl Acad. Sci. USA*, **88**, 5582.

92. Kozasa, T. and Gilman, A. G. (1995) Purification of recombinant G proteins from Sf9 cells by hexahistidine tagging of associated subunits. *J. Biol. Chem.*, **270**, 1734.

93. Singer, W. D., Miller, R. T., and Sternweis, P. C. (1994) Purification and characterization of the alpha subunit of G13. *J. Biol. Chem.*, **269**, 19796.

94. Ip, Y. T. and Davis, R. J. (1998) Signal transduction by the c-Jun N-terminal kinase (JNK)—from inflammation to development. *Curr. Opinion Cell Biol.*, **10**, 205.

95. Vara Prasad, M. V., Dermott, J. M., Heasley, L. E., Johnson, G. L., and Dhanasekaran, N. (1995) Activation of jun kinase/stress-activated protein kinase by GTPase-deficient mutants of G alpha 12 and G alpha 13. *J. Biol. Chem.*, **270**, 18655.

96. Jho, E.-H., Davis, R. J., and Malbon, C. C. (1997) c-Jun amino-terminal kinase is regulated by Galpha-12/Galpha-13 and obligate for differentiation of P19 embryonal carcinoma cells by retinoic acid. *J. Biol. Chem.*, **272**, 24468.

97. Althoefer, H., Eversole-Cire, P., and Simon, M. I. (1997) Constitutively active Galpha-q and Galpha-13 trigger apoptosis through different pathways. *J. Biol. Chem.*, **272**, 24380.

98. Berestetskaya, Y. V., Faure, M. P., Ichijo, H., and Voyno-Yasenetskaya, T. A. (1998) Regulation of apoptosis by alpha subunits of G12 and G13 proteins via apoptosis signal-regulating kinase-1. *J. Biol. Chem.*, **273**, 27816.

99. Voyno-Yasenetskaya, T., Conklin, B. R., Gilbert, R. L., Hooley, R., Bourne, H. R., and Barber, D. L. (1994) Gα13 stimulates Na-H exchange. *J. Biol. Chem.*, **269**, 4721.

100. Dhanasekaran, N., Vara Prasad, M. V. V. S., Wadsworth, S. J., Dermott, J. M., and vanRossum, G. (1994) Protein kinase C-dependent and -independent activation of Na$^+$/H$^+$ exchanger by Galpha-12 class of G proteins. *J. Biol. Chem.*, **269**, 11802.

101. Hooley, R., Yu, C.-Y., Symons, M., and Barber, D. L. (1996) Galpha-13 stimulates Na+-H+ exchange through distinct Cdc42-dependent and RhoA-dependent pathways. *J. Biol. Chem.*, **271**, 6152.

102. Wadsworth, S. J., Gebauer, G., van Rossum, G. D. V., and Dhanasekaran, N. (1997) Ras-dependent signaling by the GTPase-deficient mutant of Galpha-12. *J. Biol. Chem.*, **272**, 28829.

103. Needham, L. K. and Rozengurt, E. (1998) Galpha-12 and Galpha-13 stimulate Rho-dependent tyrosine phosphorylation of focal adhesion kinase, paxillin, and p130 Crk-associated substrate. *J. Biol. Chem.*, **273**, 14626.

104. Xu, N., Bradley, L., Ambdukar, I., and Gutkind, J. S. (1993) A mutant alpha subunit of G12 potentiates the eicosanoid pathway and is highly oncogenic in NIH 3T3 cells. *Proc. Natl Acad. Sci. USA*, **90**, 6741.

105. Ushio-Fukai, M., Griendling, K. K., Akers, M., Lyons, P. R., and Alexander, R. W. (1998) Temporal dispersion of activation of phospholipase C-beta1 and -gamma isoforms by angiotensin II in vascular smooth muscle cells. *J. Biol. Chem.*, **273**, 19772.

106. Hall, A. (1998) Rho GTPases and the actin cytoskeleton. *Science*, **279**, 509.

107. Buhl, A. M., Johnson, N. L., Dhanasekaran, N., and Johnson, G. L. (1995) G alpha 12 and G alpha 13 stimulate Rho-dependent stress fiber formation and focal adhesion assembly. *J. Biol. Chem.*, **270**, 24631.

108. Collins, L. R., Minden, A., Karin, M., and Brown, J. H. (1996) Galpha-12 stimulates c-Jun NH2-terminal kinase through the small G proteins Ras and Rac. *J. Biol. Chem.*, **271**, 17349.

109. Plonk, S. G., Park, S.-K., and Exton, J. H. (1998) The alpha-subunit of the heterotrimeric G protein G13 activates a phospholipase D isozyme by a pathway requiring Rho family GTPases. *J. Biol. Chem.*, **273**, 4823.

110. Gohla, A., Harhammer, R., and Schultz, G. (1998) The G-protein G13 but G12 mediates signaling from lysophosphatidic acid receptor via epidermal growth factor receptor to Rho. *J. Biol. Chem.*, **273**, 4653.

111. Chan, A. M.-L., Fleming, T. P., McGovern, E. S., Chedid, M., Miki, T., and Aaronson, S. A. (1993) Expression cDNA cloning of a transforming gene encoding the wild-type Galpha-12 gene product. *Mol. Cell. Biol.*, **13**, 762.

112. Jiang, H., Wu, D., and Simon, M. I. (1993) The transforming activity of activated Galpha-12. *FEBS Lett.*, **330**, 319.

113. Voyno-Yasenetskaya, T. A., Pace, A. M., and Bourne, H. R. (1994) Mutant alpha subunits of G12 and G13 proteins induce neoplastic transformation of Rat-1 fibroblasts. *Oncogene*, **9**, 2559.

114. Tolkacheva, T., Feuer, B., Lorenzi, M.V., Saez, R. and Chan, A.M. (1997) Cooperative transformation of of NIH3T3 cells by G alpha12 and Rac1. *Oncogene*, **15**, 727.

115. Hart, M. J., Sharma, S., elMasry, N., Qui, R.-G., McCabe, P., Polakis, P., and Bollag, G. (1996) Identification of a novel guanine nucleotide exchange factor for the Rho GTPase. *J. Biol. Chem.*, **271**, 25452.

116. Hart, M. J., Jiang, X., Kozasa, T., Roscoe, W., Singer, W. D., Gilman, A. G., Sternweis, P. C., and Bollag, G. (1998) Direct stimulation of the guanine nucleotide exchange activity of p115 RhoGEF by Galpha-13. *Science*, **280**, 2112.

117. Kozasa, T., Jiang, X., Hart, M. J., Sternweis, P. M., Singer, W. D., Gilman, A. G., Bollag, G., and Sternweis, P. C. (1998) p115 RhoGEF, a GTPase activating protein for Galpha-12 and Galpha-13. *Science*, **280**, 2109.

118. Jiang, Y., Ma, W., Wan, Y., Kozasa, T., Hattori, S., and Huang, X. Y. (1998) The G protein Galpha-12 stimulates Bruton's tyrosine kinase and a RasGap through a conserved PH/BM domain. *Nature*, **395**, 808.

119. Offermanns, S., Mancino, V., Revel, J. P., and Simon, M. I. (1997) Vascular system defects and impaired cell chemokinesis as a result of Galpha-13 deficiency. *Science*, **275**, 533.

120. Offermanns, S., Laugwitz, K.-L., Spicher, K., and Schultz, G. (1994) G proteins of the G12 family are activated via thromboxane A2 and thrombin receptors in human platelets. *Proc. Natl Acad. Sci. USA*, **91**, 504.

121. Parks, S. and Wieschaus, E. (1991) The *Drosophila* gastrulation gene *concertina* encodes a Galpha-like protein. *Cell*, **64**, 447.

122. Barrett, K., Leptin, M., and Settleman, J. (1997) The Rho GTPase and a putative RhoGEF mediate a signaling pathway for the cell shape changes in *Drosophila* gastrulation. *Cell*, **91**, 905.

123. Costa, M., Wilson, E. T., and Wieschaus, E. (1994) A putative cell signal encoded by the *folded gastrulation* gene coordinates cell shape changes during *Drosophila* gastrulation. *Cell*, **76**, 1075.

124. Vu, T. K., Hung, D. T., Wheaton, V. I., and Coughlin, S. R. (1991) Molecular cloning of a functional thrombin receptor reveals a novel proteolytic mechanism of receptor activation. *Cell*, **64**, 1057.

125. Matsuoka, M., Itoh, H., Kozasa, T., and Kaziro, Y. (1988) Sequence analysis of cDNA and genomic DNA for a putative pertussis toxin-insensitive guanine nucleotide-binding regulatory protein alpha subunit. *Proc. Natl Acad. Sci. USA*, **85**, 5384.

126. Fong, H. K. W., Yoshimoto, K. K., Eversole-Cire, P., and Simon, M. I. (1988) Identification of a GTP-binding protein alpha subunit that lacks an apparent ADP-ribosylation site for pertussis toxin. *Proc. Natl Acad. Sci. USA*, **85**, 3066.

127. Casey, P. J., Fong, H. K., Simon, M. I., and Gilman, A. G. (1990) Gz—A guanine nucleotide-binding protein with unique biochemical properties. *J. Biol. Chem.*, **265**, 2383.

128. Hinton, D. R., Blanks, J. C., Fong, H. K., Casey, P. J., Hildebrandt, E., and Simon, M. I. (1990) Novel localization of a G protein, Gz-alpha, in neurons of brain and retina. *J. Neurosci.*, **10**, 2763.

129. Gagnon, A. W., Manning, D. R., Catani, L., Gewirtz, A., Poncz, M., and Brass, L. F. (1991) Identification of Gz alpha as a pertussis toxin-insensitive G protein in human platelets and megakaryocytes. *Blood*, **78**, 1247.

130. Giesberts, A. N., van Ginneken, M., Gorter, G., Lapetina, E. G., Akkerman, J.-W. N., and van Willigen, G. (1997) Subcellular localization of α-subunits of trimeric G-proteins in human platelets. *Biochem. Biophys. Res. Comm.*, **234**, 439.

131. Hall, A. (1990) The cellular functions of small GTP-binding proteins. *Science*, **249**, 635.

132. Glick, J., Santoyo, G., and Casey, P. J. (1996) Arachidonate and related unsaturated fatty acids selectively inactivate the guanine nucleotide-binding regulatory protein Gz. *J. Biol. Chem.*, **271**, 2949.

133. Wong, Y. H., Conklin, B. R. and Bourne, H. R. (1992) Gz-mediated hormonal inhibition of cyclic AMP accumulation. *Science*, **255**, 339.

134. Tsu, R. C., Chan, J. S. C., and Wong, Y. H. (1995) Regulation of multiple effectors by the cloned sigma-opioid receptor: stimulation of phospholipase C and Type II adenylyl cyclase. *J. Neurochem.*, **64**, 2700.

135. Lai, H. W. L., Minami, M., Satoh, M., and Wong, Y. H. (1995) Gz coupling to the rat k-opioid receptor. *FEBS Lett.*, **360**, 97.

136. Chan, J. S. C., Chiu, T. T., and Wong, Y. H. (1995) Activation of type II adenylyl cyclase by the cloned μ-opioid receptor: coupling to multiple G proteins. *J. Neurochem.*, **65**, 2682.

137. Yung, L. Y., Tsim, S.-T., and Wong, Y. H. (1995) Stimulation of cAMP accumulation by the cloned *Xenopus* melatonin receptor through Gi and Gz proteins. *FEBS Lett.*, **372**, 99.

138. Sidhu, A., Kimura, K., Uh, M., White, B. H., and Patel, S. (1998) Multiple coupling of human D5 dopamine receptors to guanine nucleotide binding proteins Gs and Gz. *J. Neurochem.*, **70**, 2459.

139. Tsu, R. C., Allen, R. A., and Wong, Y. H. (1995) Stimulation of type II adenylyl cyclase by chemoattractant formyl peptide and C5a receptors. *Am. Soc. Pharm. Exp. Ther.*, **47**, 835.

140. Wong, Y. H., Chan, J. S., Yung, L. Y., and Bourne, H. R. (1995) Mutant alpha subunit of Gz transforms Swiss 3T3 cells. *Oncogene*, **10**, 1927.

141. Barr, A. J., Brass, L. F., and Manning, D. R. (1997) Reconstitution of receptors and GTP-binding regulatory proteins (G proteins) in Sf9 cells. *J. Biol. Chem.*, **272**, 2223.

142. Butkerait, P., Zheng, Y., and Hallak, H. (1995) Expression of the human 5-hydroxytryptamine1A receptor in Sf9 cells: reconstitution of a coupled phenotype by co-expression of mammalian G protein subunits. *J. Biol. Chem.*, **270**, 18691.

143. Nishizaki, T. and Ikeuchi, Y. (1996) Adenosine evokes potassium currents by protein kinase C activated via a novel signaling pathway in superior colliculus neurons. *FEBS Lett.*, **378**, 1.

144. Nishizaki, T. and Mori, M. (1998) Diverse signal transduction pathways mediated by endogenous P_2 receptors in cultured rat cerebral cortical neurons. *J. Neurophysiol.*, **79**, 2513.

145. Manning, D. R. and Brass, L. F. (1991) The role of GTP-binding proteins in platelet activation. *Thrombosis and Haemostasis*, 66, 393.

146. Berstein, G., Blank, J. L., Jhon, D.-Y., Exton, J. H., Rhee, S. G., and Ross, E. M. (1992) Phospholipase C-beta1 is a GTPase-activating protein for Gq/11, its physiologic regulator. *Cell*, **70**, 411.

147. Biddlecome, G. H., Berstein, G., and Ross, E. M. (1996) Regulation of phospholipase C-beta-1 and Gq and m1 muscarinic cholinergic receptor. *J. Biol. Chem.*, **271**, 7999.

148. Nakamura, F., Kato, M., Kameyama, K., Nukada, T., Haga, T., Kato, H., Takenawa, T., and Kikkawa, U. (1995) Characterization of Gq family G proteins GL1a (G14-alpha), GL2a (G11-alpha), and Gq-alpha expressed in the baculovirus-insect cell system. *J. Biol. Chem.*, **270**, 6246.

149. Dohlman, H. G. (1998) Sst2 is a GTPase-activating protein for Gpa1: purification and characterization of a cognate RGS-G protein pair in yeast. *Biochemistry*, **37**, 4815.

150. Dohlman, H. G., Song, J., Ma, D., Courchesne, W. E., and Thorner, J. (1996) Sst2, a negative regulator of pheromone signaling in the yeast *Saccharomyces cerevisiae*: expression, localization, and genetic interaction and physical association with Gpa1 (the G-protein alph a subunit). *Mol. Cell. Biol.*, **16**, 5194.

151. Druey, K. M., Blumer, K. J., Kang, V. H., and Kehrl, J. H. (1996) Inhibition of G-protein-mediated MAP kinase activation by a new mammalian gene family. *Nature*, **379**, 742.

152. Koelle, M. R. and Horvitz, H. R. (1996) EGL-10 regulates G protein signaling in the

C. elegans nervous system and shares a conserved domain with many mammalian proteins. *Cell*, **84**, 115.

153. Yu, J.-H., Wieser, J., and Adams, T. H. (1996) The *Aspergillus* FlbA RGS domain protein antagonizes G protein signaling to block proliferation and allow development. *EMBO J.*, **15**, 5184.

154. De Vries, L., Mousli, M., Wurmser, A., and Farquhar, M. G. (1995) GAIP, a protein that specifically interacts with the trimeric G protein G-alpha-i3 is a member of a protein family with a highly conserved core domain. *Proc. Natl Acad. Sci. USA*, **92**, 11916.

155. Watson, N., Linder, M. E., Druey, K. M., Kehrl, J. H., and Blumer, K. J. (1996) RGS family members: GTPase-activating proteins for heterotrimeric G-protein alpha-subunits. *Nature*, **383**, 172.

156. Berman, D. M., Wilkie, T. M., and Gilman, A. G. (1996) GAIP and RGS4 are GTPase-activating proteins for the Gi subfamily of G protein alpha subunits. *Cell*, **86**, 445.

157. Hunt, T. W., Fields, T. A., Casey, P. J., and Peralta, E. G. (1996) RGS10 is a selective activator of Galpha-i GTPase activity. *Nature*, **383**, 175.

158. Hepler, J. R., Berman, D. M., Gilman, A. G., and Kozasa, T. (1997) RGS4 and GAIP are GTPase-activating proteins for Gq alpha and block activation of phospholipase C beta by gamma-thio-GTP-Gq alpha. *Proc. Natl Acad. Sci. USA*, **94**, 428.

159. Dohlman, H. G. and Thorner, J. (1997) RGS proteins and signaling by heterotrimeric G proteins. *J. Biol. Chem.*, **272**, 3871.

160. Faurobert, E. and Hurley, J. B. (1997) The core domain of a new retina specific RGS protein stimulates the GTPase activity of transducin *in vitro*. *Proc. Natl Acad. Sci. USA*, **94**, 2945.

161. De Vries, L., Elenko, E., Hubler, L., Jones, T. L. Z., and Farquhar, M. G. (1996) GAIP is membrane-anchored by palmitoylation and interacts with the activated (GTP-bound) form of Galpha-i subunits. *Proc. Natl Acad. Sci. USA*, **93**, 15203.

162. Gundersen, C. B., Mastrogiacomo, A., Faull, K., and Umbach, J. A. (1994) Extensive lipidation of a torpedo cysteine string protein. *J. Biol. Chem.*, **269**, 19197.

163. Pupier, S., Leveque, C., Marqueze, B., Kataoka, M., Takahashi, M., and Seager, M. J. (1997) Cysteine string proteins associated with secretory granules of the rat neurohyphophysis. *J. Neurosci.*, **17**, 2722.

164. Ingi, T., Krumins, A. M., Chidiac, P., Brothers, G. M., Chung, S., Snow, B. E., Barnes, C. A., Lanahan, A. A., Siderovski, D. P., Ross, E. M., Gilman, A. G., and Worley, P. F. (1998) Dynamic regulation of RGS2 suggests a novel mechanism in G-protein signaling and neuronal plasticity. *J. Neurosci.*, **18**, 7178.

165. Yan, Y., Chi, P. P., and Bourne, H. R. (1997) RGS4 inhibits Gz-mediated activation of mitogen-activated protein kinase and phosphoinositide synthesis. *J. Biol. Chem.*, **272**, 11924.

166. Huang, C., Hepler, J. R., Gilman, A. G., and Mumby, S. M. (1997) Attenuation of Gi- and and Gq-mediated signaling by expression of RGS4 or GAIP in mammalian cells. *Proc. Natl Acad. Sci. USA*, **94**, 6159.

167. Heximer, S. P., Watson, N., Linder, M. E., Blumer, K. J., and Hepler, J. R. (1997) RGS2/GOS8 is a selective inhibitor of Gq-alpha function. *Proc. Natl Acad. Sci. USA*, **94**, 14389.

168. Tu, Y., Wang, J., and Ross, E. M. (1997) Inhibition of brain Gz GAP and other RGS proteins by palmitoylation of G protein alpha subunits. *Science*, **278**, 1132.

169. Popov, S., Yu, K., Kozasa, T., and Wilkie, T. M. (1997) The regulators of G protein signaling (RGS) domains of RGS4, RGS10, and GAIP retain GTPase activating protein activity in vitro. *Proc. Natl Acad. Sci. USA*, **94**, 7216.

170. Sanchez-Blazquez, P., Garcia-Espana, A., and Garzon, J. (1995) *In vivo* injection of antisense oligodeoxynucleotides to Galpha subunits and supraspinal analgesia evoked by mu and delta opioid agonists. *J. Pharm. Exp. Ther.*, **275**, 1590.

171. Al-Aoukaty, A., Rolstad, B., and Maghazachi, A. A. (1997) Functional coupling of NKR-P1 receptors to various heterotrimeric G proteins in rat interleukin-2-activated natural killer cells. *J. Biol. Chem.*, **272**, 31604.

172. Nagata, K., Okano, Y., and Nozawa, Y. (1995) Identification of heterotrimeric GTP-binding proteins in human megakaryoblastic leukemia cell line, MEG-01, and their alteration during cellular differentiation. *Life Sci.*, **57**, 1675.

173. Ammer, H. and Schulz, R. (1994) Retinoic acid-induced differentiation of human neuroblastoma SH-SY5Y cells is associated with changes in the abundance of G proteins. *J. Neurochem.*, **62**, 1310.

174. Glassner, M., Jones, J., Kligman, I., Woolkalis, M. J., Gerton, G. L., and Kopf, G. S. (1991) Immunocytochemical and biochemical characterization of guanine nucleotide-binding regulatory proteins in mammalian spermatozoa. *Developmental Biol.*, *146*, 438.

175. Paulssen, E. J., Paulssen, R. H., Haugen, T. B., Gautvik, K. M., and Gordeladze, J. O. (1991) Cell specific distribution of guanine nucleotide-binding regulatory proteins in rat pituitary tumour cell lines. *Mol. Cell. Endocrinol.*, **76**, 45.

176. Matsuoka, M., Kaziro, Y., Asano, S., and Ogata, E. (1993) Analysis of the expression of seven G protein alpha subunit genes in hematopoetic cells. *Am. J. Med. Sci.*, **306**, 89.

177. Paulssen, E. J., Paulssen, R. H., Haugen, T. B., Gautvik, K. M., and Gordeladze, J. O. (1991) Regulation of G protein mRNA levels by thyroliberin, vasoactive intestinal peptide and somatostain in prolactin-producing rat pituitary adenoma cells. *Acta Physiol. Scand.*, **143**, 195.

178. Bargmann, C. I. (1998) Neurobiology of the *Caenorhabditis elegans* genome. *Science*, **282**, 2028.

3 | Ras

JOHANNES L. BOS

1. Introduction

Ras is a molecular switch that controls the transduction of signals from cell-surface receptors to intracellular targets (Fig. 1a). Like all GTPases, the switch mechanism consists of activation by exchange of bound GDP for GTP, and inactivation by hydrolysis of GTP into GDP. Ras achieved notoriety in the 1980s with the discovery that 15% of all human tumours contain a point mutation in one of the three *Ras* genes. The resulting mutant protein is unable to hydrolyse bound GTP and therefore remains constitutively in the active conformation. As a consequence, cells experience a continuous growth-promoting signal in the absence of extracellular cues. Ras has been studied extensively and much is known about mechanisms of activation and of downstream signalling. In most cells, Ras is activated in response to a variety of extracellular signals, which stimulate tyrosine kinases either directly or indirectly. In its GTP-bound state, Ras associates with one of several effector molecules, most notably members of the Raf family, the RalGDS family, or PI3K. The biological consequences of these interactions depend greatly on the cell type and on the context of other signalling events. In addition, poorly understood parameters, such as the timing and intensity of the signal, may determine the ultimate response. Interestingly, Ras is highly conserved both in structure and function during evolution. In the yeast *Saccharomyces cerevisiae*, Ras is involved in nutrient response, in *Schizosaccharomyces pombe* Ras regulates the pheromone response in mating, and in the nematode *Caenorhabditis elegans* and the fruitfly *Drosophila* Ras is a key component of numerous developmental processes. This chapter will provide a background to Ras and present some current ideas about the molecular mechanisms of Ras signalling.

2. Background

2.1 Ras is the product of proto-oncogenes

Ras was first identified in two acutely transforming RNA tumour viruses isolated from a rat sarcoma. During passage through animals, both Harvey rat sarcoma virus and Kirsten rat sarcoma virus acquired related cellular genes that were responsible for their acutely transforming properties. The transforming viral genes turned out to encode fusion proteins derived from the viral *gag* gene and either the H-*ras* or Ki-*ras*

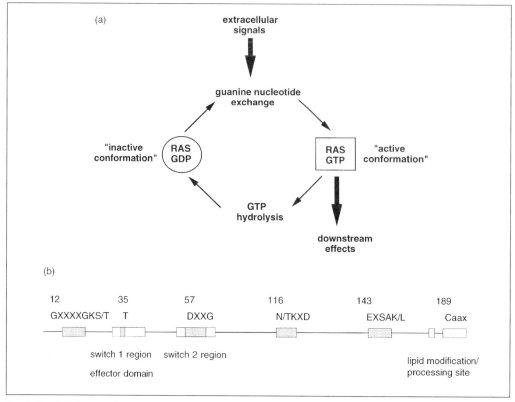

Fig. 1 Small GTPases of the Ras family. (a) Ras and other small GTPases in the Ras family cycle between an inactive (GDP-bound) and an active (GTP-bound) conformation. Activation occurs in response to extracellular signals that promote guanine nucleotide exchange. Once activated, RasGTP interacts with cellular targets to generate downstream effects and a cellular response. The molecular switch is inactivated through GTP hydrolysis. (b) The five conserved regions involved in guanine nucleotide binding are indicated by grey boxes (for more detail, see Chapter 9). The consensus sequence is indicated as well as the position of the first residue of the sequence in Ras. The switch 1 and 2 regions are the regions that show significant conformational differences in the GDP- and GTP-bound states. The lipid modification or processing site is a CaaX box (a, aliphatic; X, any residue) and a cysteine or a polybasic region close to the C terminus. Fully processed GTPases lack the last three amino acids and have a carboxymethylated C-terminal cysteine, attached to the farnesyl lipid. Other members of the Ras family have a geranylgeranyl group instead of the farnesyl group, in combination with either a neighbouring polybasic region or palmitoylation of a neighbouring cysteine.

gene picked up from the rat cells. Further analysis revealed that the v-H-*ras* and v-Ki-*ras* genes differed from their mammalian counterparts by mutations at codons 12 and 59. The huge, world-wide interest in Ras, however, came from experiments designed to identify human 'oncogenes'. NIH-3T3 cells (an immortalized, but non-transformed cell line) were transfected with DNA isolated from human tumour cells and screened for the appearance of transformed cells (visualized as foci growing on a flat monolayer). Foci induced with DNA isolated from a human bladder carcinoma cell line (T24/EJ) were found to contain the human H-*ras* gene and, moreover, the transfected gene had a single point mutation in codon 12 that was crucial for its

Table 1 Frequency of Ras mutations in common types of human cancer (from (2))

Tumour type	Approximate frequency (%)
Lung adenocarcinoma	30
Pancreas adenocarcinoma	90
Colon adenocarcinoma	50
Breast adenocarcinoma	0
Myeloid leukaemia	30
Lymphoid leukaemia	0

transforming properties (1). This was the first demonstration of a mutated cellular gene in a human tumour and caused a great deal of excitement. Using a similar approach, mutations were later found in the cellular Ki-*ras* gene and in a third *ras* gene, N-*ras* (N from neuroblastoma). Mutations were invariably found in codons 12, 13 or 61 of the three homologous *Ras* genes. We now know that these mutations abolish the ability of Ras to hydrolyse bound GTP, and thus these versions of Ras are constitutively in the GTP-bound, active conformation (Fig. 1a)(2).

Subsequent analysis of a large number of human tumours using more direct approaches to detect mutations revealed that around 15% of all human tumours contain a mutated *Ras* gene (Table 1). The most common types of human cancers in which Ras mutations are found are adenocarcinomas of the colon (50%) and the pancreas (90%), whereas mutations in Ras are rarely found in the commonly occurring adenocarcinoma of the breast or in squamous cell carcinomas of the lung (2). Apparently, the effect of mutant Ras proteins on the process of carcinogenesis is restricted to certain tumour types.

2.2 Ras is involved in signal transduction

The first indication that Ras is involved in signal transduction came from the observation that Ras binds GDP and GTP with high affinity (3). Furthermore, micro-injection of quiescent NIH3T3 fibroblasts with a neutralizing monoclonal antibody, Y13–259, raised against the Ras protein completely inhibited serum-induced DNA synthesis (4). These results, in combination with the transforming properties of mutant Ras, have led to the hypothesis that Ras is a molecular switch that controls a signal-transduction pathway leading from extracellular signals acting on membrane receptors to the promotion of cell proliferation. Direct proof for this hypothesis came with the finding that extracellular growth factors rapidly activate Ras, leading to an increase in the GTP-bound state (5, 6). In general, most ligands that activate receptor tyrosine kinases or receptors that associate with non-receptor tyrosine kinases induce Ras activation. However, activation of G protein-coupled receptors, such as those for LPA and thrombin, can induce Ras activation, through a pathway that also appears to be dependent on tyrosine kinase activity (7). More recently, it was found

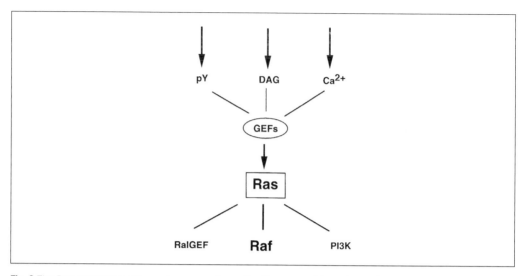

Fig. 2 The Ras signalling pathways in mammalian cells. Three possible pathways by which Ras can be activated are indicated, along with the three currently well-established downstream effectors of Ras. Raf is the best-characterized Ras target. pY, Phosphotyrosine; DAG, diacylglycerol.

that second messengers such as calcium and diacylglycerol can activate Ras in some cell types (8, 9) (Fig. 2).

3. Biochemical properties of Ras

Each of the H-*ras* and N-*ras* genes encodes a single protein, whereas the Ki-*ras* gene potentially encodes two proteins, which differ in their C termini as a result of alternative splicing. Only one of these, Ki-ras4B, has been detected to date. All Ras proteins are post-translationally modified by:

- farnesylation of a cysteine in the C-terminal 'CAAX' motif (A, aliphatic; X, undefined) (see Fig. 1b);
- proteolysis of the last three residues (AAX);
- methylation of the exposed C-terminal carboxylic acid group.

In addition, H-ras and N-ras are palmitoylated at the penultimate C-terminal cysteine. Both the farnesyl and the palmitoyl groups serve to anchor Ras on the inside of the plasma membrane. Ki-ras4B is not palmitoylated; instead, a polybasic region close to the C terminus is required, along with the farnesyl group, for plasma membrane localization. All four proteins bind GDP and GTP with equally high affinity (K_d = 10 pM) and exhibit a relatively slow spontaneous dissociation rate. The intrinsic GTPase rate is also rather slow, but both dissociation and hydrolysis are greatly enhanced by guanine nucleotide exchange factors (GEFs) and GTPase-activating proteins (GAPs) (10).

4. Regulation of Ras activity

4.1 Guanine nucleotide exchange factors (GEFs)

Currently, five mammalian GEFs for Ras are known (Fig. 3a). Sos 1 and 2 are very closely related and ubiquitously expressed GEFs that mediate Ras activation induced by tyrosine kinase activity (11). Recently the crystal structure of Ras with the catalytic domain of Sos has been solved, showing that Sos binds to the switch 1 and switch 2 regions of GDP-bound Ras (see Chapter 9) (12). As a consequence, a loop of the Sos protein inserts into the groove that binds GDP and displaces the switch 1 region. GDP dissociates from Ras, and GTP, which is much more abundant in mammalian cytosol than GDP, binds (for further details, see Chapter 9). Receptor tyrosine kinases can recruit Sos to the plasma membrane through an SH2- and SH3-domain-containing adaptor protein, Grb2. In addition, membrane lipid interactions via the pleckstrin homology (PH) domain of Sos may be important (13). G protein-coupled receptors activate Ras indirectly by activation of receptor tyrosine kinases, such as the EGF receptor, or by activation of non-receptor tyrosine kinases, such as Src (14). Two other GEFs, RasGRF1 (Cdc25Mm) and RasGRF2, are highly expressed in brain. These GEFs have a calmodulin-binding (IQ) domain and are activated by calcium (9). In addition, GRF1 has been implicated in the activation of Ras by $\beta\gamma$ subunits of heterotrimeric G proteins (15). Recently, RasGRP has been described; this GEF is expressed predominantly in brain and contains a diacylglycerol-binding site and a Ca^{2+}-binding 'EF-hand' (8).

4.2 GTPase-activating proteins (GAPs)

At least two mammalian GTPase-activating proteins (GAPs), p120GAP and neurofibromin, are responsible for the downregulation of Ras (Fig. 3b). As shown for p120GAP, these proteins function by providing an essential arginine (arginine finger, see Chapter 9) for the GTPase reaction (16). It is thought that association of GAPs with Ras is tightly regulated, but the details are unclear. p120GAP has an N-terminal region containing an SH3 domain flanked by two SH2 domains and a Ca^{2+}-dependent lipid-binding motif. These domains mediate the interaction of p120GAP with several other signalling molecules. For instance, after receptor tyrosine kinase stimulation, p120GAP associates with several tyrosine phosphorylated proteins, including growth factor receptors, p190RhoGAP (17) and p62[dok] (18, 19). The functional consequences of these associations are still unclear, but the association with p190RhoGAP indicates a potential connection with Rho GTPases (20). Neurofibromin is a large protein that is mutated in patients with the disease neurofibromatosis type I. These patients frequently acquire benign neurofibromas (21). GAP1[M] represents a family of related GAPs, including the IP4 receptor GAP[IP4], that act on Ras *in vitro*. It is unclear, however, whether Ras is the true physiological target of these proteins (22, 23). More recently, p135SynGAP has been cloned. This GAP protein is expressed pre-

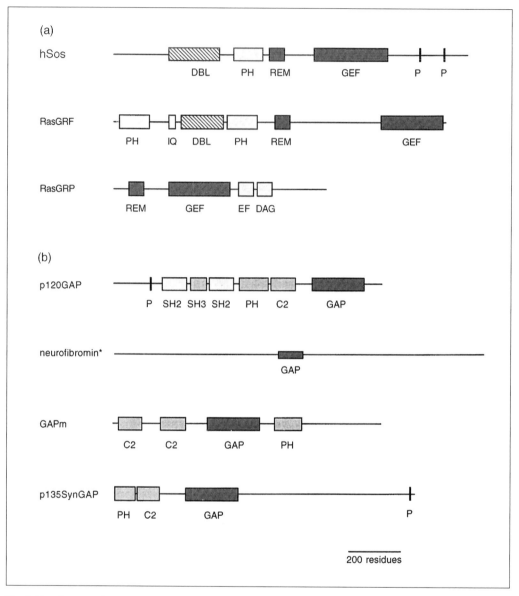

Fig. 3 The Ras guanine nucleotide exchange factors (a) and Ras GTPase activating proteins (b). DBL, Dbl homology domain (guanine nucleotide exchange factor for Rho-like GTPases); PH, pleckstrin homology domain; REM, Ras exchange motif (necessary for stabilization of the GEF domain); GEF, catalytic domain; P, proline-rich sequences; IQ, calmodulin-binding motif; EF, calcium-binding motif with 'EF hands'; DAG, diacylglycerol-binding motif; SH2, Src homology region 2 (binds tyrosine phosphorylated peptides); SH3, Src homology region 3 (binds proline-rich sequences); C2, calcium-dependent lipid-binding motif; GAP, catalytic region. * Neurofibromin is indicated in a twofold smaller scale than the other proteins.

dominantly in synapses of hippocampal neurons and is negatively regulated by CaM-kinase II (24).

One of the major functions of GAPs is obviously the downregulation of Ras activity, although it has been suggested that they may have additional functions and contribute to downstream signalling. However, these additional functions are poorly characterized.

5. Effectors of Ras

5.1 The serine–threonine kinases of the Raf family

The best characterized downstream effector of Ras is the serine–threonine kinase, Raf1, and its close relatives, B-raf and A-raf. Although the complete mechanism of activation of Raf appears to be rather complex, the initial event involves recruitment of cytosolic Raf to the plasma membrane by its direct interaction with the active conformation of Ras. Further details on the interaction of Ras with Raf can be found in Chapter 9, Section 9.2. In addition, covalent modification by phosphorylation and the interaction of Raf with auxilliary proteins such as 14-3-3 may be involved in the activation process (25). Raf1 can also be activated by additional pathways, including protein kinase C (PKC) (26, 27), although interaction with RasGTP may still be necessary (28). In certain cell types Ras-induced Raf1 activation is under negative control by cAMP, which may induce phosphorylation and inactivation of Raf1 (29).

Raf1 phosphorylates and activates the dual-specificity kinase, MEK (a MAP kinase kinase), which in turn phosphorylates and activates ERK (a MAP kinase), thereby activating the so-called 'classical' MAP kinase (mitogen activated protein kinase) cascade. One of the functions of activated ERK is to translocate to the nucleus and to phosphorylate and activate transcription factors, such as ternary complex factors.

5.2 The p110 catalytic subunit of phosphatidylinositol 3-kinase

A second effector of Ras is thought to be the p110 catalytic subunit of phosphatidylinositol 3-kinase (PI3K), a kinase that generates the lipids phosphotidylinositol 3,4-bisphosphate (PI3,4P$_2$) and phosphatidylinositol 3,4,5-triphosphate (PI3,4,5P$_3$)(30). PI3K mediates the activation of protein kinase B (PKB) (31, 32), which is responsible for the inactivation GSK3 (33), and is involved in protecting cells against apoptotic signals (34, 35). Through a distinct pathway, PI3K also activates the small GTPase, Rac, leading to cell-surface membrane protrusions (Chapter 4) (36). The mechanism by which Ras activates PI3K is still unclear but, like Raf, a large contribution to activation may come from membrane targeting. Indeed, addition of membrane-targeting sequences directly to the p110 subunit is sufficient to induce constitutive PI3K activity (37). The catalytic subunit of PI3K is always found intimately associated with a regulatory, SH2-containing subunit, p85. This allows association with receptor phosphotyrosines after receptor stimulation, inducing PI3K activation independently of Ras. However, recent mutational analysis of the platelet-derived growth factor (PDGF) receptor suggests that both the association of

PI3K with the receptor and with Ras may be necessary for an increase in PI3K products (38–40).

5.3 Ral guanine nucleotide exchange factors (Ral GEFs)

A third group of downstream targets of Ras, a family of GEFs for the small GTPase Ral, was identified through yeast two-hybrid screening of cDNA libraries with Ras as a bait (41–43). The prototype, RalGDS, is related in sequence to Ras GEFs (41). Two distant relatives with similar structural and binding properties, Rgl (44) and Rlf (45), were subsequently identified. Both RalGDS and Rlf have been shown to exhibit exchange activity for Ral *in vitro* and *in vivo*. Importantly, in cotransfection experiments in COS-7 cells, active Ras induces both RalGDS and Rlf activity, suggesting that both are true effectors for Ras (46). The possible consequences of Ral activation (via Ral GEF) in Ras-mediated signal transduction are discussed below.

5.4 Other effectors

A number of other proteins have been identified, primarily through yeast two-hybrid analysis, that may serve as effectors of Ras (25, 47). These include Rin, Rsb, PKC-ζ, MEKK-1, Nore, and AF6. However, for most of these proteins it has not been shown that an interaction occurs in the cell with endogenous proteins, and the physiological relevance of these interactions remains unclear.

6. Biological responses

6.1 Oncogenic Ras induces proliferation and differentiation

Ras is involved in many different biological responses, including cell proliferation, differentiation, migration, and apoptosis, depending on the cell type and the surrounding conditions. The most notable responses induced by expression of a constitutively active Ras protein in immortalized rodent fibroblasts such as NIH3T3, are the induction of serum-independent cell proliferation, anchorage-independent growth, loss of cell–cell contact, and malignant transformation. In a very elegant approach using various effector mutants of Ras with amino-acid substitutions in the effector region (switch 1), it was shown that Ras must interact with multiple downstream targets to induce efficient cell transformation. These effector-region mutants have been analysed in more detail (48, 49). Thus RasV12S35 and RasV12E38 only interact with Raf and not with RalGDS or PI3K; RasV12G37 only binds RalGDS, and RasV12C40 only binds PI3K. None of these mutants alone can induce transformation. From these results it can be concluded that for full oncogenic transformation, Ras needs to activate several pathways (Fig. 4). Interestingly, not only Ral but also several other small GTPases seem to be important for Ras-induced proliferation and oncogenic transformation, in particular the Rho GTPases Cdc42, Rac, and Rho.

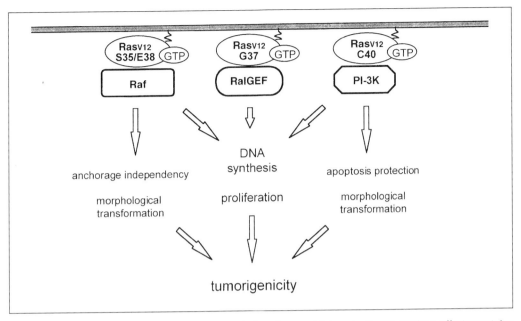

Fig. 4 Behaviour of effector-site mutants of oncogenic Ras. Ras must interact with at least three effector proteins to induce the various features associated with oncogenic transformation.

Expression of active Ras does not always lead to the induction of proliferation. For instance, in PC12 cells, Ras activation induces the cessation of DNA synthesis and the outgrowth of neurites, illustrating the induction of terminal differentiation. Clearly, the biological response of cells to oncogenic Ras is determined by the cell type.

6.2 Function of endogenous Ras

Many of the studies described above involve overexpression of wild-type or mutant Ras, and this may not reflect the real function of endogenous Ras in a particular cell type. A nice example of this is PC12 cells. In these cells EGF induces proliferation, whereas nerve growth factor (NGF) induces differentiation. For both events, however, Ras-mediated activation of the ERK MAP kinase cascade is essential. Interestingly, EGF induces ERK (and Ras) only transiently, whereas NGF induces a sustained activation of ERK (and Ras). By using cell lines overexpressing the EGF receptor, it could be demonstrated that quantitative differences in ERK activation are responsible for the different biological responses of the PC12 cells (50).

Other elegant experiments have been carried out to probe the function of endogenous Ras in mammalian cells. Stacey and co-workers (51) were the first to micro-inject neutralizing Ras antibodies into NIH3T3 fibroblasts. This resulted in the inhibition of serum-induced DNA synthesis, demonstrating the essential role of Ras in G_1 cell cycle progression. Interestingly, only inhibition of Ras during the first 12 hours or so of the G_1 phase affected the cell cycle (51). This agrees with the observation that in proliferating cells Ras is activated in mid-G_1 (52) and in cells with a disrupted G_1–S control, such as Rb-negative cells, Ras inhibition has little effect (53, 54). One of the proteins regulated by Ras is the cell-cycle inhibitor p21Waf1/Cip1 (55–57). Although it is not exactly clear how Ras affects cell-cycle inhibitor activities, it was recently shown that when Rho is inhibited, activated Ras induces p21Waf1/Cip1 expression and, as a consequence, entry into the cell cycle is blocked. When Rho is active, induction of p21Waf1/Cip1 by Ras is suppressed and Ras promotes S-phase entry. In line with this observation, cells that lack p21Waf1/Cip1 do not require Rho signalling for the induction of DNA synthesis by activated Ras (55). These observations point to the multiple activities required to ensure cell-cycle progression.

Some *in vivo* effects of Ras inhibition have also been determined. A dominant negative Ras, under the control of the T-cell-specific lck promoter, was introduced into transgenic mice. This resulted in the inhibition of T-cell receptor-induced positive selection in the thymus, but it did not affect either negative selection by self-antigens or proliferation during early T-cell development (58). This appears to be more similar to the observations made by the genetic analysis of the Ras pathway in the lower eukaryotes *C. elegans* and *Drosophila* (see below), where Ras is involved in the determination of cell fate, rather than cell proliferation.

The generation of knockout mice has thus far revealed only limited information on the function of Ras (59). Mice with homozygous deletions of either H-*ras* or N-*ras* are completely viable, whereas Ki-*ras*-deficient mice die at day 12–14 of embryogenesis due to fetal liver failure. These results suggest a redundancy in function of at least two of the three Ras proteins, but also support the idea of some unique function for Ki-*ras* in the developing liver.

7.   Ras in lower eukaryotes

The Ras protein is highly conserved in all eukaryotes, including yeast, and genetic studies in *Drosophila* and C. *elegans* have provided a great deal of information concerning the nature of Ras-regulated signal-transduction pathways. Most notably, Ras is required for the development of the vulva in C. *elegans* and for the induction of R7 photoreceptor cells in the compound eye of *Drosophila*.

7.1   Ras in vulva development in C. *elegans*

The vulva in C. *elegans* develops from six potential vulval precursor cells (60, 61). An inductive signal generated by an anchor cell determines the fate of three of these precursor cells to form the vulva. C. *elegans* Ras (let-60) plays an essential role in determining the fate of the precursor cells. A gain-of-function mutant of Ras (too much signal) results in a multi-vulva phenotype, while a loss-of-function mutant (too little signal) results in a vulva-less phenotype. The inductive signal provided by the

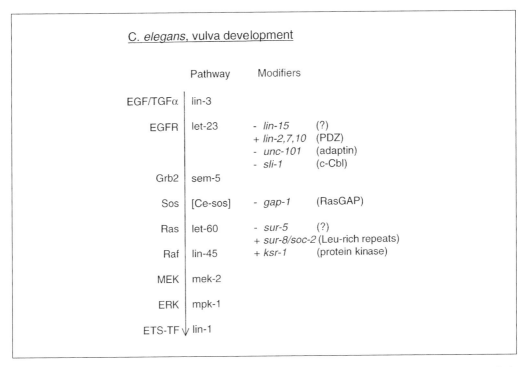

Fig. 5 The Ras signalling pathway involved in vulva development in *C. elegans*. The gene products that most likely represent a linear pathway from the cell surface to the nucleus, as determined mainly by genetic analysis, are indicated. *Ce-sos* has thus far not been identified in genetic screens, but has been retrieved from the genomic sequence of *C. elegans*. In addition, a number of proteins are indicated that either positively (+) or negatively (-) influence the pathway. (Adapted from (60).)

anchor cell is lin-3, an EGF-like growth factor that activates let-23, a EGF receptor homologue in the vulva precursor cells (Fig. 5). By screening for mutations that complement the 'too little signal' effect (enhancer screen) or the 'too much signal' (suppressor screen) of let-23 or let-60 several intermediates in the developmental pathway of the precursor cells have been identified (Fig. 5). The first to be identified was the *C. elegans* Ras protein itself (let-60). Later, upstream components linking Ras to the membrane receptor (sem-5) and downstream components linking Ras to transcription factors (Raf, MEK, and ERK) were found. In addition, a number of proteins have been identified that modify the let-23- or let-60-induced vulva phenotype. These factors interact directly with intermediates of the Ras pathway or operate in an essential parallel pathway. Many, though not yet all, of the components in the C. *elegans* pathway have been identified in mammalian cells. In fact, several were identified first in *C. elegans* (e.g. sem-5 and ksr-1). Ras not only plays a role in vulva development, but also in many other developmental pathways in the worm. Interestingly though, Ras appears to be involved primarily in the determination of cell fate, rather than in cell proliferation (60).

7.2   Ras in eye development in *Drosophila*

The ommatidia of the *Drosophila* eye is the basic structure for photoreception. This ommatidia consists of eight photoreceptor cells (R1–8), each developing from its respective precursor. The R7 precursor cell only develops into a photoreceptor cell when induced to do so by a cell-surface molecule on a neighbouring R8 cell. This surface molecule, Boss, is a membrane-anchored, EGF-like molecule that activates Sevenless, a receptor tyrosine kinase related to the EGF receptor, present on the R7 precursor cell. This signal is subsequently transduced to the nucleus of the R7 precursor cell by a pathway involving Ras (D-ras) (Fig. 6) (62). Again, as described for *C. elegans*, enhancer and suppressor screens have revealed a number of inter-mediates in the pathway. The most notable example was the discovery of Sos as a Ras GEF (63). Also in *Drosophila,* a number of modifier proteins have been identified, although their precise role in the signalling pathway is still under intense investigation.

7.3   Ras in yeast

In the yeast *S. cerevisiae*, Ras is involved predominantly in the control of cAMP levels during vegetative growth by direct binding and activation of adenylate cyclase. This

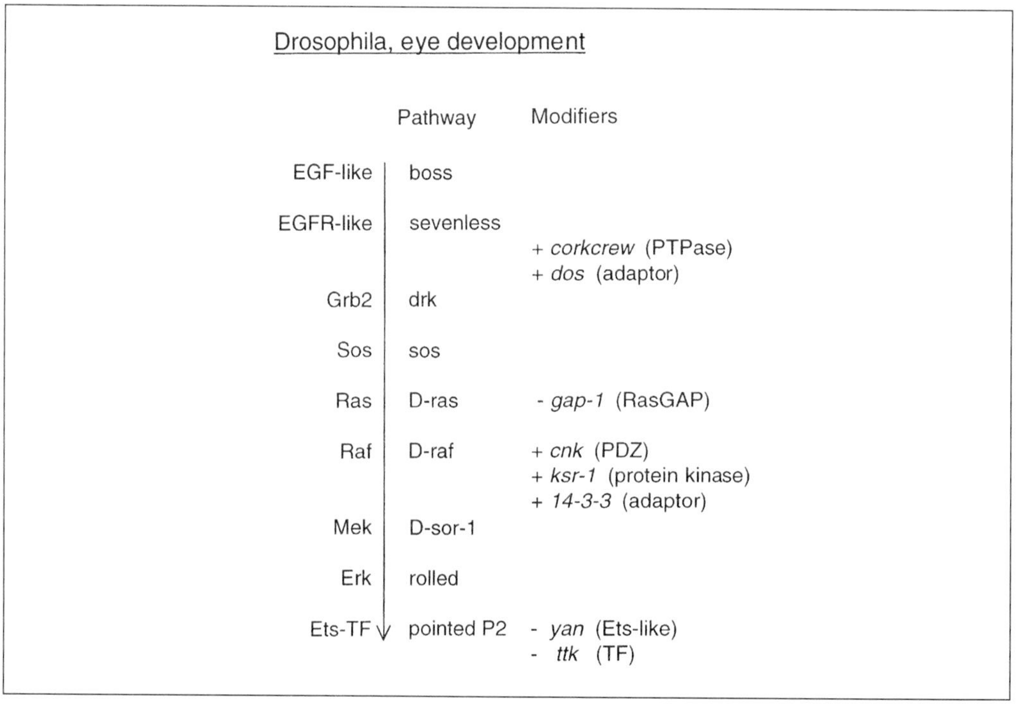

Drosophila, eye development

	Pathway	Modifiers
EGF-like	boss	
EGFR-like	sevenless	+ *corkcrew* (PTPase)
		+ *dos* (adaptor)
Grb2	drk	
Sos	sos	
Ras	D-ras	- *gap-1* (RasGAP)
Raf	D-raf	+ *cnk* (PDZ)
		+ *ksr-1* (protein kinase)
		+ *14-3-3* (adaptor)
Mek	D-sor-1	
Erk	rolled	
Ets-TF	pointed P2	- *yan* (Ets-like)
		- *ttk* (TF)

Fig. 6 The Ras signalling pathway involved in the development of the *Drosophila* eye. Sos was originally identified in the *Drosophila* genetic screen and later identified in mammalian cells.

function of Ras is not conserved in the yeast *S. pombe*, or in other higher eukaryotes. In *S. pombe* Ras is a part of the pheromone response pathway and upon activation it interacts directly with, and activates, the MAP kinase kinase kinase (MAPKKK), Byr2, leading to stimulation of a MAP kinase cascade. In addition, *S. pombe* Ras also binds to Scd1, a putative GEF for Cdc42sp, thereby contributing to the organization of the actin cytoskeleton (25).

8. Other Ras-like GTPases

Ras is the paradigm of a family of closely related proteins that share considerable sequence identity, particularly in their effector regions (Fig. 7). As a consequence, these proteins may interact with some of the same effector proteins as Ras. At the moment, however, the function of most of these Ras-like GTPases is poorly understood.

8.1 Ral

RalA and B are two very homologous proteins and have six out of nine amino residues in the core effector region identical to those of Ras (region 32–40 of Ras) (64). Ral is ubiquitously expressed, but is particularly abundant in brain, testis, and platelets and is mainly localized on endocytic and exocytic vesicles. The intrinsic exchange rate and GTPase activity of Ral are similar to those of Ras. Different Ral

```
                        G37  C40
                    S35 |E38|
            FVDEYDPTIEDSYRKQVV        H-ras
            -----------------         K-ras4A/B
            -----------------         N-ras

            --EK-----------E          Rap1a
            --EK-----------E          Rap1b
            -IEK-------F---EIE         Rap2a
            -IEK-------F---EIE         Rap2b

            --SD---------T-ICS         R-ras
            --TD---------T--C-         TC21
            --PD---------L-HTE         M-ras

            --ED-E--KA-----K--         RalA
            --ED-E--KA-----K--         RalAB

            ---S------NTFT-LIT         Rheb
```

Fig. 7 Effector regions of Ras family members. The core region, usually taken as residues 32–40, is identical in Ras, Rap1, R-ras, TC21, and M-Ras. Indicated are some of the Ras effector mutations (T35S, E37G, D38E, and Y40C) that have been shown to selectively disrupt binding to effectors (see Fig. 4).

GEFs have been identified, as discussed above (RalGDS, Rlf, and Rgl) and a Ral GAP activity has been identified, but the corresponding protein has not been characterized (65).

Two possible functions have been proposed for Ral. First, Ral may link Ras to Rho GTPases and thereby affect the actin cytoskeleton. This was suggested by the observation that a GAP for Cdc42, RalBP (RLIP, RIP), binds directly to the GTP-bound form of Ral and hence is a candidate effector (65). The N terminus of Ral associates with a phospholipase D (PLD) activity, suggesting a second role of Ral, and expression of a dominant negative Ral has been shown to inhibit Src- and Ras-induced PLD activity (66).

8.2 Rap1

Rap1 is found as two isoforms, Rap1a and Rap1b, which differ in only a few amino acids at the C terminus. A characteristic feature of Rap1 is a threonine residue at position 61; in most other GTPases this residue is glutamine. Substitution of threonine for glutamine in Ras results in weak oncogenic activation and, as expected, Rap1 has a tenfold lower intrinsic GTPase activity than Ras. Rap1 is not sensitive to p120RasGAP or neurofibromin; however, a distinct protein with Rap GAP activity has been characterized (67). In contrast to Ras, Rap1 does not transform cells, in fact it was originally discovered as a suppressor of Ki-*ras* transformed cells. This has led to the suggestion that Rap1 might act as an antagonist of Ras. To date, however, there is little evidence that this is the role of endogenous Rap1.

Significant progress has been made in the elucidation of mechanisms of Rap1 activation. Rap1 is a very early sensor of a number of second messengers, including diacylglycerol, calcium, and cAMP (68, 69), and, as a consequence, many receptors can activate Rap1. Several GEFs have been identified recently that may mediate the activation of Rap1; two of these are particularly interesting, since they are activated directly by second messengers. CalDAG-GEF I, a very close relative of RasGRP (see Fig. 3) (70), is responsive to both diacylglycerol and calcium, while Epac/cAMP-GEF I and II contain a cAMP-binding site that is very similar to that found in the regulatory subunit of protein kinase A (Fig. 8) (71, 72). Epac (exchange protein activated by cAMP) responds to cAMP both *in vivo* and *in vitro*. Interestingly, like PKA, the cAMP-binding region of Epac functions as an inhibitory sequence and inhibition of exchange activity is relieved by binding of cAMP (71). In this respect, the regulation of Epac differs conceptually from the regulation of the RasGEF Sos, for which membrane translocation is the most likely mechanism of activation. The identification of RasGRP, CalDAG-GEF, and Epac demonstrates that GEFs for small GTPases can be regulated directly by second messengers without kinases acting as mediators. An additional form of regulation of Rap1 may be the phosphorylation of a serine in the polybasic domain close to the C terminus. This phosphorylation most likely regulates membrane attachment of the GTPase (73). Several Rap1-specific GAPs have been identified, but their function in the regulation of Rap1 is still elusive (68).

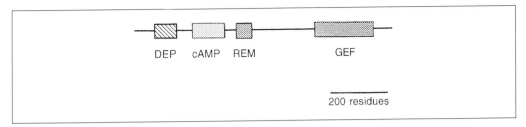

Fig. 8 Schematic representation of Epac, a Rap1-specific GEF that binds directly to and is responsive to cAMP. The indicated DEP domain is a protein motif that may mediate membrane attachment (68). Epac is identical to cAMP-GEF I and is related to cAMP-GEF II (69).

The similarity with Ras suggests that Rap1 may have a similar function, but thus far no clear biochemical downstream effects of Rap1 have been established (68). Interestingly, Rap1 is very abundant in platelets and is rapidly activated after platelets have been stimulated to induce aggregation and secretion. This suggests that Rap1 might be involved in one of these two processes. A further possible clue about the function of Rap1 comes from the analysis Bud1 (Rsr1), a Rap1 orthologue in the budding yeast *S. cerevisiae*. Bud1, and the products of two other genes, Bud5 (a GEF for Bud1) and Bud2 (a GAP for Bud1) are involved in bud site selection (74). Interestingly, Bud5 and Bud2 null cells have the same phenotype as Bud1 null cells, strongly suggesting that Bud1 needs to cycle between the GTP- and GDP-bound states to perform its role (75, 76). The function of Bud1 is to recruit polarity estab-lishment factors to the bud site (77–79); these include Cdc24 (a GEF for Cdc42), Cdc42 itself, and Bem1 (an SH3-domain-containing, scaffold protein). Cdc24 interacts directly with the GTP-bound form of Bud1, whereas Bem1 appears to associate specifically with the GDP-bound form of Bud1 (79). Thus both the GDP- and the GTP-bound conformations of Bud1 may affect complex formation. Multiple cycles of Bud1GDP–GTP exchange are thought to be essential to get sufficient amounts of Bem1 and Cdc24 to the target site. Cdc24 activates Cdc42, which, together with Bem1, coordinates actin filament organization to form the future bud. Perhaps Rap1 has a similar function and regulates some aspect of polarity establishment in mammalian cells.

8.3 Rap2

Rap2 differs from Rap1 by a single amino acid residue in its effector region and a number of differences outside this region. Like Rap1, Rap2 has the characteristic threonine at position 61. In its constitutively activated form Rap2 does not transform cells, but in contrast to Rap1, it does not suppress Ras-induced transformation (80). Little is known about the mechanism of regulation of Rap2. In T-cells, Rap2 was found to be activated by the phorbol ester TPA, but the mechanism of activation and the downstream effects are unknown (81). In platelets Rap2, but not Rap1, was found in a complex containing the major integrin $\alpha_{IIb}\beta_3$, suggesting that Rap2 may be involved in integrin signalling (82).

8.4 R-ras

R-Ras is more than 50% identical in amino-acid sequence to Ras, although it has a 26 amino acid N-terminal extension (83). The similarity between R-ras and TC21 is even higher (70%). Thus far, no R-ras-specific guanine nucleotide exchange factors have been reported. Several GAPs active on R-ras have been identified (84–87). p120RasGAP and neurofibromin interact with R-ras and stimulate GTP hydrolysis (86, 87). GAP^{IP4}, a protein previously identified as the inositol 1,3,4,5-tetraphosphate (IP$_4$) receptor in platelets, acts as a GAP for Ras, R-ras, and Rap1 (84, 85). This may indicate a link between R-ras, IP$_4$, and calcium metabolism.

Recently it was shown that R-ras is activated by thrombin in platelets and during differentiation of myoblasts, but the function of R-ras is still not established. Over-expression of R-ras does induce cell transformation in some cell types, suggesting that R-ras may be involved in the control of cell proliferation, but a role in the regulation of apoptosis has also been suggested (88). Constitutively activated R-ras has been reported to stimulate cell adhesion, suggesting that R-ras is involved in the activation of integrins (89), but the mechanisms are unclear.

8.5 TC21

TC21 (R-ras-3) shares 55% sequence similarity to Ras and is the only small GTPase, other than Ras, that has been found mutated in human tumour cell lines. In an ovarian carcinoma cell line, a single point mutation substituting glutamine for leucine at position 72 was found (90). This mutation is equivalent to a mutation in glutamine 61 in Ras, which is frequently found in human tumours. In a leiomyo-sarcoma cell line, an insertion of three amino acid residues at position 24 was observed (91). Both mutants are highly transforming in the NIH3T3 focus formation assay, as is TC21-V23, a mutation equivalent to RasV12 (92).

Recently R-ras-3 (M-Ras), a new member of the R-ras family, has been identified. This GTPase may function in the control of the actin cytoskeleton (93, 94).

8.6 Rheb

Rheb is a ubiquitously expressed small GTPase, with six out of nine identical and two similar amino-acid residues to the effector region of Ras (95, 96). Rheb interacts with Raf1 *in vitro* in a largely GTP-dependent manner (97), but its *in vivo* function is unknown.

9. Conclusions

The mammalian family of Ras GTPases currently contains 13 members, which share considerable homology with their family founder, Ras. Ras is one of the best studied of all signalling molecules and much is known about the proteins that interact with Ras, and how they transduce signals to generate a biological response. More recently, other members of the Ras family have begun to be analysed. *In vitro*, many of these interact with the same effector proteins as Ras; however, their biological effects are clearly different from Ras, and their real targets are likely to be distinct. Ras and Ras family GTPases remain therefore an exciting area for future research.

Acknowledgement

I thank Dr B. M.Th. Burgering for reading and commenting on the manuscript.

References

1. Parada, L. F., Tabin, C. J., Shih, C., and Weinberg, R. A. (1982) Human EJ bladder carcinoma oncogene is homologue of Harvey sarcoma virus *ras* gene. *Nature,* **297**, 474.
2. Bos, J. L. (1989) *ras* oncogenes in human cancer: a review. *Cancer Res.,* **49**, 4682.
3. Shih, T. Y., Papageorge, A. G., Stokes, P. E., Weeks, M. O., and Scolnick, E. M. (1980) Guanine nucleotide-binding and autophosphorylating activities associated with the p21src protein of Harvey murine sarcoma virus. *Nature,* **287**, 686.
4. Smith, M. R., DeGudicibus, S. J., and Stacey, D. W. (1986) Requirement for c-*ras* proteins during viral oncogene transformation. *Nature,* **320**, 540.
5. Satoh, T., Endo, M., Nakafuku, M., Nakamura, S., and Kaziro, Y. (1990) Platelet-derived growth factor stimulates formation of active p21*ras*-GTP complex in Swiss mouse 3T3 cells. *Proc. Natl Acad. Sci. USA,* **87**, 5993.
6. Downward, J., Graves, J. D., Warne, P. H., Rayter, S., and Cantrell, D. A. (1990) Stimulation of p21*ras* upon T-cell activation. *Nature,* **346**, 719.
7. van Corven, E. J., Hordijk, P. L., Medema, R. H., Bos, J. L., and Moolenaar, W. H. (1993) Pertussis toxin-sensitive activation of p21*ras* by G protein-coupled receptor agonists. *Proc. Natl Acad. Sci. USA,* **90**, 1257.
8. Ebinu, J. O., Bottorff, D. A., Chan, E. Y., Stang, S. L., Dunn, R. J., and Stone, J. C. (1998) RasGRP, a Ras guanyl nucleotide- releasing protein with calcium- and diacylglycerol-binding motifs. *Science,* **280**, 1082.
9. Farnsworth, C. L., Freshney, N. W., Rosen, L. B., Ghosh, A., Greenberg, M. E., and Feig, L. A. (1995) Calcium activation of Ras mediated by neuronal exchange factor Ras-GRF. *Nature,* **376**, 524.
10. Barbacid, M. (1987) *ras* Genes. *Annu. Rev. Biochem.,* **56**, 779.
11. Simon, M. A., Dodson, G. S., and Rubin, G. M. (1993) An SH3-SH2-SH3 protein is required for p21*ras1* activation and binds to Sevenless and SOS proteins *in vitro*. *Cell,* **73**, 169.
12. Boriack-Sjodin, P. A., Margarit, S. M., Bar-Sagi, D., and Kuriyan, J. (1998) The structural basis of the activation of Ras by Sos. *Nature,* **394**, 337.
13. Byrne, J. L., Paterson, H. F., and Marshall, C. J. (1996) p21Ras activation by the guanine nucleotide exchange factor Sos, requires the Sos/Grb2 interaction and a second ligand-dependent signal involving the Sos N-terminus. *Oncogene,* **13**, 2055.
14. van Biesen, T. *et al.* (1995) Receptor-tyrosine-kinase- and G beta gamma-mediated MAP kinase activation by a common signalling pathway. *Nature,* **376**, 781.
15. Mattingly, R. R., Sorisky, A., Brann, M. R., and Macara, I. G. (1994) Muscarinic receptors transform NIH 3T3 cells through a Ras-dependent signalling pathway inhibited by the Ras-GTPase-activating protein SH3 domain. *Mol. Cell. Biol.,* **14**, 7943.

16. Scheffzek, K., Ahmadian, M. R., Kabsch, W., Wiesmuller, L., Lautwein, A., Schmitz, F., and Wittinghofer, A. (1997) The Ras-RasGAP complex: structural basis for GTPase activation and its loss in oncogenic Ras mutants [see comments]. *Science*, **277**, 333.

17. Settleman, J., Narasimhan, V., Foster, L. C., and Weinberg, R. A. (1992) Molecular cloning of cDNAs encoding the GAP-associated protein p190: implications for a signaling pathway from ras to the nucleus. *Cell*, **69**, 539.

18. Carpino, N., Wisniewski, D., Strife, A., Marshak, D., Kobayashi, R., Stillman, B., and Clarkson, B. (1997) p62(dok): a constitutively tyrosine-phosphorylated, GAP-associated protein in chronic myelogenous leukemia progenitor cells. *Cell*, **88**, 197.

19. Yamanashi, Y. and Baltimore, D. (1997) Identification of the Abl- and rasGAP-associated 62 kDa protein as a docking protein, Dok. *Cell*, **88**, 205.

20. Settleman, J., Albright, C. F., Foster, L. C., and Weinberg, R. A. (1992) Association between GTPase activators for Rho and Ras families. *Nature*, **359**, 153.

21. Ballester, R., Marchuk, D., Boguski, M., Saulino, A., Letcher, R., Wigler, M., and Collins, F. (1990) The *NF1* locus encodes a protein functionally related to mammalian GAP and yeast *IRA* proteins. *Cell*, **63**, 851.

22. Hattori, M., Tsukamoto, N., Nur-e-Kamal, M. S., Rubinfeld, B., Iwai, K., Kubota, H., Maruta, H., and Minato, N. (1995) Molecular cloning of a novel mitogen-inducible nuclear protein with a Ran GTPase-activating domain that affects cell cycle progression. *Mol. Cell. Biol.*, **15**, 552.

23. Cullen, P. J. (1998) Bridging the GAP in inositol 1,3,4,5-tetrakisphoaphate signalling. *Biophys. Biochim. Acta*, **1436**, 35.

24. Chen, H. J., Rojas-Soto, M., Oguni, A., and Kennedy, M. B. (1998) A synaptic Ras-GTPase activating protein (p135 SynGAP) inhibited by CaM kinase II. *Neuron*, **20**, 895.

25. Marshall, C. J. (1996) Ras effectors. *Curr. Opinion Cell Biol.*, **8**, 197.

26. Burgering, B. M., de Vries-Smits, A. M., Medema, R. H., van Weeren, P. C., Tertoolen, L. G., and Bos, J. L. (1993) Epidermal growth factor induces phosphorylation of extracellular signal- regulated kinase 2 via multiple pathways. *Mol. Cell. Biol.*, **13**, 7248.

27. Sozeri, O., Vollmer, K., Liyanage, M., Frith, D., Kour, G., Mark, G. D., and Stabel, S. (1992) Activation of the c-Raf protein kinase by protein kinase C phosphorylation. *Oncogene*, **7**, 2259.

28. Nori, M., L'Allemain, G., and Weber, M. J. (1992) Regulation of tetradecanoyl phorbol acetate-induced responses in NIH3T3 cells by GAP, the GTPase-activating protein associated with p21c-ras. *Mol. Cell. Biol.*, **12**, 936.

29. Burgering, B. M. T., Pronk, G. J., van Weeren, P., Chardin, P., and Bos, J. L. (1993) cAMP antagonizes p21ras-directed activation of extracellular-signal regulated kinase 2 and phosphorylation of mSos nucleotide exchange factor. *EMBO J.*, **12**, 4211.

30. Rodriguez-Viciana, P., Warne, P. H., Dhand, R., Vanhaesebroeck, B., Gout, I., Fry, M. J., Waterfield, M. D., and Downward, J. (1994) Phosphatidylinositol 3-OH kinase as a direct target of Ras. *Nature*, **370**, 527.

31. Burgering, B. M. T. and Coffer, P. J. (1995) Protein kinase B (c-Akt) in phosphatidylinositol-3-OH kinase signal transduction. *Nature*, **376**, 599.

32. Franke, T. F., Yang, S. I., Chan, T. O., Datta, K., Kazlauskas, A., Morrison, D. K., Kaplan, D. R., and Tsichlis, P. N. (1995) The protein kinase encoded by the proto-oncogene Akt is a target of the PDGF-activated phosphatidylinositol 3-kinase. *Cell*, **81**, 727.

33. van Weeren, P. C., de Bruyn, K. M., de Vries-Smits, A. M., van Lint, J., and Burgering, B. M. (1998) Essential role for protein kinase B (PKB) in insulin-induced glycogen synthase kinase 3 inactivation. Characterization of dominant-negative mutant of PKB. *J. Biol. Chem.*, **273**, 13150.

34. Dudek, H. *et al.* (1997) Regulation of neuronal survival by the serine-threonine protein kinase Akt. *Science,* **275**, 661.

35. Kauffmann-Zeh, A., Rodriguez-Viciana, P., Ulrich, E., Gilbert, C., Coffer, P., Downward, J., and Evan, G. (1997) Suppression of c-Myc-induced apoptosis by Ras signalling through PI(3)K and PKB. *Nature,* **385**, 544.

36. Hawkins, P. T. *et al.* (1995) PDGF stimulates an increase in GTP-Rac via activation of phosphoinositide 3-kinase. *Curr. Biol.,* **5**, 393.

37. Hu, Q., Klippel, A., Muslin, A. J., Fantl, W., J., and Williams, L. T. (1995) Ras-dependent induction of cellular responses by constitutively active phosphatidylinositol-3 kinase. *Science,* **268**, 100.

38. Klinghoffer, R. A., Duckworth, B., Valius, M., Cantley, L., and Kauzlauskas, A. (1996) Platelet-derived growth factor-dependent activation of phospatidylinositol 3-kinase is regulated by receptor binding of SH2 domain containing proteins which influence Ras activaty. *Mol. Cell. Biol.,* **16**, 5905.

39. Rodriguez-Viciana, P., Warne, P. H., Vanhaesebroeck, B., Waterfield, M. D., and Downward, J. (1996) Activation of phosphoinositide 3-kinase by interaction with Ras and by point mutation. *EMBO J.,* **15**, 2442.

40. van Weering, D. H., de Rooij, J., Marte, B., Downward, J., Bos, J. L., and Burgering, B. M. (1998) Protein kinase B activation and lamellipodium formation are independent phosphoinositide 3-kinase-mediated events differentially regulated by endogenous Ras. *Mol. Cell. Biol.,* **18**, 1802.

41. Albright, C. F., Giddings, B. W., Liu, J., Vito, M., and Weinberg, R. A. (1993) Characterization of a guanine nucleotide dissociation stimulator for a ras-related GTPase. *EMBO J.,* **12**, 339.

42. Hofer, F., Fields, S., Schneider, C., and Martin, G. S. (1994) Activated Ras interacts with the Ral guanine nucleotide dissociation stimulator. *Proc. Natl Acad. Sci. USA,* **91**, 11089.

43. Spaargaren, M. and Bischoff, J. R. (1994) Identification of the guanine nucleotide dissociation stimulator for Ral as a putative effector molecule of R-ras, H-ras, K-ras, and Rap. *Proc. Nat. Acad. Sci. USA,* **91**, 12609.

44. Kikuchi, A., Demo, S. D., Ye, Z. H., Chen, Y. W., and Williams, L. T. (1994) ralGDS family members interact with the effector loop of ras p21. *Mol. Cell. Biol.,* **14**, 7483.

45. Wolthuis, R. M. *et al.* (1996) RalGDS-like factor (Rlf) is a novel Ras and Rap 1A-associating protein. *Oncogene,* **13**, 353.

46. Urano, T., Emkey, R., and Feig, L. A. (1996) Ral-GTPases mediate a distinct downstream signaling pathway from Ras that facilitates cellular transformation. *EMBO J.,* **15**, 810.

47. Katz, M. E. and McCormick, F. (1997) Signal transduction from multiple Ras effectors. *Curr. Biol.,* **7**, 75.

48. White, M. A., Nicolette, C., Minden, A., Polverino, A., Van Aelst, L., Karin, M., and Wigler, M. H. (1995) Multiple Ras functions can contribute to mammalian cell transformation. *Cell,* **80**, 533.

49. Rodriguez-Viciana, P. *et al.* (1997) Role of phosphoinositide 3-OH kinase in cell transformation and control of the actin cytoskeleton by Ras. *Cell,* **89**, 457.

50. Marshall, C. J. (1995) Specificity of receptor tyrosine kinase signaling: transient versus sustained extracellular signal-regulated kinase activation. *Cell,* **80**, 179.

51. Mulcahy, L. S., Smith, M. R., and Stacey, D. W. (1985) Requirements for *ras* proto-oncogene function during serum-stimulated growth of NIH 3T3 cells. *Nature,* **313**, 241.

52. Taylor, S. J. and Shalloway, D. (1996) Cell cycle-dependent activation of Ras. *Curr. Biol.,* **6**, 1621.

53. Peeper, D. S., Upton, T. M., Ladha, M. H., Neuman, E., Zalvide, J., Bernards, R., DeCaprio,

J. A., and Ewen, M. E. (1997) Ras signalling linked to the cell-cycle machinery by the retinoblastoma gene. *Nature,* **386**, 177.

54. Mittnacht, S., Paterson, H., Olson, M. F., and Marshall, C. J. (1997) Ras signalling is required for inactivation of the tumour suppressor pRb cell-cycle control protein. *Curr. Biol.,* **3**, 219.

55. Olson, M. F., Paterson, H. F., and Marshall, C. J. (1998) Signals from Ras and Rho GTPases interact to regulate expression of p21Waf1/Cip1. *Nature,* **394**, 295.

56. Sewing, A., Wiseman, B., Lloyd, A. C., and Land, H. (1997) High-intensity Raf signal causes cell cycle arrest mediated by p21Cip1. *Mol. Cell. Biol.,* **17**, 5588.

57. Woods, D., Parry, D., Cherwinski, H., Bosch, E., Lees, E., and McMahon, M. (1997) Raf-induced proliferation or cell cycle arrest is determined by the level of Raf activity with arrest mediated by p21Cip1. *Mol. Cell. Biol.,* **17**, 5598.

58. Swan, K. A., Alberola-Ila, J., Gross, J. A., Appleby, M. W., Forbush, K. A., Thomas, J. F., and Perlmutter, R. M. (1995) Involvement of p21ras distinguishes positive and negative selection in thymocytes. *EMBO J.,* **14**, 276.

59. Johnson, L. *et al.* (1997) K-ras is an essential gene in the mouse with partial functional overlap with N-ras. *Genes Dev.,* **11**, 2468.

60. Sternberg, P. W. and Han, M. (1998) Genetics of Ras signaling in C. *elegans. Trends Genet.,* **14**, 466.

61. Sternberg, P. W. and Alberola-Ila, J. (1998) Conspiracy theory: Ras and Raf do not act alone. *Cell,* **95**, 447.

62. Rommel, C. and Hafen, E. (1998) *Curr. Opin. Genet. Dev.,* **8**, 412.

63. Pronk, G. J. and Bos, J. L. (1994) The role of p21ras in receptor tyrosine kinase signalling. *Biochim. Biophys. Acta,* **1198**, 131.

64. Chardin, P. and Tavitian, A. (1986) The Ral gene: a new Ras-related gene isolated by the use of a synthetic probe. *EMBO J.,* **5**, 2203.

65. Feig, L. A., Urano, T., and Cantor, S. (1996) Evidence for a Ras/Ral signaling cascade. *Trends Biochem. Sci.,* **21**, 438.

66. Jiang, H., Luo, J. Q., Urano, T., Frankel, P., Lu, Z., Foster, D. A., and Feig, L. A. (1995) Involvement of Ral GTPase in v-Src-induced phospholipase D activation. *Nature,* **378**, 409.

67. Noda, M. (1993) Structure and function of Krev-1 transformation suppressor gene and its relatives. *Biochim. Biophys. Acta,* **1155**, 97.

68. Bos, J. L. (1998) All in the family? New insights and questions regarding interconnectivity of Ras, Rap1 and Ral. *EMBO J.,* **17**, 6776.

69. Zwartkruis, F. J., Wolthuis, R. M., Nabben, N. M., Franke, B., and Bos, J. L. (1998) Extra-cellular signal-regulated activation of Rap1 fails to interfere in Ras effector signalling. *EMBO J.,* **17**, 5905.

70. Kawasaki, H. *et al.* (1998) A Rap guanine nucleotide exchange factor enriched highly in the basal ganglia. *Proc. Natl Acad. Sci. USA,* **95**, 13278.

71. de Rooij, J., Zwartkruis, F. J., Verheijen, M. H., Cool, R. H., Nijman, S. M., Wittinghofer, A., and Bos, J. L. (1998) Epac is a Rap1 guanine-nucleotide-exchange factor directly activated by cyclic AMP. *Nature,* **396**, 474.

72. Kawasaki, H., Springett, G. M., Mochizuki Shinichiro Toki, N., Nakaya, M., Matsuda, M., Housman, D. E., and Graybiel, A. M. (1998) A family of cAMP-binding proteins that directly activate rap1. *Science,* **282**, 2275.

73. Altschuler, D. and Lapetina, E. G. (1993) Mutational analysis of the cAMP-dependent protein kinase-mediated phosphorylation site of Rap1b. *J. Biol. Chem.,* **268**, 7527.

74. Chant, J., Corrado, K., Pringle, J. R., and Herskowitz, I. (1991) Yeast BUD5, encoding a

putative GDP-GTP exchange factor, is necessary for bud site selection and interacts with bud formation gene BEM1. *Cell,* **65**, 1213.

75. Bender, A. (1993) Genetic evidence for the roles of the bud-site-selection genes BUD5 and BUD2 in control of the Rsr1p (Bud1p) GTPase in yeast. *Proc. Natl Acad. Sci. USA,* **90**, 9926.

76. Park, H. O., Chant, J., and Herskowitz, I. (1993) BUD2 encodes a GTPase-activating protein for Bud1/Rsr1 necessary for proper bud-site selection in yeast. *Nature,* **365**, 269.

77. Chenevert, J., Corrado, K., Bender, A., Pringle, J., and Herskowitz, I. (1992) A yeast gene (BEM1) necessary for cell polarization whose product contains two SH3 domains. *Nature,* **356**, 77.

78. Michelitch, M. and Chant, J. (1996) A mechanism of Bud1p GTPase action suggested by mutational analysis and immunolocalization. *Curr. Biol.,* **6**, 446.

79. Park, H. O., Bi, E., Pringle, J. R., and Herskowitz, I. (1997) Two active states of the Ras-related Bud1/Rsr1 protein bind to different effectors to determine yeast cell polarity. *Proc. Natl Acad. Sci. USA,* **94**, 4463.

80. Jimenez, B., Pizon, V., Lerosey, I., Beranger, F., Tavitian, A., and de Gunzburg, J. (1991) Effects of the ras-related rap2 protein on cellular proliferation. *Int. J. Cancer,* **49**, 471.

81. Reedquist, K. A. and Bos, J. L. (1998) Costimulation through CD28 suppresses T cell receptor-dependent activation of the Ras-like small GTPase Rap1 in human T lymphocytes. *J. Biol. Chem.,* **273**, 4944.

82. Torti, M., Ramaschi, G., Sinigaglia, F., Lapetina, E. G., and Balduini, C. (1994) Glycoprotein IIb-IIIa and the translocation of Rap2B to the platelet cytoskeleton. *Proc. Natl Acad. Sci. USA,* **91**, 4239.

83. Lowe, D. G., Capon, D. J., Delwart, E., Sakaguchi, A. Y., Naylor, S. L., and Goeddel, D. V. (1987) Structure of the human and murine R-ras genes, novel genes closely related to ras proto-oncogenes. *Cell,* **48**, 137.

84. Yamamoto, T., Matsui, T., Nakafuku, M., Iwamatsu, A., and Kaibuchi, K. (1995) A novel GTPase-activating protein for R-Ras. *J. Biol. Chem.,* **270**, 30557.

85. Cullen, P. J., Hsuan, J. J., Truong, O., Letcher, A. J., Jackson, T. R., Dawson, A. P., and Irvine, R. F. (1995) Identification of a specific Ins(1,3,4,5)P4-binding protein as a member of the GAP1 family. *Nature,* **376**, 527.

86. Garrett, M. D., Self, A. J., van Oers, C., and Hall, A. (1989) Identification of distinct cytoplasmic targets for ras/R-ras and rho regulatory proteins. *J. Biol. Chem.,* **264**, 10.

87. Rey, I., Taylor-Harris, P., van Erp, H., and Hall, A. (1994) R-ras interacts with rasGAP, neurofibromin and c-raf but does not regulate cell growth or differentiation. *Oncogene,* **9**, 685.

88. Wang, H. G., Millan, J. A., Cox, A. D., Der, C. J., Rapp, U. R., Beck, T., Zha, H., and Reed, J. C. (1995) R-Ras promotes apoptosis caused by growth factor deprivation via a Bcl-2 suppressible mechanism. *J. Cell. Biol.,* **129**, 1103.

89. Zhang, Z., Vuori, K., Wang, H., Reed, J. C., and Ruoslahti, E. (1996) Integrin activation by R-ras. *Cell,* **85**, 61.

90. Chan, A. M., Miki, T., Meyers, K. A., and Aaronson, S. A. (1994) A human oncogene of the RAS superfamily unmasked by expression cDNA cloning. *Proc. Natl Acad. Sci. USA,* **91**, 7558.

91. Huang, Y., Saez, R., Chao, L., Santos, E., Aaronson, S. A., and Chan, A. M. (1995) A novel insertional mutation in the TC21 gene activates its transforming activity in a human leiomyosarcoma cell line. *Oncogene,* **11**, 1255.

92. Graham, S. M., Cox, A. D., Drivas, G., Rush, M. G., D'Eustachio, P., and Der, C. J. (1994)

Aberrant function of the Ras-related protein TC21/R-Ras2 triggers malignant transformation. *Mol. Cell. Biol.,* **14**, 4108.

93. Kimmelman, A., Tolkacheva, T., Lorenzi, M. V., Osada, M., and Chan, A. M. (1997) Identification and characterization of R-ras3: a novel member of the RAS gene family with a non-ubiquitous pattern of tissue distribution. *Oncogene,* **15**, 2675.

94. Matsumoto, K., Asuno, T., and Endo, T. (1997) Novel small GTPase M-Ras participates in reorganization of actin cytoskeleton. *Oncogene,* **15**, 2409.

95. Yamagata, K., Sanders, L. K., Kaufmann, W. E., Yee, W., Barnes, C. A., Nathans, D., and Worley, P. F. (1994) rheb, a growth factor- and synaptic activity-regulated gene, encodes a novel Ras-related protein. *J. Biol. Chem.,* **269**, 16333.

96. Gromov, P. S., Madsen, P., Tomerup, N., and Celis, J. E. (1995) A novel approach for expression cloning of small GTPases: identification, tissue distribution and chromosome mapping of the human homolog of rheb. *FEBS Lett.,* **377**, 221.

97. Yee, W. M. and Worley, P. F. (1997) Rheb interacts with Raf-1 kinase and may function to integrate growth factor- and protein kinase A-dependent signals. *Mol. Cell. Biol.,* **17**, 921.

4 | Rho

ANNE J. RIDLEY

1. Introduction

Rho was first identified in 1985 as a cDNA encoding a small GTP-binding protein related to Ras. Later, Rho was identified as the target of the *Clostridium botulinum* exoenzyme C3 transferase, which had been shown previously to induce changes to the morphology of mammalian tissue-culture cells. Subsequently, Rho and then a closely related protein, Rac, were shown to be major regulators of the actin cytoskeleton in all eukaryotic cells. A major surprise at that time was that Rac was also required for activation of the NADPH oxidase enzyme complex in phagocytic cells, a process seemingly independent of the actin cytoskeleton. Since then, additional members of the Rho family have been identified and many other cellular responses have been linked to them, including activation of MAPK cascades and regulation of transcription factors, secretion, endocytosis, cell polarity, and the cell cycle. Members of the Rho family have emerged as important coordinators of signal transduction processes activated by a wide variety of both extracellular and intrinsic cell cycle linked signals. Accordingly, they play crucial roles in the development and behaviour of multicellular organisms.

2. Structure and expression of Rho GTPases

2.1 Sequence and structure

Rho GTPases have been found in all eukaryotic species examined, from yeast and worms to plants and mammals (Table 1). Different mammalian Rho GTPases are at least 40% identical to each other at the amino-acid level, whereas they are approximately 25% identical to Ras. To date, only Rho, Rac, and Cdc42 have been characterized extensively. *Rho* was the first Rho family gene to be cloned, initially from the sea slug, *Aplysia*, and subsequently from humans (1). In mammals, there are three highly homologous isoforms of Rho, known as RhoA, RhoB, and RhoC, which are over 85% identical at the amino-acid level. The majority of differences lie within the last 15 amino acids of the carboxy terminus. Similarly, Rac1, Rac2, and Rac1B/Rac3 are over 88% identical, and differ primarily within the carboxy-terminal 13 amino acids (2, 3). The *Cdc42* gene was initially identified in *S. cerevisiae* as a cell cycle mutant defective in budding (4). The two mammalian isoforms of Cdc42 are

Table 1 Rho GTPases. Members of the Rho family are listed for mammals and for selected model organisms where their function has been analysed in most detail. Some of these are grouped into subfamilies based on their homology to mammalian Rho, Rac, or Cdc42. Other members (Others) are not organized into subfamilies as homologues of the mammalian proteins have so far not been identified in other species. Rho2 in *S. cerevisiae* and *S. pombe* are homologues, however.

	Mammals	*S. cerevisiae*	*S. pombe*	*Drosophila*	*Dictyostelium*	*C. elegans*
Rho	Rho (A, B, C)	Rho1	Rho1	Rho1		RhoA
Rac	Rac (1, 2, 3)			Rac (1, 2)	Rac (1A, 1B, B)	Rac (1, 2)
Cdc42	Cdc42, G25K	Cdc42	Cdc42	Cdc42		
Others	RhoD	Rho2	Rho2	RhoL	RacA	mig2
	RhoE/Rnd1	Rho3			RacC	
	Rnd2	Rho4				
	Rnd3					
	RhoG				RacD	
	TC10				RacE	
	TTF					

actually two alternatively spliced variants of the same gene, with different carboxy-terminal sequences (5, 6). Subsequently, other members of the Rho family have been identified either through deliberate searches for Ras-homologous genes, or by serendipity, for example as a serum-inducible gene (RhoG) (7).

When the sequences of Rho GTPases are aligned with Ras, it is apparent that most show conservation of specific amino acids critically required for nucleotide binding. However, four members of the family, TTF, RhoE (also known as Rnd3), Rnd1, and Rnd2, have amino-acid substitutions that are predicted to inhibit GTP hydrolysis (Table 2) (8–10). In addition, members of the Rho family have a stretch of 13 additional amino acids (compared to Ras) encompassing amino acids 123–135 (Rac numbering) and referred to as the 'insert region' (Fig. 1). Exceptions to this are TTF,

Table 2 Amino-acid substitutions in Rho GTPase. Amino acids 12, 59, and 61 (Ras/Rac numbering) are identical in all Ras and Rho family members, with the exception of Rnd1, Rnd2, RhoE/Rnd3, and TTF, which have the substitutions shown. These substitutions in Ras convert it to an oncogenic protein (see Chapter 3).

Amino acid

	12	59	61
Ras/Rho	G	A	Q
Ras oncogene	Any except P	T	Any except Q/P
Rnd1	V	S	S
Rnd2	A	S	S
RhoE/Rnd3	S	S	S
TTF	S	A	N

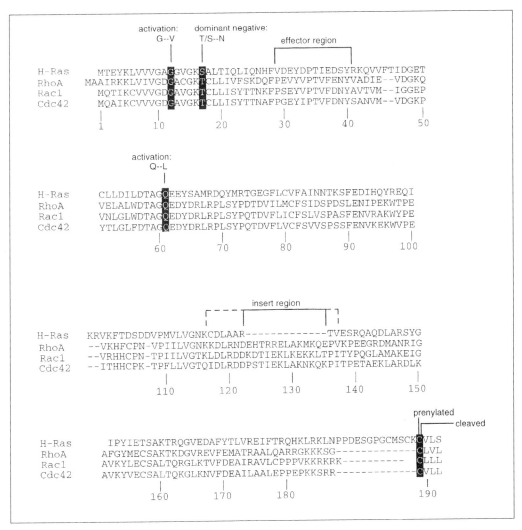

Fig. 1 Amino-acid sequence alignment of human H-Ras, RhoA, Rac1, and Cdc42 proteins. Structurally important regions of the proteins involved in interaction with other proteins are indicated. The effector region is required for interaction with GAPs and downstream target proteins; the insert region, present in Rho family members but not in other Ras superfamily members, is required for interaction with guanine nucleotide dissociation inhibitors (GDIs) and some downstream targets; and the C-terminal cysteine is prenylated, allowing interaction with membranes and GDIs. Amino acids that are commonly mutated to produce proteins with altered behaviour are also indicated (white lettering): mutation of amino acids 12 and 61 generates GTPase-defective mutants, and mutation of amino acid 17 generates dominant-negative mutants. Numbers refer to Rac/Cdc42.

which has only eight additional amino acids in this region (Fig. 2), and the *Drosophila* RhoL protein, which lacks the insert region altogether. Several members of the Rho family also have extensions either at their amino or carboxy terminus, compared to Rac and Cdc42, which are the shortest members of the family.

The crystal structures of RhoA complexed with GDP (11) and of Rac1 complexed

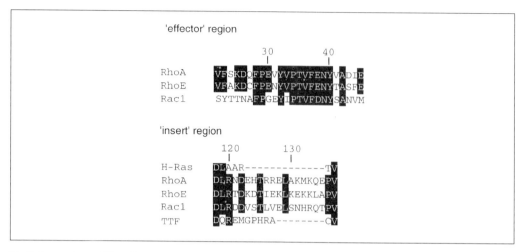

Fig. 2 Comparison of the effector and insert regions of Rho proteins. Alignment of the amino acid sequences of effector and insert regions, highlighting (white lettering) identical amino acids. Note that TTF has a shorter insert region and does not have any of the conserved amino acids found in all other Rho family proteins that are important for formation of the loop structure of this region (D, T, L, P). Numbers refer to Rac/Cdc42.

with a GTP analogue (12) have been solved, showing that their overall structure is very similar to Ras but that the insert region (around residues 123–135) forms part of an extra α-helical domain, not present in Ras, encompassing amino acids 117–137 (Rac numbering) (see Chapter 9). Although the conformation of this insert region is not altered in the GDP- and GTP-bound forms (11–13), it is clear that it is important for some functions of Rho GTPases (14, 15).

2.2 Post-translational modifications

As with other Ras superfamily members, Rho GTPases are modified by prenylation and carboxymethylation of a conserved cysteine four amino acids from the C terminus, followed by removal of the last three amino acids (Fig. 1) (16, 17). RhoA, RhoC, Rac1, Rac2, and Cdc42 proteins are all prenylated by a 20-carbon-chain geranylgeranyl group, while RhoD and RhoE are prenylated by a 15-carbon-chain farnesyl group (9, 18). RhoB is unique in that it can either be geranylgeranylated or farnesylated (16). Prenylation has been shown to be important for at least some, although not all, of the functions of Rho proteins (19, 20), and facilitates their association with membranes (16, 21). The farnesylation of RhoB is important for its localization to an early/late endosomal compartment (22, 23). In addition to prenylation, RhoA can be phosphorylated by protein kinase A (PKA) *in vitro* on serine at residue 188 (not present in RhoB or RhoC), and this has been suggested to regulate its activity (24).

Rho GTPases are covalently modified by a number of bacterial toxins and exoenzymes (see Chapter 10) (25). The most extensively used is C3 transferase, an exoenzyme produced by *Clostridium botulinum* that ADP-ribosylates Rho at amino acid 41 and thereby inactivates it. The enzyme is highly specific for Rho and shows at least 50-fold less activity towards Rac or Cdc42. *Clostridium difficile* toxins A and B, on the other hand, glucosylate and inactivate Rho, Rac, and Cdc42 (26). These toxins have been very useful in determining whether Rho proteins are required for a particular response. Yet other toxins activate Rho proteins, including *E. coli* cytotoxic necrotizing factor-1, which deamidates Gln63 in Rho (Gln61 in Rac/Cdc42) to Glu, creating constitutively active versions of the proteins (27, 28). These may also prove to be useful in elucidating the function of Rho GTPases.

2.3   Expression of Rho GTPases

In most cases, Rho proteins are widely expressed in different tissues or in mammalian cell lines (6, 9, 29, 30), although *Rac*2 is only expressed in haematopoietic cells (31), and in chick embryos the *Rac*1B/*Rac*3 gene appears to be expressed selectively in the developing nervous system (3). *RhoB*, *RhoG*, and *Rac*1B/*Rac*3 are serum/growth factor-inducible genes (2, 7, 32), but *RhoB* transcription is also upregulated by damage-inducing agents such as UV irradiation (33, 34). In addition, RhoB protein accumulates during the G_1 phase of the cell cycle, and in response to TGF-β (transforming growth factor β) (35). So far, however, it is unclear what roles changes in protein levels play in signalling or cell cycle control.

3.   Mechanisms of activating and inactivating Rho GTPases

In general, Rho proteins can bind GDP and GTP, and hydrolyse GTP. They also bind to proteins in the cytoplasm known as guanine nucleotide dissociation inhibitors (GDIs), which are thought to sequester them in an inactive form. They are generally assumed to be active when bound to GTP, and inactive when bound to GDP (Fig. 3). Mutant forms of the proteins that are defective for GTP hydrolysis generally induce a well-defined cellular response, although it is also likely that cycling between GTP- and GDP-bound forms is important for the normal function of these proteins. Proving that extracellular signals 'activate' Rho family proteins by increasing the level of GTP-bound protein has been difficult, but there is now good evidence that this does indeed occur in response to different signals (36, 37).

An exception to the pattern of GTP/GDP cycling is found in RhoE and Rnd1, which do not detectably bind GDP and appear to be unable to hydrolyse GTP (9, 10, 38). This is a consequence of specific amino-acid substitutions in these proteins, which in other Ras superfamily members are known to inhibit intrinsic GTP hydrolysis rates (Table 2). Indeed, mutation of these amino acids in RhoE to residues found at the equivalent positions in Ras convert the protein to an effective GTPase (9). Two other Rho family proteins, Rnd2 and TTF, have similar amino-acid substitutions to RhoE and Rnd1 (Table 2), and are therefore also presumed to have very

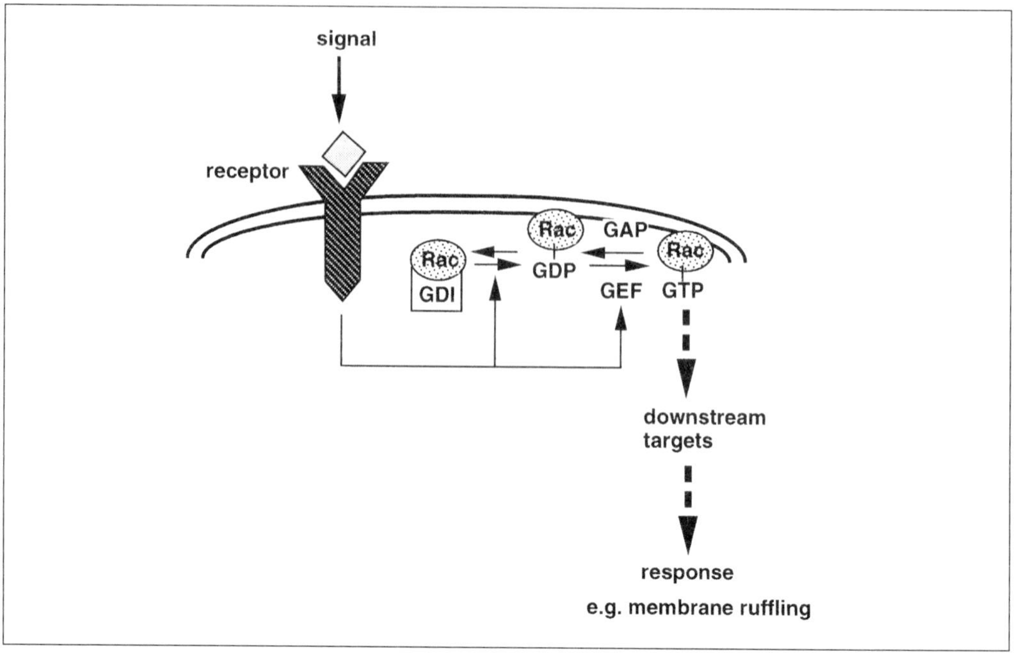

Fig. 3 Model for activation of Rho GTPases by extracellular signals. Rho proteins are held in an inactive complex with GDIs in the cytoplasm. Extracellular signals induce an increase in active, GTP-bound protein by stimulating the release of GDI and the translocation of the GTPase to membranes. Exchange factors stimulate release of GDP and allow GTP to bind, and the active conformation of the protein interacts with downstream targets, or effectors, to produce a response. The signal is terminated by GTPase-activating proteins that stimulate hydrolysis of bound GTP.

low rates of GTP hydrolysis (8, 10). This suggests that the regulation of these proteins will be significantly different from that of other Ras superfamily members.

3.1 Nucleotide exchange and exchange factors

The intrinsic rate of nucleotide exchange on Rho proteins is normally very slow in the presence of physiological concentrations of Mg^{2+} (39). A number of amino-acid substitutions can alter this intrinsic nucleotide exchange rate: for example, mutation of amino acid 28 from Phe to Leu in Cdc42 speeds up nucleotide exchange, leading to accumulation of the protein predominantly in a GTP-bound form in cells (40). However, nucleotide exchange is normally catalysed in cells by exchange factors, and well over 30 potential exchange factors (GEFs) for Rho GTPases have been identified (41, 42). All possess a conserved exchange factor domain, known as the Dbl homology (DH) domain, adjacent to a pleckstrin homology (PH) domain. They are referred to as the Dbl family, after the first identified member of the Rho GEF family, Dbl, which was isolated through its ability to transform NIH3T3 fibroblasts onco-genically (43). Its DH domain was subsequently found to be essential for cellular transformation, and also to be sufficient by itself to catalyse nucleotide exchange on

Rho and Cdc42 (44). Several other members of the Dbl family were similarly identified in fibroblast transformation assays (42), although no Rho GEF has yet been found to be mutated or amplified in any human cancer.

The function of the PH domain is less clear but it may, in part at least, be required for correct localization and orientation of the DH domain at the plasma membrane (45–47), where many GEFs and their target Rho GTPases are presumed to act. In the case of Dbl, the PH domain is not required for exchange factor activity *in vitro*, but is required for its transforming activity (see Chapter 9) (41, 45, 48). Many Dbl family proteins are large proteins and contain additional recognizable domains, such as SH2 and SH3 domains, and thus may act to bring GEFs into a complex with other signalling molecules (41).

The majority of Dbl family GEFs are active on several Rho GTPases *in vitro*, although they normally show preference for one or a subgroup of proteins. Over-expression of a GEF in cells usually leads predominantly to one form of actin rearrangement, reflecting the preference of the GEF for Rho, Rac, or Cdc42 *in vitro* (49, 50), but some effects of GEF expression do not appear to correlate with activation of a particular Rho GTPase (51, 52). This raises the possibility that other domains of the GEF may dictate the pattern of signalling. Interestingly, the Ras exchange factors Sos and RasGRF also have DH/PH domains (see Chapter 9), and appear to co-ordinate activation of Rac by Ras (53, 54), while Trio has two DH/PH domains, one active on Rho and the other on Rac, and this could therefore coordinately activate two Rho GTPases (55).

Several Dbl family GEFs show restricted expression patterns: Vav, for example, is expressed only in haematopoeitic cells, whereas Tiam-1 shows highest expression in the brain, although it is expressed elsewhere (42, 56). This may explain partly why there are far more GEFs than Rho proteins. It is also possible that different GEFs are activated in response to different extracellular stimuli, or that each GEF acts as a scaffold protein, bringing Rho GTPases together with a subset of signalling proteins. This could explain why certain Rac/Cdc42 GEFs stimulate activation of PAK1 but not JNK, although constitutively active Rac or Cdc42 can activate both these kinases (57).

Several mechanisms have been suggested for how GEF activity might be regulated in cells. Oncogenic activation of Dbl family members normally involves truncation of the N-terminal region (41), suggesting that this region negatively regulates exchange factor activity, and that activation involves a conformational change. In general, activation of Rac-dependent responses by tyrosine kinase receptors is dependent on phosphatidylinositide 3-kinase (PI3K) activity, which stimulates production of phos-phatidylinositol 3,4,5-trisphosphate (PI3,4,5P$_3$) (58). It is postulated that PI3,4,5P$_3$ regulates GEFs in some way, and an attractive hypothesis is that it binds to their PH domains. In the case of Vav, there is good evidence that it is activated by tyrosine phosphorylation in response to extracellular signals (by the non-receptor tyrosine kinase, lck) (59), and that binding of PI3,4,5P$_3$ enhances tyrosine phosphorylation and therefore Vav exchange factor activity (60). Relocalization of some GEFs in response to extracellular stimuli is likely to be important for their action and may also reflect binding to phosphoinositides. Tiam-1, for example, relocalizes from the cytoplasm to the plasma membrane following stimulation of cells with growth factors, and this is required for activation of Rac-dependent responses (61). Interestingly, Tiam-1 has an extra N-terminal PH domain in addition to the one adjacent to the DH domain, and it is the N-terminal domain that is required for membrane localization (61). Sos also accumulates at the plasma membrane following serum stimulation, but here the single PH domain adjacent to the DH domain facilitates interaction with the plasma membrane (47).

Dominant negative forms of Rho proteins have been particularly useful in analysing their function, and are believed to inhibit the activity of their respective endogenous GTPases by competing for binding to exchange factors (62). The most widely used dominant-negative proteins are created by substituting amino acid 17 (Rac numbering) from Thr to Asn (Fig. 1). This amino acid coordinates to an essential Mg^{2+} ion required for guanine nucleotide binding in all Ras superfamily GTP-binding proteins (see Chapter 9) (11, 63). Substitution of this amino acid is therefore predicted to interfere with both Mg^{2+} and nucleotide binding, and indeed Rac and Cdc42 proteins with this mutation have a much lower affinity for GTP/GDP than their wild-type counterparts (39).

3.2   GTP hydrolysis and GAPs

Once Rho proteins are activated and bound to GTP, their activity is terminated by GTP hydrolysis, yielding bound GDP and free phosphate. Some Rho proteins, for example Rac and Cdc42, have a relatively high rate of spontaneous GTP hydrolysis *in vitro* in comparison to Ras (39). However, as with both Ras (see Chapter 3) and heterotrimeric G proteins (see Chapters 1 and 2), the intrinsic GTP hydrolysis rate can be stimulated by GTPase-activating proteins (GAPs). Mutation of a number of conserved amino acids in Ras superfamily proteins decreases the intrinsic and GAP-stimulated GTPase activity, allowing these proteins to remain predominantly in the GTP-bound, active form in cells (63, 64). Proteins with mutations of amino acid 12 (Rac numbering) from Gly to Val, or of amino acid 61 from Glu to Leu, are the most commonly used constitutively activated forms of Rho GTPases (Fig. 1) with reduced GTPase activity. The crystal structures show that these amino acids are positioned very similarly in Rho, Rac, and Ras, and are therefore predicted to function similarly in facilitating GTP hydrolysis (11, 12) (see Chapter 9).

GAPs for Rho family proteins all share a related 140-amino-acid domain, known as the RhoGAP domain, which is sufficient to confer GAP activity. As with GEFs, most GAP domains are actually part of much larger, multidomain proteins (65). Bcr, for example, not only has a GAP domain, but also a DH/PH, GEF-like domain. Micro-injection or expression of isolated GAP domains in cells can inhibit Rho-, Rac-, and Cdc42-mediated actin reorganization (66, 67), but the other domains of the proteins can alter their signalling function. For example, N-chimerin shows GAP activity for Rac, but overexpression of full-length N-chimerin has been reported to stimulate rather than inhibit the formation of lamellipodia and filopodia (68). So far, little is known about how GAPs are regulated in cells, although, as with GEFs, changes in subcellular localization and their incorporation into multimolecular signalling complexes are likely to be important (69).

3.3 GDIs

GDIs bind with high affinity to prenylated Rho proteins and, although there is little direct evidence, it is believed that they maintain inactive Rho GTPases in the cytoplasm, until an appropriate stimulus induces dissociation of the complex and concomitant binding of the Rho protein to membranes (70). *In vitro*, binding of GDIs to Rho proteins both prevents nucleotide exchange (thereby preventing activation) and prevents intrinsic and GAP-stimulated GTP hydrolysis (thereby preventing deactivation). GDIs can also extract Rho proteins from membranes. Overexpression of GDIs in cells generally inhibits signalling pathways normally regulated by Rho proteins, including actin reorganization and activation of JNK (71, 72). These observations are consistent with GDIs acting as negative regulators of Rho signalling.

Three mammalian GDIs specific for the Rho family have been characterized: RhoGDI, D4/LyGDI, and GDIγ/RhoGDI-3. RhoGDI is ubiquitously expressed, whereas D4/LyGDI is expressed only in haematopoietic cells, and GDIγ/RhoGDI-3 is expressed preferentially in the brain and pancreas (73, 74). A RhoGDI has also been identified and characterized in *S. cerevisiae* (75).

C-terminal prenylation of Rho proteins is essential for efficient binding of GDIs, and structures of RhoGDI determined by X-ray crystallography and NMR spectroscopy indicate that it contains a hydrophobic cleft for binding to isoprenes (see Chapter 9) (76, 77). This explains how GDIs keep Rho proteins in a soluble complex in the cytoplasm, by masking the membrane-binding prenyl group. The isoprene-binding cleft of RhoGDI is found in the highly structured C-terminal portion, whereas the N-terminal portion of RhoGDI is unstructured in solution, but is necessary for inhibition of nucleotide dissociation from Cdc42 (76, 77). In addition to the prenylated C terminus of Rho proteins, it is likely that other regions also interact with GDIs. Interestingly, the insert region of Cdc42 (Figs 1 and 2) is not required for RhoGDI binding but is required for GDI-dependent inhibition of GDP dissociation and GTP hydrolysis (15), suggesting that this region might interact with RhoGDI.

The three mammalian GDIs show differences in their activities and specificities, which are presumably reflected in different structures. D4/LyGDI is ten- to twenty-fold less active than RhoGDI in inhibiting GDP dissociation, and has a similarly lower affinity for RhoA, Rac1, and Cdc42, explaining why it is not detectably complexed to these proteins in cells. RhoGDI, on the other hand, can be purified in a complex with all three GTPases (78). A few amino-acid differences near the C terminus of D4/LyGDI appear to be responsible for its reduced activity relative to RhoGDI, and in particular residue 177, which forms part of the isoprene-binding pocket, is critical for the difference in affinity between the two GDIs (76, 79). RhoGDIγ shows preferential binding to RhoB and RhoG over RhoA and Rac1 (80), suggesting that it may have a higher affinity for farnesylated over geranylgeranylated proteins (see Section 2.2 above). In addition, it is unique among GDIs in being membrane associated rather than cytoplasmic (80, 81).

As well as keeping Rho family members in an inactive complex in the cytoplasm, GDIs may play a role in regulating translocation of the proteins to the plasma membrane. RhoGDI interacts with ezrin, radixin, and moesin (ERM) proteins, which in turn interact with the transmembrane protein CD44 and the cell-cell adhesion molecule, ICAM-1 and also with actin (82). The amino-terminal region of radixin can inhibit RhoGDI activity *in vitro* and induce release of Rho from RhoGDI (83). As ERM proteins can adopt either a 'closed' or 'open' conformation, stimuli that induce unfolding of ERM proteins to expose the amino-terminal region could lead to release of Rho proteins from RhoGDI and their subsequent availability for activation. Interestingly, ERM proteins can also bind to the Rho GEF, Dbl (84), suggesting that they may coordinately induce Rho release from GDIs and enhance nucleotide exchange.

Genetic studies of the *in vivo* function of GDIs have been carried out in *S. cerevisiae*, where deletion of the RhoGDI gene has no apparent phenotype (75). Disruption of the D4/LyGDI gene in mouse embryonic stem cells had only a small effect on superoxide production in macrophages derived from these stem cells (85), suggesting that D4/LyGDI does not play a major role in regulating Rac delivery to the superoxide complex in these cells. D4/LyGDI-deficient mice also show no striking abnormalities but do have a subtle defect in the apoptosis of lymph node cells following interleukin-2 (IL-2) withdrawal (86). Interestingly, D4/LyGDI is a substrate for apoptotic caspases, and is inactivated following cleavage (87, 88), although whether this is physiologically relevant for the process of apoptosis is not known.

4. Downstream interacting proteins

Rho GTPases interact directly with a large number of different target or effector proteins, which are believed to mediate downstream effects. Target proteins can be divided into groups depending on the presence of conserved sequences required for interaction with each GTPase. The majority of target proteins recognize preferentially the GTP-bound conformation of Rho proteins, but some, such as PIP-5K (phosphoinositol (4) phosphate-5 kinase) and the MAP kinase kinase kinase, MEKK4, do not appear to distinguish between the GTP- and GDP-bound conformations (42, 89–91). Many targets for Rho, Rac, and Cdc42 have been characterized, some of which are specific for only one of these three proteins, whereas others recognize two or more (Figs 4 and 5). Little is known about targets for other Rho family members, although TC10 is most closely homologous to Cdc42 and shares some, although not all, of its targets (92).

4.1 Rho targets

Many Rho-binding proteins fall into two classes based on sequence homology in the Rho-binding regions, although others do not fit into either class (Fig. 4) (93, 94). There is no significant homology between the REM-1 motif (Rho effector motif, class 1) or

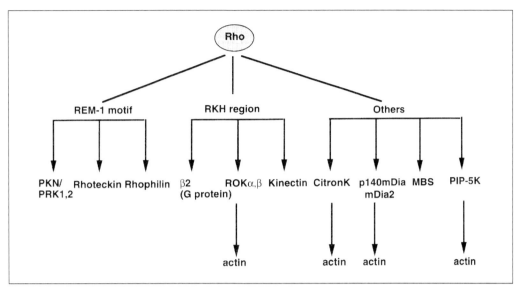

Fig. 4 Downstream targets for Rho. Proteins that have been shown to interact directly with GTP-bound Rho are subdivided into three groups, based upon sequence similarities in the Rho-binding region of each target. Those targets that are thought to affect actin organization are indicated.

the RKH region (ROK-kinectin homology region, class 2). This suggests that they recognize different subsets of amino acids on Rho, and indeed Rhophilin (REM-1 motif) binding requires amino acids 72–90 of Rho, whereas ROK (RKH region) requires both this region and a region encompassing amino acids 23–40 (94). Extensive mutagenesis of amino acids 37–42 has shown that even in this small region of Rho, different amino acids are important for interaction with specific subsets of targets (95). In addition, even different REM-1 motifs have evolved to prefer distinct regions of Rho, one REM-1 motif in PRK1 (HR1a) binds only to GTP-bound RhoA, whereas an adjacent REM-1 motif (HR1b) binds to both GTP- and GDP-bound RhoA (96). Other Rho-binding proteins that do not have REM-1 motifs or RKH regions, such as PLD and CitronK, again require different regions of Rho for efficient interaction (94, 97).

4.2 Rac and Cdc42 targets

The largest class of target proteins for Rac and Cdc42 are the CRIB (Cdc42/Rac-interactive binding) motif-containing proteins (Fig. 5). The first identified members of the CRIB family were the protein kinases PAK1 (also known as α-PAK) and ACK-1 (98, 99). PAK1 has the additional distinction of being the first kinase shown to be activated directly by a Ras superfamily GTP-binding protein (98). In mammalian cells, three PAK isoforms have been described so far: PAK1, PAK2, and PAK3. Homologues of PAK exist in yeast, and there is also a growing family of PAK-related kinases that lack the CRIB motif (100). PAK kinase activity is activated by binding to Rac or Cdc42, and many extracellular signals activate PAKs (100). PAK2 can also be cleaved by caspases during apoptosis, leading to a constitutively active kinase

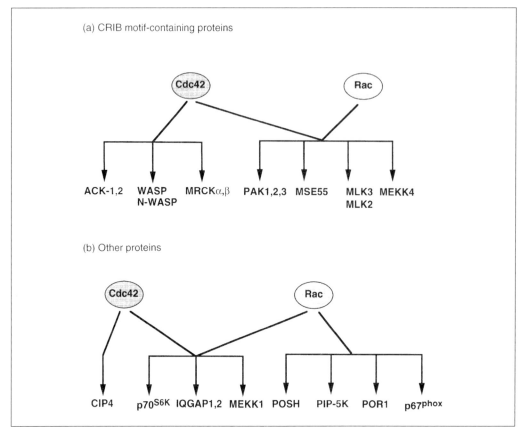

Fig 5 Downstream targets for Cdc42 and Rac. Proteins that have been shown to interact directly with GTP-bound Cdc42 and/or Rac are subdivided into two groups, based on whether or not they contain a Cdc42/Rac-interactive binding (CRIB) motif.

domain fragment (101). In contrast to PAKs, the tyrosine kinases ACK-1 and ACK-2 interact only with Cdc42 and not with Rac, and Cdc42 binding does not stimulate their kinase activity *in vitro* (99, 102). The tyrosine kinase activity of ACK-2 is, however, activated in cells when co-expressed with Cdc42, and ACKs are also activated in response to epidermal growth factor and bradykinin, as well as various stresses (102, 103).

CRIB-containing proteins interact with the effector regions of Rac and Cdc42 (Fig. 1), and mutation of specific amino acids within this region strongly reduces binding (104, 105). It is clear, however, that other regions of Rac and Cdc42, in addition to the effector region, are important for interaction with PAKs, and conversely, that regions of PAK in addition to the CRIB motif are important for optimal binding to Rac (106).

Apart from the CRIB motif, no other conserved sequences have been identified in Rac and Cdc42 targets which might identify further interacting domains (Fig. 5). Some targets for Rho, including ROK and mDia2, can also interact with Rac or Cdc42 in yeast two-hybrid or dot-blot analysis (93, 104), although whether this interaction is physiologically relevant is unclear as ROK, at least, appears to be involved principally in Rho-dependent responses (see Section 5.1 below). Little is known about the regions of Rac or Cdc42 involved in interacting with non-CRIB-containing proteins, with the exception of p67[phox], a component of the phagocytic NADPH oxidase enzyme complex, where several regions of Rac are required for optimal interaction (see Section 9.1 below). In particular, amino acids in both the effector and insert regions are involved in p67[phox] binding. As the sequences of insert regions vary greatly between Rho family members (Figs 1 and 2), it is likely that, in general, this region is important for conferring specificity to their interactions with downstream targets (11).

5. Rho GTPases and the actin cytoskeleton

Changes in the organization of filamentous actin were the first characterized responses to Rho GTPases in mammalian cells (Fig. 6), and these are now frequently used as a readout to monitor the activity of Rho GTPases, as well as the behaviour of GEFs, GAPs, GDIs, and target proteins.

5.1 Rho

5.1.1 Rho-induced actin reorganization

Most cultured mammalian fibroblasts, endothelial cells, and epithelial cells contain stress fibres, which consist of contractile bundles of actin filaments associated with myosin filaments and other proteins. The presence of stress fibres is modulated both by the type of extracellular matrix protein and by the medium in which the cells are cultured. An increase in the level of active Rho stimulates the accumulation of stress fibres (Fig. 6), and Rho is required for the formation of stress fibres induced by both soluble factors and integrin engagement (25, 107). The contractile nature of Rho-

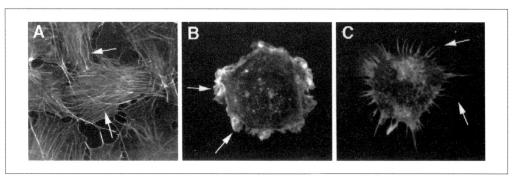

Fig. 6 Three distinct actin cytoskeletal structures regulated by Rho, Rac, and Cdc42. The actin cytoskeleton is shown in mouse fibroblasts (A) or macrophages (B, C). Arrows indicate examples of stress fibres regulated by Rho (A), lamellipodia and membrane ruffles regulated by Rac (B), and filopodia regulated by Cdc42 (C). Filamentous actin is visualized by immunofluorescence after staining with fluorescent phalloidin.

induced stress fibres in fibroblasts is clearly observed when the cells are plated on deformable substrata (108), and Rho is also able to enhance contraction where contractile fibres are already present, for example in smooth muscle cells (109, 110). Stress fibres are linked at the plasma membrane to multi-protein complexes known as focal adhesions, where transmembrane integrins are associated directly or indirectly with a large number of structural and signalling proteins (111). The formation of focal adhesions is regulated by Rho and is intimately linked with stress-fibre formation, although the two responses are to a certain extent separable (see Section 6.1, below).

Many observations on Rho function are consistent with its ability to stimulate the formation of actomyosin-based structures and to regulate their contractility. These structures are not always stress fibres. For example, Rho is required *in vivo* for the healing of small wounds in chick embryos, which are closed through the contraction of an actin-based purse string (112). This purse string appears morphologically similar to the Rho-regulated peripheral bundles of actin filaments observed in isolated epithelial-cell colonies (113). Rho is required for the integrity of a perijunctional ring of F-actin and myosin II that forms the apical pole of polarized epithelial cells (114). In macrophages, microinjection of activated Rho protein rapidly stimulates cell contraction, and although macrophages do not have stress fibres, they possess fine, Rho-regulated actin cables in the cytoplasm, which are presumably contractile (115). In neuronal cell lines, Rho also stimulates cell contraction and mediates neurite retraction in response to a variety of extracellular agents (116, 117).

As well as stimulating actomyosin-based contractility, Rho can stimulate an overall increase in F-actin in cells (118, 119), which could be due to stabilization of actin-containing structures, but in the case of mast cells appears to involve new actin polymerization, as it is blocked by cytochalasin D (118). Consistent with this, Rho is required for *Shigella* entry into HeLa cells, and appears to mediate actin polymerization from new actin nucleation sites, although it does not seem to be involved in the formation of nucleation sites (120). In Swiss 3T3 fibroblasts, however, Rho does not significantly induce new actin polymerization as an early response, as measured by the incorporation of fluorescently labelled actin monomers into stress fibres (121). The extent to which Rho stimulates increased actin polymerization may therefore vary between cell types and the status of cells.

In *S. cerevisiae* and in *S. pombe*, there is evidence that the RhoA homologue, Rho1p, regulates actin organization (122–124). Rho1p localizes to the bud tip and to the site of cytokinesis (123, 125), and appears to coordinate cell wall synthesis (126) with polarized actin organization to allow growth of the presumptive bud. In *Drosophila*, Rho1 and an exchange factor, DRhoGEF2, appear to play an essential role in the cell shape changes occurring during gastrulation, consistent with a role in regulating actin organization (127, 128).

5.1.2 Targets of Rho involved in actin reorganization

A number of target proteins interacting with Rho have been implicated in mediating actin reorganization, and it seems likely that Rho interacts with several different effector proteins to allow the coordinated assembly of actin, leading to the formation of stress fibres (Fig. 4).

ROKs (Rho-associated kinase)

The most intensively studied Rho targets are the two highly related Ser/Thr kinases, ROKα and ROKβ (also known as Rho-kinases or p160ROCK), which are involved in regulating actomyosin-based contractility via increased phosphorylation of myosin light chain (111, 129). ROK phosphorylates and inhibits the activity of the myosin-binding subunit of myosin phosphatase, and can also directly phosphorylate myosin light chain (MLC) *in vitro* (130). In addition, expression of activated RhoA or ROK increases the level of MLC phosphorylation in cells (111, 131). Since MLC phosphory-lation stimulates the interaction between myosin II and actin filaments, as well as the ATPase activity of myosin, ROK is thereby able to regulate actomyosin-based con-tractility (131). This is postulated to lead to stress fibre formation (111), and indeed expression of activated ROK mutants stimulates the formation of actin cables in various cell types (132). In addition, studies with Rho mutants indicate that ROK is essential for Rho-dependent stress-fibre formation in fibroblasts. However, it appears that ROK activation alone is not sufficient for the correct assembly of stress fibres, and that at least one other Rho target is also involved (95).

In addition to regulating myosin light chains, ROKs can phosphorylate a number of other proteins that may be relevant to the function of Rho in regulating cell morphology. These include α-adducin, a protein that promotes the binding of spectrin to actin filaments (133); glial fibrillary acidic protein (GFAP), an intermediate fila-ment protein (134); the Na^+/H^+ exchanger NHE1, which has been implicated in stress-fibre formation (135); and ezrin/radixin/moesin (ERM) proteins (136) (see Section 6.2 below). The mechanism underlying ROK activation is unclear, although it is likely to be regulated by phosphorylation, and indeed increased ROK phosphory-lation is observed in response to thrombin (137). In addition, it may relocalize following cell stimulation: in platelets integrin activation coincides with translo-cation of ROK from a detergent-soluble to a detergent-insoluble fraction (137).

p140mDia/Diaphanous

A second target for Rho that is likely to play an important role in actin reorganization is p140mDia, a structural protein that can bind to the actin-binding protein profilin and which, when overexpressed in COS cells, induces actin filament formation (138). *p140mDia* is the mammalian homologue of a *Drosophila* gene, *diaphanous,* which is required for cytokinesis (139). A second mammalian protein related to Diaphanous, mDia2, has been identified recently (93). Interestingly, p140mDia is mutated in an inherited form of progressive deafness (140), and it has therefore been suggested that it regulates actin organization in the hair cells of the inner ear. p140mDia and Diaphanous are members of a larger family of formin-related proteins (138), which includes the Bni1p protein in *S. cerevisiae*. Bni1p also binds profilin, and is involved in cytokinesis and establishment of cell polarity. It interacts with both Rho1p and

Cdc42p, and may therefore be a common target for these proteins involved in regulating actin organization (122, 141).

PIP 5-kinase

Rho may also stimulate actin reorganization at the plasma membrane by enhancing synthesis of phosphatidylinositol 4,5-bisphosphate (PI4,5P$_2$), a lipid that has been shown to interact with a number of actin-binding proteins (e.g. profilin, vinculin, and gelsolin) involved in regulating actin filament turnover and organization (107, 142). In cells, the barbed ends of actin filaments, which are the preferred end for actin polymerization, are normally masked by capping proteins to prevent spontaneous polymerization. Phosphoinositides are able to induce dissociation of capping proteins from actin filaments *in vitro* (143, 144), and therefore one model for the stimulation of actin polymerization by extracellular signals is through changes in phosphoinositide levels leading to filament uncapping and subsequent addition of actin monomers. It should be noted, however, that most actin-associated PI4,5P$_2$-binding proteins do not show complete specificity for PI4,5P$_2$, but also bind to related phosphoinositides, such as PI4P, PI3,4P$_2$, and PI3,4,5P$_3$ (142, 144, 145), and thus it is not clear precisely how the activities of these proteins are regulated during cell signalling. PI4,5P$_2$ is generated from PI through the sequential actions of a PI4 kinase and a PIP5-kinase (146). Consistent with a model where Rho enhances generation of PI4,5P$_2$, there has been one report that recombinant Rho can associate with PIP 5-kinase activity in cell lysates, and that an increase in PIP 5-kinase activity, seen following integrin-mediated cell adhesion, is dependent on Rho (147). So far, however, a direct interaction between Rho and any PIP 5-kinase isoform has not been reported.

Other targets

Other Rho targets may also influence actin organization either directly or indirectly. Overexpression of kinase-deficient PRK2 (aPKC-related kinase) induces loss of stress fibres (148), but this may be a consequence of titrating out endogenous Rho protein. Citron kinase is involved in regulating actin reorganization at cytokinesis (see Section 6 below).

5.2 Rac and Cdc42

5.2.1 Rac-induced lamellipodium formation

Microinjection and transfection studies have revealed that Rac regulates the formation of lamellipodia and membrane ruffles in a variety of cell types (107). Lamellipodia are plasma membrane protrusions containing a meshwork of actin filaments (Fig. 6). They extend over the substratum to form new adhesive contacts known as focal complexes (149) and are commonly found at the leading edge of migrating cells, where they drive forward extension. Membrane ruffles are similar in structure to lamellipodia, but protrude upwards from the dorsal surface of adherent cells. Both lamellipodium extension and membrane ruffling require active actin polymerization at the plasma membrane.

Lamellipodium extension is believed to be driven by *de novo* actin polymerization at the plasma membrane (150), and, in agreement with this, Rac stimulates the incorporation of fluorescently labelled actin monomers into filaments in lamellipodia (121). The Arp2/3 protein complex, which is able to nucleate new actin filaments and is localized to lamellipodia (151), is a good candidate for mediating Rac's effects, but this has not yet been demonstrated directly. As well as stimulating new actin poly- merization, actin filament turnover is required to provide a motile lamellipodium, and this may be regulated by cofilin/ADF, a factor that can induce actin depolymer- ization. Cofilin/ADF is inactivated by phosphorylation, and recently a kinase that specifically phosphorylates cofilin, LIMK, has been identified (152, 153). This kinase may be regulated by Rac, and consistent with this the level of cofilin phosphorylation is slightly increased in cells expressing activated Rac. It is possible that Rac activates LIMK locally and thus inhibits cofilin/ADF activity adjacent to the plasma membrane, allowing filament elongation here. Cofilin/ADF activity would then be restricted to regions away from the plasma membrane, where filament depolymer- ization would be expected to occur.

In contrast to its effects in other cell types, activated Rac protein does not stimulate lamellipodium formation in epithelial cells where intercellular junctions are present: instead, it appears to enhance actin filament accumulation at cell–cell junctions (113, 154, 155; and see Section 6.2 below). Dominant negative Rac does, however, inhibit lamellipodium formation induced after addition of HGF (hepatocyte growth factor) to MDCK epithelial cells (113). This suggests that the function of Rac in epithelial cells is more complicated than in fibroblasts, and may be dependent on other incoming signals.

No Rac homologue has been reported in *S. cerevisiae*, but studies on the localization and expression of *Rac*-related genes in *Drosophila*, *C. elegans,* and *Dictyostelium* suggest that they play key roles in regulating actin dynamics during morphogenesis. In *Dictyostelium*, seven different Rac-like genes have been identified, and although little is known about the function of each isoform, it is interesting that the expression of some isoforms is induced when *Dictyostelium* cells transit from a growth phase to a highly motile stage (156). In *C. elegans*, CeRac1 and Cdc42Ce co-localize with a target, CePAK, at hypodermal cell boundaries during embryonic body elongation, a process involving cytoskeletal reorganization (157). Finally, in *Drosophila*, regulated ex- pression of dominant-negative Rac protein during embryonic development inhibits processes involving cell-shape changes and disrupts actin organization (158).

5.2.2 Cdc42 and filopodium formation

Microinjection and transfection studies have demonstrated that Cdc42 induces the formation of filopodia in several mammalian cell types, including fibroblasts and macrophages (107, 115) (Fig. 6). As with lamellipodia, filopodia in fibroblasts and macrophages are associated with focal complexes located approximately at the base of each filopodium (115, 159). Filopodia are fine, finger-like plasma-membrane pro- trusions, containing bundles of actin filaments cross-linked by actin-binding proteins, that extend from the leading edge of migrating cells, forming new contacts with the substratum. It has been suggested that they play a sensory role, rather than being a driving force per se for migration. In agreement with this, correct guidance of extending nerve growth cones is dependent on filopodia, and in a neuroblastoma cell line, Cdc42 has been shown to promote filopodium formation on growth cones (117) (see also Section 5.2.4 below). Under different conditions or in other cell types, however, Cdc42 does not detectably induce filopodium formation and instead other actin-containing structures are observed. For example, in confluent fibroblasts Cdc42 stimulates some stress-fibre formation but no filopodia are observed (159), perhaps due to cell–cell contacts.

Recombinant Cdc42 can stimulate actin polymerization directly in neutrophil or *Xenopus* oocyte extracts, providing an ideal *in vitro* system for dissecting the molecular mechanisms underlying Cdc42-induced actin responses (160, 161). To date, it has not been possible to induce actin polymerization in cell lysates using Rac or Rho. Interestingly, the requirement for phosphoinositides appears to vary, depending on the type of extract and the conditions used. In one report, stimulation of actin polymerization did not correlate with $PI4,5P_2$ or $PI3,4,5P_3$ synthesis, nor was it inhibited by an antibody to polyphosphoinositides, suggesting that neither $PI4,5P_2$ nor other polyphosphoinositides act downstream of Cdc42 in this response (160). In another report, vesicles containing phosphoinositides appear to be required for actin polymerization, and may have promoted GTP exchange on Cdc42 as well as the interaction between Cdc42 and actin nucleating activities (161).

5.2.3 Cdc42 in yeast

In the yeast *S. cerevisiae*, Cdc42 is essential for coordinating polarized cell surface growth and reorganization of the actin cytoskeleton during both budding and pheromone-induced mating (162, 163). It is also involved in controlling polarized cell growth in the fission yeast *Schizosaccharomyces pombe* (164). The *S. cerevisiae* actin cytoskeleton consists of cortical actin patches associated with the plasma membrane, and actin cables running through the body of the cell. These actin cables become polarized along the axis of the cell as a new bud forms, and they are believed to be important for the directed delivery of membrane components to the bud site or mating projection (165). Cdc42 is involved in the polarization of the actin cytoskeleton and is localized to bud sites, although actual bud-site selection is dependent on many other gene products (162).

5.2.4 Rac, Cdc42 and neuritogenesis

In addition to regulating lamellipodium and filopodium formation in a variety of cultured cell types, a growing body of work has shown that Rac and Cdc42 regulate neuritogenesis in cultured neuronal cells, as well as affecting neuronal development *in vivo*. Rac1 microinjection stimulates lamellipodium formation on growth cones and along neurites of N1E-115 neuroblastoma cells (117). Transfection of activated Rac1, Cdc42, or the Rac exchange factor Tiam-1 also stimulates neurite outgrowth in these cells (166). In PC12 cells, Rac is required for NGF-induced neurite elongation, and is hypothesized to be required for the adhesive and motile function of growth cones rather than for the actual assembly process (167). *In vivo*, targeted expression of dominant-negative or constitutively active forms of Rho, Rac, and Cdc42 in specific populations of neuronal cells in mice affects dendritic development and axonal outgrowth (168, 169). Different Rac isoforms may have distinct functions in neuronal cells, as they each show a unique pattern of expression in the developing chick brain (3). In addition, overexpression of Rac1B/Rac3 but, surprisingly, not Rac1 in primary chick neurites induces an increase in the number of neurites per neuron, and promotes neurite branching, suggesting that Rac1B/Rac3 may specifically regulate the ability of neuronal cells to extend multiple neurites. The specificity of Rac1B/Rac3 resides in its C terminus, which is the only region that differs significantly from Rac1 (170).

In *Drosophila* and *C. elegans*, there is also abundant evidence supporting a central role for Rho proteins in regulating neuronal cell behaviour and axonal outgrowth. In *C. elegans*, a protein known as UNC-73 is required for axon guidance, and encodes a large multidomain protein with two DH/PH elements, the first of which acts as an exchange factor for Rac (171). Similarly, the still life protein in *Drosophila* has a DH/PH element, is localized in synaptic terminals, and regulates neuronal morphology (172). In addition, selective expression of constitutively active or dominant-negative versions of Rac1 and Cdc42 in neuronal cells leads to defects in axon outgrowth and targetting in *Drosophila* (173).

5.2.5 Targets for Rac and Cdc42 involved in actin reorganization

A number of Rac and Cdc42 targets have been implicated as regulators of actin organization.

PAKs

Expression of PAK1 mutants has been reported to induce actin reorganization, leading to extension of filopodia and lamellipodia in fibroblasts (174) and neurite outgrowth in neuronal cells (175, 176). PAK1 localizes to areas of membrane ruffling and cell extension in fibroblasts (177) and also to Rac-regulated focal complexes (175). However, neither the kinase activity of PAK1 nor a functional Rac/Cdc42-binding domain (CRIB motif) is required for PAK1-induced actin reorganization or neurite outgrowth (174, 178), suggesting that another region of PAK1 mediates these responses. One possibility is the N-terminal proline-rich domain, which interacts with the adaptor protein Nck and with a DH/PH domain-containing GEF protein known as PIX (179, 180). By recruiting PIX, PAK is postulated to activate an exchange factor for Rac, leading to lamellipodium formation (180). This model fits with the observation that PAK-induced neurite outgrowth is dependent on Rac (178). It would also provide a molecular explanation for the observation that Cdc42 can act upstream of Rac (181), if Cdc42 induces PAK association with PIX. Another complication to the Rac/PAK story is provided by the observation that the cyclin-dependent kinase 5 (Cdk5) in complex with its neuron-specific regulator p35 binds Rac and is able to phosphorylate PAK1, thereby inhibiting its kinase activity (182). Again, this suggests that the kinase activity of PAKs may not be important for morphological changes.

The contribution of endogenous PAKs to Cdc42- and Rac-mediated actin reorganization in mammalian cells is therefore unclear. To further confuse the issue, Rac mutants with greatly reduced affinity for PAK can still induce lamellipodium formation when introduced into fibroblasts (104, 105). In *S. cerevisiae*, however, mutants lacking the PAK homologues Ste20 and Cla4 show actin cytoskeletal defects resembling those in *cdc42* mutants, and Ste20 can correct defects in actin assembly in *cdc42* mutant cells (183, 184). These observations argue that the yeast PAKs do participate in Cdc42 pathways regulating actin organization.

IQGAPs

IQGAPs were originally thought to possess GAP activity for Ras because of the homology of the 'GAP' domain with known Ras GAPs. However, biochemical studies together with structural analysis of the RasGAP domain show that this domain of IQGAP is not a GAP (185). Instead, this domain is required for interaction with Cdc42 and, more weakly, with Rac. In addition, IQGAPs are also known to interact with actin filaments (186) and calmodulin (187), and presumably have the potential to alter actin organization. As they are large, multidomain proteins (185), they may act as scaffolds to bring together a number of molecules involved in actin reorganization close to the actin cytoskeleton. Interestingly, IQGAP1 localizes to intercellular junctions in epithelial cells and when overexpressed induces loss of cell–cell interaction (188), suggesting that it plays a role in organizing the actin cytoskeleton at these junctions. The best evidence that IQGAPs are involved in regulating actin organization, however, comes from studies on cytokinesis in *Dictyostelium* and *S. pombe* (see Section 5.3 below).

WASP/N-WASP

The *WASP* gene is mutated in patients with Wiskott–Aldrich syndrome (WAS), an inherited disease characterized by defects in immune responses (189). WASP expression is restricted to haematopoietic cells, whereas the related protein N-WASP shows more widespread expression (190, 191). When WASP was first identified as a Cdc42-interacting protein, it was shown that its overexpression induced the formation of actin filament clusters in several cell types, and that this effect was dependent on Cdc42 function (192). Subsequently, N-WASP was found to interact directly with actin filaments (191), and overexpression of N-WASP has been shown to enhance Cdc42-induced filopodium formation (193). In *S. cerevisiae*, a WASP-related protein, Las17p/Bee1p, is required for assembly of cortical actin patches (194, 195).

Although the mechanism whereby WASPs contribute to Cdc42-induced filopodium formation is not clear, it is interesting that N-WASP and WASP have a highly proline-rich region, and that this region of N-WASP can interact with profilins (196). In addition, WASP can recruit another profilin-binding protein, WIP, via its N-terminal domain (197). A similar interaction is observed between the *S. cerevisiae* homologues of WASP and WIP, Las17p and End5p (198). These observations suggest that WASPs could recruit actin monomers bound to profilin directly or indirectly to sites of actin polymerization. N-WASP and WASP can also interact via their proline-rich sequences with the SH3 domains of a number of proteins, including tyrosine kinases such as Btk (199), and the adaptor protein Grb2 (191). These interactions might serve to regulate the localization of WASPs or their ability to recruit further proteins.

PIP 5-kinase

As observed with Rho, recombinant Rac can associate with PIP 5-kinase activity in cell lysates (89), and a complex of Rac with a type I PIP 5-kinase and diacylglycerol kinase has been purified from cells (200). A model for Rac-induced actin reorganization involving $PI4,5P_2$ has been proposed, based on the observation that in permeabilized platelets Rac stimulates actin polymerization and concomitantly augments incorporation of phosphate into $PI4,5P_2$ (201). Addition of a peptide that binds to polyphosphoinositides inhibits Rac-induced actin polymerization, suggesting that this response to Rac may involve $PI4,5P_2$. Addition of $PI4,5P_2$-containing lipid micelles can also stimulate actin polymerization, presumably by removing phosphoinositide-regulated capping proteins from actin filaments, as described above (Section 5.1.2). It remains to be determined whether this model for Rac-induced actin polymerization is applicable to other cell types in addition to platelets.

Other Rac and Cdc42 targets

Two other Rac/Cdc42 targets, CIP4 and POR1, have been identified in yeast two-hybrid screens, and although much less is known about them than the other targets described above, the data so far indicate that they are involved in regulating actin organization. CIP4 interacts specifically with Cdc42 and not with Rac, and CIP4 overexpression has been shown to induce actin reorganization and loss of stress fibres (202). Conversely, POR1 interacts with Rac but not Cdc42, and has been implicated in potentiating Ras- but not Rac-induced membrane ruffling (203).

5.3 Rho and Cdc42 in cytokinesis

Dramatic changes in actin organization occur in cells during mitosis and cytokinesis (204, 205). At mitosis, stress fibres disperse in adherent cultured cells and lamellipodia and ruffles are lost. As a consequence of these changes, cells round up and become very loosely attached to the substratum. Subsequently the actomyosin-based contractile ring assembles and then constricts to form the cleavage furrow, finally separating each cell into its two daughter cells.

Both Rho and Cdc42 have been shown to play a role during cytokinesis. Rho is apparently localized to the cleavage furrow, and is required for actin filament assembly and constriction of the contractile ring in *Xenopus* eggs and embryos, and in sand dollar eggs (206, 207). *Saccharomyces cerevisiae* Rho1p localizes to the mother–daughter neck at cytokinesis (126), suggesting that it is involved in this process. The Rho target citron kinase may be a key mediator of Rho-induced responses, as it localizes to the contractile ring, and expression of various mutants either enhances constriction of the contractile ring or inhibits cytokinesis (208). The kinase domains of citron kinase and ROK are related, and it is therefore possible that citron kinase, like ROK, is able to phosphorylate the intermediate filament protein GFAP at the same site during cytokinesis (134). As GFAP is localized to the cleavage furrow, these results suggest that Rho could act via citron kinase or ROK to coordinate changes in the actomyosin cytoskeleton and intermediate filament network during cytokinesis.

Cdc42 plays a role in cleavage furrow formation in *Xenopus* embryos (207), and mammalian cells expressing activated Cdc42 accumulate multiple nuclei, suggesting that cytokinesis is inhibited (209). Genetic and cell biological analysis in *Dictyostelium* and in fission yeast (*S. pombe*) has shown that IQGAPs, which are targets for Cdc42/Rac, are important in cytokinesis. In *S. pombe*, the IQGAP homologue Rng2p is a component of the actomyosin contractile ring, and is required for construction of the ring (210). In *D. discoideum* , correct cytokinesis is dependent on a Rac/Cdc42-related protein, RacE, and on two members of the IQGAP family (211–214). Interestingly, the effects of disrupting each *IQGAP* gene on cytokinesis are phenotypically distinguishable, suggesting that cytokinesis may involve several distinctly regulated steps (214).

5.4   Other Rho GTPases

So far, the abilities of Rho, Rac, and Cdc42 to regulate actin organization have been most thoroughly characterized, but there is increasing evidence that other Rho family members can influence cell morphology and cytoskeletal arrangements. Expression of activated RhoD leads to actin rearrangement in several different mammalian cell types, and concomitantly alters the motility and distribution of early endosomes, suggesting a link between vesicular transport and the actin cytoskeleton (18).

RhoE (also known as Rnd3) and the related protein Rnd1 act antagonistically to Rho and induce a loss of stress fibres (10, 38). In fibroblasts, Rnd1 induces cell detachment from the substratum, whereas epithelial cells remain attached to the substratum but stress fibres are lost. Interestingly, the effector regions of RhoE, Rnd1, and Rnd2 are almost identical to RhoA, B, and C (Fig. 2), although the remainder of the proteins only show 43% overall homology to Rho. This suggests that they may compete with Rho for binding to some common targets, preventing Rho from activating these targets.

RhoG can act via Cdc42 and Rac1 to induce filopodium formation and lamellipodium formation or membrane ruffling respectively, but despite its homology to Rac1 and Cdc42, it does not appear to interact with targets such as PAK1, POR1, or WASP (215). TC10, which is most closely homologous to Cdc42, can also induce filopodium formation when expressed in fibroblasts, but in contrast to RhoG appears to interact with a subset of Cdc42 targets (92). Finally, in *Drosophila*, a novel member of the Rho family, RhoL, has been identified (216). Expression of activated RhoL induces breakdown of cortical actin filaments in nurse cells associated with oocytes.

6. Rho GTPases and cell adhesion

The actin cytoskeleton is structurally linked to sites of cell–cell and cell–matrix adhesion, at integrin-mediated adhesion sites, cadherin-based adherens junctions, and tight junctions. In altering actin organization, Rho family proteins inevitably affect these linked adhesion sites. It is also likely that they act directly to coordinate changes in the organization or distribution of adhesion sites with actin cytoskeletal remodelling.

6.1 Adhesion to the extracellular matrix

In addition to regulating responses to cytokines and growth factors, Rho is involved in signalling induced by activation of integrins. When fibroblasts adhere to the extra-cellular matrix protein fibronectin, they can form stress fibres and focal adhesions in the absence of growth factors. In addition, tyrosine phosphorylation of a number of focal adhesion proteins is stimulated, and in some cell types activation of MAPK pathways and gene transcription is induced (111, 217). Rho is required for the fibronectin-induced assembly of stress fibres and focal adhesions in fibroblasts, and for the tyrosine phosphorylation of focal adhesion components (218, 219). Similarly, in activated platelets Rho inactivation decreases the formation of vinculin-rich focal adhesions, and also decreases their adhesion to fibrinogen (220). Constitutively active Rho stimulates actin stress-fibre formation in cells plated on fibronectin but not in cells plated on the non-specific substrate, poly-L-lysine, showing that integrin engagement is important for Rho-induced organization of the actin cytoskeleton into actomyosin bundles (221, 222). Overexpression of Dbl, an exchange factor for Rho and Cdc42, acts like Rho in stimulating stress-fibre formation in cells adhering to fibronectin, but interestingly actually inhibits stress-fibre formation in cells plated on gelatin (223). This suggests that, on gelatin, Dbl acts preferentially on Cdc42, which under appropriate conditions induces loss of stress fibres (224), rather than on Rho. One possible downstream target involved in Rho-mediated signalling from integrins is the Na^+/H^+ exchanger NHE1 (225), which appears to play a role in regulating Rho-induced assembly of stress fibres and focal adhesions (226), and is itself localized to focal adhesions. The Rho target ROK can phosphorylate NHE1, suggesting that it provides a link between Rho and this protein (135).

Rac and Cdc42 also mediate the formation of adhesion sites to the extracellular matrix. In fibroblasts, these sites are much smaller than Rho-mediated focal adhesions, and are predominantly localized in lamellipodia or associated with filopodia (159). In macrophages, which do not have focal adhesions, Cdc42 acts upstream of Rac to regulate the formation of small focal complexes, which contain integrins and other associated proteins (115). It is presumed that these structures are linked to the actin cytoskeleton, although this has not formally been proven. Integrin-mediated cell spreading on fibronectin is dependent on Rac and Cdc42 (219, 227), and the Rac/Cdc42 target PAK is rapidly activated following adhesion, implying that Rac and/or Cdc42 are activated in response to integrin engagement (227). Expression of activated Rac in T lymphocytes enhances integrin-mediated adhesion and spreading (228), and in T lymphoma cells it stimulates invasion, possibly as a consequence of increased adhesions (49).

6.2 Intercellular adhesions

As well as regulating cell–extracellular matrix adhesions, Rho family proteins affect cell–cell junctions between epithelial cells. These junctions include adherens junctions, where the transmembrane components are cadherins (229), and tight junctions, which act as impermeable barriers preventing the movement of macromolecules between cells (230). Both of these junctions are linked to the actin cytoskeleton. Several Rho family proteins localize to epithelial cell junctions, including Rac, RhoE, and Rnd3 (10, 38, 155, 231, 232). Of these, only Rac has so far been shown to be involved in regulating junction integrity: it is required for the formation and maintenance of adherens junctions between keratinocytes (233), and in kidney-derived MDCK epithelial cells expression of activated Rac1 or its exchange factor Tiam-1 enhances the formation of adherens junctions (155, 234). In contrast, Rac is also required for the loss of adherens junctions induced by scatter factor/hepatocyte growth factor (SF/HGF) (235), which reversibly stimulates the conversion of epithelial cells to a highly motile, fibroblastic, phenotype. In this case, it is probable that other signals initiated by SF/HGF override Rac and induce weakening of adherens junctions. As a consequence Rac can now stimulate lamellipodium extension, leading to cell migration and the pulling apart of intercellular junctions.

Rho has also been implicated in the formation of adherens junctions (155, 231, 233) but whether this reflects a direct role for Rho in maintaining junctions or an indirect consequence of its effects on actin cytoskeletal organization and adhesion to the extracellular matrix is not known. Similarly, altering Rho and Rac activity affects tight junctions (114, 232), although in MDCK cells only the activated forms of Rho and Rac visibly cause disruption of tight-junction morphology, suggesting that more subtle effects are responsible for the loss of functionally intact junctions induced by dominant-negative Rho or Rac (232).

Little is known of the signalling pathways linking Rho and Rac to intercellular junctions, although one possible target for Rac involved in regulating adherens junctions is IQGAP1, which is localized to intercellular junctions and interacts with E-cadherin and β-catenin (188). In addition, overexpression of IQGAP1 results in a decrease in E-cadherin-mediated adhesion. Rho, on the other hand, may affect junctions via its connections with the ezrin/moesin/radixin (ERM) family of actin-binding proteins. ERM proteins link actin filaments to the plasma membrane via transmembrane binding partners such as CD44 or ICAM-1 (82). They are located primarily in microvilli on the apical surface of epithelial cells, although they have also been observed at intercellular boundaries (236). Rho can co-localize to the plasma membrane with ERM proteins, and Rho inactivation prevents the localization of ERM proteins to the plasma membrane (237). RhoGDI apparently binds in a complex with CD44 and ERM proteins, and could therefore act as an intermediary to target Rho to ERM proteins, or alternatively ERM proteins could increase Rho activity by titrating out RhoGDI (83). Finally, ERM protein interaction with actin is required for Rho and Rac to induce actin reorganization and focal complex assembly in permeabilized fibroblasts (238). Although the functional consequences of all these links between Rho and ERM proteins are far from clear, they do suggest that ERM proteins may act as a focal point both for Rho activation and subsequent signalling.

7. Rho GTPases, cell polarity, and cell migration

The ability of Rho GTPases to coordinate changes in actin organization and cell adhesion suggests that they are likely to play a role in cell migration (239) (Fig. 7). Migrating cells extend lamellipodia at their leading edge and simultaneously make new contacts with the substratum. Not surprisingly, given its role in lamellipodium formation, Rac is absolutely required for the migration of a variety of cell types, including epithelial cells and macrophages (113, 240). Studies using C3 transferase to block Rho function also point to a requirement for Rho activity in cell migration (113, 240, 241). As Rho stimulates actomyosin-mediated contractility, it is likely that Rho is required for contraction of the cell body, leading to detachment of the rear of the cell (239, 240). Interestingly, although many cell types extend filopodia at their leading edge when migrating, Cdc42 is not essential for cell migration but instead is required for cell polarization and the ability to sense a gradient of chemoattractant (240) (Fig. 7). In T cells, Cdc42 is also required for polarization of the cytoskeleton towards antigen-presenting cells (242). The concept that Cdc42 plays a role more generally in

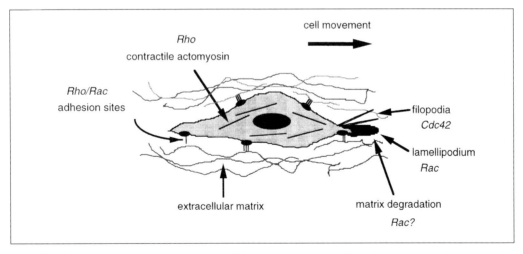

Fig. 7 Model for the roles of Rho, Rac, and Cdc42 in cell migration. A motile cell is represented schematically, and actin-filament-containing structures regulated by Rho, Rac, and Cdc42 are indicated. Rac and Cdc42 act at the leading edge of the cell to stimulate forward protrusion via actin polymerization, and concomitantly induce the formation of new adhesions to the substratum. Rac may also stimulate production of matrix-degrading proteases indirectly by regulating their expression levels. Rho stimulates actomyosin-based contractility within the body of the cell, allowing the cell body to move forward towards the leading edge.

establishing cell polarity is supported by studies in yeast and *Drosophila*. In *S. cerevisiae*, Cdc42 is required for polarized growth of buds and also for polarization of cells towards mating partners (162). In the *Drosophila* wing-disc epithelium, Cdc42 is required for epithelial cell elongation, which in turn is essential for the generation of apico-basal polarity (154).

Cell invasion is another assay for assessing the migratory phenotype of cells. Invasion *in vivo* is dependent not only on cell motility but also on the strength of cell adhesions and on secretion of matrix-degrading proteases. Using T-lymphoma cells as a model system, it has been shown that activated Rac1 and an exchange factor for Rac, Tiam-1, promote their invasion into a fibroblast monolayer (49, 243, 244). In addition, Cdc42 can stimulate invasion, and Rho is required for invasion although it is not by itself sufficient (244). Rac is also required for invasion of breast carcinoma cells into Matrigel, a mixture of extracellular matrix proteins (245), and activated forms of Rac and Cdc42 enhance the migration of breast epithelial cells through collagen-coated filters (246). It is not clear which of the cellular processes involved in invasion is enhanced by Rac and Cdc42, but as mentioned above, increased cell adhesion could contribute to the responses observed. In this context, it is interesting that Rac and its exchange factor Tiam-1 can either inhibit or enhance the migratory capacity of MDCK epithelial cells, depending on the matrix protein on which the cells are growing (234, 247). It is also possible that Rac may contribute to invasion by stimulating production of matrix-degrading proteases (248).

8. Regulation of transcription, cell cycle, and transformation

Unlike Ras, there is no evidence so far that any of the Rho GTPases are mutated in human cancers. Overexpression of RhoC has been found to correlate with progression in pancreatic tumours, but whether it plays any role in tumour progression is not known (249). Cancer development involves both increased proliferative potential and changes in cell adhesion and migration leading to invasion and metastasis. As described above, there is good reason to believe that Rho family proteins will play a central role in the latter processes. In addition, evidence is accumulating to support an involvement of these proteins in regulating cell proliferation. Cdc42, Rac, and Rho are required for transformation of fibroblast cell lines induced by a variety of oncogenes, and in these cells they either directly or indirectly influence the transcription of many genes. Some of these genes are relevant to cell cycle progression, and could thereby contribute to transformation. Whether Rho family proteins play a similar role in the epithelial cell types that form the majority of human cancers remains to be determined.

8.1 Rho GTPases in oncogenic transformation

In some rodent fibroblast lines, activated RhoA, RhoB, RhoG, Rac1, and Cdc42 are themselves weakly oncogenic, but in addition are required for malignant transformation by the Ras oncogene (20, 42, 250–252). Each of these proteins appears to make a different contribution to the transformed phenotype. For example, RhoG and Rac1 but not Cdc42 increase the saturation density of NIH3T3 fibroblasts at confluence (250–252).

In addition to the effects of Rho proteins themselves, deregulated expression of many Dbl family exchange factors in NIH3T3 fibroblasts has been shown to induce transformation (42), with variable morphological and growth alterations depending on the specific exchange factor (253, 254). Indeed, several Dbl family members were originally isolated as oncogenes in NIH3T3 fibroblast transformation assays (42).

Consistent with their transforming potential in fibroblasts, microinjection of Rho, Rac, and Cdc42 proteins into quiescent Swiss 3T3 fibroblasts can stimulate DNA synthesis (255). However, Cdc42 can also inhibit DNA synthesis (256), and when expressed in the thymus of mice, Rac enhances apoptosis (257). In these cases, Rac and Cdc42 may act by stimulating the serine/threonine kinase JNK, which has been implicated in inducing apoptosis (258). In fact, it seems likely that the long-term effects of activating these proteins vary considerably, depending on both the cell type and the status of the cells, as has also been observed for Ras (259).

Various point mutants and deletion mutants of Rho, Rac, or Cdc42 have been tested to determine whether cell transformation correlates with any of the other known responses induced by these proteins. This approach has shown that the contributions of Rac and Cdc42 to cell transformation are separable from their effects on the actin cytoskeleton and cell adhesion, whereas with Rho there is a correlation between actin reorganization and transformation (95, 105). It appears that proteins binding to the insert region of Cdc42 and Rac are important, as deletion of this region inhibits their ability to stimulate transformation but does not alter their effects on the actin cytoskeleton (14, 15). Interestingly, the ability of Rac to contribute to transformation correlates with its ability to activate superoxide production (14), suggesting that superoxide production itself may be important for mitogenesis.

8.2 Transcriptional regulation and links with cell cycle control

The first indication that Rho proteins could regulate the activity of transcription factors came from studies on the serum response element of the c-*fos* gene promotor. Rho, Rac, and Cdc42 are all able to activate serum response factor (SRF)-dependent reporter constructs, although only Rho is required for lysophosphatidic acid (LPA)-induced activation of these constructs (260). In cells, however, only Cdc42 is able to activate transcription of the endogenous c-*fos* gene on a chromosomally integrated reporter construct (261). Rho is able to induce this response when a second signal leading to histone H4 hyperacetylation is provided, indicating that alterations in chromatin organization are required for efficient activation of chromosomal genes (261). Using Rho effector domain mutants, the ability of Rho to activate SRF has been separated from its ability to transform cells, suggesting that the contribution of Rho to the transformed phenotype is not dependent on this transcription factor (95). It appears more likely that the transforming potential of Rho lies in its effects on cell

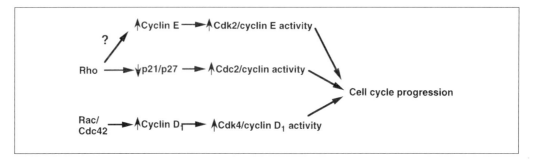

Fig. 8 Model for how Rho, Rac, and Cdc42 may induce cell cycle progression. Expression of activated Rho, Rac, and Cdc42 can increase the expression levels of cyclin D_1 or cyclin E. These cyclins interact with and are essential for the activation of cyclin-dependent protein kinases (Cdk2 and Cdk4), which stimulate cell cycle progression through G_1 into S phase. Alternatively, Rho can affect cell cycle progression by decreasing the expression level of the Cdk inhibitor p21/p27, which normally acts to repress cell cycle progression by inhibiting the activity of the Cdk, Cdc2.

cycle inhibitors (Fig. 8). Rho is required for Ras-induced DNA synthesis in fibroblasts by virtue of its ability to repress Ras-induced transcription of the cyclin-dependent kinase inhibitor p21[cip] (262). C3 transferase also inhibits p27[kip] degradation and cyclin E upregulation during cell cycle progression, suggesting that Rho may coordinately affect several components of the cell cycle machinery (263, 264).

In addition to their effects on SRF, expression of constitutively activated Rac and Cdc42 in cells leads to activation of c-*Jun* N-terminal kinase (JNK) and in some cells the MAPK p38, which can then phosphorylate and activate a number of transcription factors, including Jun (265). Activation of JNK is required for Cdc42 to induce transcription of chromosomally integrated SRF-dependent reporter genes (261). However, the ability of various Rac and Cdc42 mutants, or different Dbl proteins, to induce transformation or cell cycle progression does not correlate with activation of JNK (15, 104, 105, 254), which tallies with the observation that JNK activation is normally associated primarily with signals that are not mitogenic (265). Rac and Cdc42 have also been implicated in regulating transcription in other signalling pathways not linked to proliferation, including TGF-β-induced gene transcription (37) and NF-κB-dependent transcription (266, 267). In the latter case, however, it appears that this is an indirect consequence of superoxide production in response to Rac (266, 248).

Recent evidence suggests that the ability of Rac and Cdc42 to contribute to transformation is linked to expression of cyclin D_1, a key regulator of cell cycle progression that is often overexpressed in cancers (268). Rac and Cdc42 can stimulate transcription mediated by the transcription factor E2F, leading to increased expression of cyclin D_1 (269). The transforming potential of Rac mutants and various Dbl family proteins correlates with their ability to stimulate transcription of cyclin D1 (105, 254), implying that this activity is critical to the involvement of Rac and/or Cdc42 in stimulating both cell cycle progression and transformation (Fig. 8).

9. Other activities of Rho GTPases

9.1 Activation of the NADPH oxidase

At the same time that Rac was shown to regulate actin organization, it was identified as a key component of the NADPH oxidase enzyme complex of phagocytes. The cytoplasmic proteins Rac, p67[phox], and p47[phox] together are required *in vitro* for activation of the membrane-bound cytochrome that produces superoxide ions (270). Rac interacts directly with p67[phox] (271, 272), although it is still not clear precisely how this induces activation of the oxidase. Extensive mutagenesis of Rac, as well as construction of chimeras between Rac and related proteins, has been carried out in order to identify the amino acids that are important for activation of the NADPH oxidase and for interaction with p67[phox]. Phagocytic cells express both Rac1 and Rac2, and both of these are equally able to activate the NADPH oxidase when post-translationally modified at the C terminus, because prenylation greatly increases their association with membranes. However, as recombinant, non-prenylated proteins, Rac1 is more efficacious than Rac2, primarily because it has a more positively charged carboxy terminus that allows it to interact with negatively charged phospholipids in membranes (19, 275). However, in a yeast two-hybrid analysis p67[phox] interacts better with Rac2 than with Rac1 (273). Specific amino acids in the effector region of Rac are required for NADPH oxidase activation, and in addition several other regions are involved, including amino acid 103, the Rac insert region (Fig. 1), and a region encompassing amino acids 163–169 (271, 273–276). These regions are all exposed on the surface of Rac (12), and are therefore likely to be involved directly in interaction with other NADPH oxidase components. Interestingly, the Rac target PAK can phosphorylate p67[phox] adjacent to the Rac binding site on this protein (277), suggesting that PAK may itself regulate Rac's interaction with p67[phox].

Superoxide ions are highly toxic and thus activation of the NADPH oxidase is normally restricted to phagosomes, in order to limit the effects of the ions to the phagosome contents. As Rac is also required for phagocytosis (see below), it may be that Rac plays a dual role in coordinating sealing of the phagosome with activation of the NADPH oxidase.

9.2 Phagocytosis and entry of bacteria into cells

Since actin polymerization is necessary for phagocytosis, and actin filaments are localized around growing phagocytic extensions (278), it was expected that Rho family proteins would regulate this process. Indeed, Rac and Cdc42 are both required for Fcγ-receptor-mediated phagocytosis (279–281). Cdc42 appears to mediate the extension of actin-rich protrusions around particles, whereas Rac is required for the final stages of particle engulfment and phagosome sealing. It is intriguing that PI3 kinase has also been shown to play a role in the final membrane fusion step of phagocytosis (282), given that PI3 kinase can act upstream of Rac in signalling pathways activated by tyrosine kinases (58). Rho is not required for Fcγ-mediated phagocytosis, but it is essential for complement-mediated phagocytosis through the CR3 integrin receptor (281).

Many pathogenic bacteria use a variety of different approaches to enter host cells. In several cases bacterial entry has been shown to involve actin reorganization, suggesting that Rho proteins may be activated. Entry of *Shigella* into cells requires Rho proteins, which localize directly to the membrane folds surrounding the bacterium (120). Interestingly, RhoA shows a slightly different localization to RhoB and RhoC within the folds, suggesting that the isoforms may carry out distinct functions. In contrast, entry of *Salmonella typhimurium* requires Cdc42 but not Rac or Rho (283, 284). Binding of *S. typhimurium* to epithelial cells rapidly stimulates actin reorganization and bacteria are internalized by subsequent macropinocytosis.

9.3 Secretion and endocytosis

Trafficking of vesicles within cells involves interactions with the actin cytoskeleton as well as with microtubules (285). Rho family proteins can affect several aspects of vesicle trafficking, and this may, at least in part, reflect their ability to induce actin reorganization.

In permeabilized mast cells, recombinant Rho, Rac, and Cdc42 proteins enhance secretion induced by various agents, although they cannot by themselves stimulate secretion (286–288). Conversely, C3 transferase and dominant-negative Rac or Cdc42 inhibit secretion induced by calcium and non-hydrolysable GTP analogues (286, 287). Rac was also purified from mast cells as a factor that can enhance secretion (287), and in natural killer cells, expression of dominant-negative Rac1 inhibits granule exocytosis (289). Mast-cell secretion is accompanied by actin reorganization that is also mediated by Rac and Rho, although it is not clear that these changes in the actin cytoskeleton play any role in the secretory process (290).

Rho and Rac have also been implicated in the regulation of both constitutive and receptor-mediated endocytosis (291, 292). In *Xenopus* oocytes, Rho can enhance constitutive endocytosis/pinocytosis, whereas C3 transferase inhibits the process (292). In contrast, in mammalian cells and in a cell-free endocytosis system, Rho and Rac appear to inhibit receptor-mediated endocytosis of clathrin-coated vesicles (291). In this case, however, the effects may be due to changes in the actin cytoskeleton indirectly affecting endocytosis (293). Interestingly, one of the three Rho isoforms, RhoB, is localized to early endosomes, suggesting a role in endocytosis (22). RhoD also localizes to early endosomes, and overexpression of GTPase-defective RhoD decreases organelle motility, impeding movement and fusion of endosomes (18). Cycling of RhoD between a GTP- and GDP-bound form is therefore likely to be important for endosomal motility.

The mechanisms underlying the involvement of Rho and Rac in endocytosis and secretion have not been established, but in addition to altering the actin cytoskeleton it is also possible that they act by altering membrane phospholipid composition, for example through activation of PIP 5-kinase, PI3 kinase, or phospholipase D (PLD)

(294). These enzymes are potential targets for Rho family proteins (see above) and are known to play a role in vesicle trafficking, where it is likely that regulated changes in phosphoinositide composition of vesicles is important for vesicle budding or fusion (295).

10. Conclusions

Rho family proteins have emerged as central players in cell signalling and, through their ability to interact with a diverse range of upstream regulators and downstream targets, they appear to act as key coordinators of cellular responses (summarized in Table 3). Depending on the cell type and the incoming signals, activation of any particular Rho family member will have different consequences. During wound healing, for example, it may be that Rac coordinately activates signals leading to cell migration across the wound with signals leading to cell proliferation, in order to replace damaged cells. During phagocytosis, on the other hand, Rac may co-ordinately regulate the formation of phagosomes with activation of the NADPH oxidase (281). The apparently diverse array of responses regulated by each Rho family member may therefore reflect their role in ensuring that specific responses are linked in time and place.

Table 3 Summary of cellular responses involving Rho, Rac, and Cdc42. Each protein has been shown to be involved in the indicated responses, by introducing or expressing constitutively activated mutants and/or dominant-negative mutants

	Actin cytoskeleton	NADPH oxidase	Secretion	Phagocytosis	JNK activation	NF-κB activation	c-fos transcription	Cell cycle progression
Rho	+	-	+	+	-	+	+	+
Rac	+	+	+	+	+	+	-	+
Cdc42	+	-	+	+	+	+	+	+

We now have a concept of how Rho GTPases act, but much molecular detail remains to be resolved. Important issues for the future include determining how the activity and specificity of exchange factors, GAPs, and GDIs are regulated, investigating more fully the functions of other members of the family besides Rho, Rac, and Cdc42, and elucidating how the various downstream targets mediate all the described cellular responses involving Rho family members.

References

1. Hall, A. (1994) Small GTP-binding proteins and the regulation of the actin cytoskeleton. *Annu. Rev. Cell Biol.,* **10**, 31.
2. Haataja, L., Groffen, J., and Heisterkamp, N. (1997) Characterization of Rac3, a novel member of the Rho family. *J. Biol. Chem.,* **272**, 20384.
3. Malosio, M. L., Gilardelli, D., Paris, S., Albertinazzi, C., and de Curtis, I. (1997) Differential expression of distinct members of Rho family GTP-binding proteins during neuronal

development: identification of Rac1B, a new neural-specific member of the family. *J. Neurosci.*, **17**, 6717.

4. Johnson, D. I. and Pringle, J. R. (1990) Molecular characterization of CDC42, a *Saccharomyces cerevisiae* gene involved in the development of cell polarity. *J. Cell Biol.*, **111**, 143.

5. Shinjo, K., Koland, J. G., Hart, M. J., Narasimhan, V., Johnson, D. I., Evans, T., and Cerione, R. A. (1990) Molecular cloning of the gene for the human placental GTP-binding protein Gp (G25K): identification of this GTP-binding protein as the human homolog of the yeast cell-division-cycle protein CDC42. *Proc. Natl Acad. Sci., USA*, **87**, 9853.

6. Marks, P. W. and Kwiatkowski, D. J. (1996) Genomic organization and chromosomal location of murine *Cdc42*. *Genomics*, **38**, 13.

7. Vincent, S., Jeanteur, P., and Fort, P. (1992) Growth-regulated expression of *rhoG*, a new member of the ras homolog gene family. *Mol. Cell. Biol.*, **12**, 3138.

8. Dallery, E., Galiegue-Zouitina, S., Collyn-d'Hooghe, M., Quief, S., Denis, C., Hildebrand, M.-P., Lantoine, D., Deweindt, C., Tilly, H., Bastard, C., and Kerckaert, J.-P. (1995) TTF, a gene encoding a novel small G protein, fuses to the lymphoma-associated LAZ3 gene by t(3;4) chromosomal translocation. *Oncogene*, **10**, 2171.

9. Foster, R., Hu, K.-Q., Lu, Y., Nolan, K. M., Thissen, J., and Settleman, J. (1996) Identification of a novel human Rho protein with unusual properties: GTPase deficiency and in vivo farnesylation. *Mol. Cell. Biol.*, **16**, 2689.

10. Nobes, C. D., Lauritzen, I., Mattei, M.-G., Paris, S., Hall, A., and Chardin, P. (1998) A new member of the Rho family, Rnd1, promotes disassembly of actin filament structures and loss of cell adhesion. *J. Cell Biol.*, **141**, 187.

11. Wei, Y., Zhang, Y., Derewenda, U., Liu, X., Minor, W., Nakamoto, R. K., Somlyo, A. V., Somlyo, A. P., and Derewenda, Z. S. (1997) Crystal structure of RhoA-GDP and its functional implications. *Nature Struct. Biol.*, **4**, 699.

12. Hirshberg, M., Stockley, R. W., Dodson, G., and Webb, M. R. (1997) The crystal structure of human rac1, a member of the rho-family complexed with a GTP analogue. *Nature Struct. Biol.*, **4**, 147.

13. Rittinger, K., Walker, P. A., Eccleston, J. F., Smerdon, S. J., and Gamblin, S. J. (1997) Structure at 1.65 A of RhoA and its GTPase-activating protein in complex with a transition-state analogue. *Nature*, **389**, 758.

14. Joneson, T. and Bar-Sagi, D. (1998) A Rac1 effector site controlling mitogenesis through superoxide production. *J. Biol. Chem.*, **273**, 17991.

15. Wu, W. J., Lin, R., Cerione, R. A., and Manor, D. (1998) Transformation activity of Cdc42 requires a region unique to Rho-related proteins. *J. Biol. Chem.*, **273**, 16655.

16. Adamson, P., Marshall, C. J., Hall, A., and Tilbrook, P. A. (1992a) Post-translational modifications of p21[rho] proteins. *J. Biol. Chem.*, **267**, 20033.

17. Marshall, C. J. (1993) Protein prenylation: a mediator of protein–protein interactions. *Science*, **259**,1865.

18. Murphy, C., Saffrich, R., Grummt, M., Gournier, H., Rybin, V., Rubino, M., Auvinen, P., Lütcke, A., Parton, R. G., and Zerial, M. (1996) Endosomal dynamics regulated by a Rho protein. *Nature*, **384**, 427.

19. Kreck, M. L., Freeman, J. L., Abo, A., and Lambeth, J. D. (1996) Membrane association of Rac is required for high activity of the respiratory burst oxidase. *Biochemistry*, **35**, 15683.

20. Lebowitz, P. F., Du, W., and Prendergast, G. C. (1997) Prenylation of RhoB is required for its cell transforming function but not its ability to activate serum response element-dependent transcription. *J. Biol. Chem.*, **272**, 16093.

21. Cox, A. D. and Der, C. J. (1992) Protein prenylation: more than just glue? *Curr. Opinion Cell Biol.,* **4**, 1008.

22. Adamson, P., Paterson, H. F., and Hall, A. (1992) Intracellular localization of the p21[rho] proteins. *J. Cell Biol.,* **119**, 617.

23. Lebowitz, P., Davide, J., and Prendergast, G. C. (1995) Evidence that farnesyltransferase inhibitors suppress Ras transformation by interfering with Rho activity. *Mol. Cell. Biol.,* **15**, 6613.

24. Schmidt, G., Sehr, P., Wilm, M., Selzer, J., Mann, M., and Aktories, K. (1997) Gln63 of Rho is deaminated by *E. coli* cytotoxic necrotizing factor-1. *Nature,* **387**, 725.

25. Machesky, L. M. and Hall, A. (1996) Rho: a connection between membrane receptor signalling and the actin cytoskeleton. *Trends Cell Biol.,* **6**, 304.

26. Just, I., Selzer, J., Wilm, M., von Eichel Streiber, C., Mann, M., and Aktories, K. (1995) Glucosylation of Rho proteins by *Clostridium difficile* toxin B. *Nature,* **375**, 500.

27. Lang, P., Gesbert, F., Carmagnat, M., Stancour, R., Pouchelet, M., and Bertoglio, J. (1996) PKA phosphorylation of RhoA mediates morphological and functional effects of cAMP in lymphocytes. *EMBO J.,* **15**, 510.

28. Flatau, G., Lemichez, E., Gauthier, M., Chardin, P., Paris, S., Florentini, C., and Boquet, P. (1997) Toxin-induced activation of the G protein p21 Rho by deamidation of glutamine. *Nature,* **387**, 729.

29. Olofsson, B., Chardin, P., Touchot, N., Zahraoui, A., and Tavitian, A. (1988) Expression of the ras-related ralA, rho12 and rab genes in adult mouse tissues. *Oncogene,* **3**, 231.

30. Moll, J., Sansig, G., Fattori, E. , and van der Putten, H. (1991) The murine *rac*1 gene: cDNA cloning, tissue distribution and regulated expression of *rac*1 mRNA by disassembly of actin microfilaments. *Oncogene,* **6**, 863.

31. Shirsat, N. V., Pignolo, R. J., Kreider, B. L., and Rovera, G. (1990) A member of the *ras* gene superfamily is specifically expressed in T, B and myeloid hemopoietic cells. *Oncogene,* **5**, 769.

32. Jähner, D. and Hunter, T. (1991) The ras-related gene rhoB is an immediate-early gene inducible by v-fps, epidermal growth factor and platelet-derived growth factor in rat fibroblasts. *Mol. Cell. Biol.* **11**, 3682.

33. Fritz, G., Kaina, B., and Aktories, K. (1995) The ras-related small GTP-binding protein RhoB is immediate-early inducible by DNA damaging treatments. *J. Biol. Chem.,* **270**, 25172.

34. Fritz, G. and Kaina, B. (1997) rhoB encoding a UV-inducible ras-related small GTP-binding protein is regulated by GTPases of the rho family and independent of JNK, ERK, and p38 MAP kinase. *J. Biol. Chem.,* **272**, 30637.

35. Engel, M. E., Datta, P. K., and Moses, H. L. (1998) RhoB is stabilized by transforming growth factor β and antagonizes transcriptional activation. *J. Biol. Chem.,* **272**, 9921.

36. Hawkins, P. T., Eguinoa, A., Qiu, R. G., Stokoe, D., Cooke, F. T., Walters, R., Wennström, S., Claesson-Welsh, L., Evans, T., Symons, M., *et al.* (1995) PDGF stimulates an increase in GTP-Rac via activation of phosphoinositide 3-kinase. *Curr. Biol.,* **5**, 393.

37. Mucsi, I., Skorecki, K. L., and Goldberg, H. J. (1996) Extracellular signal-regulated kinase and the small GTP-binding protein, Rac, contribute to the effects of transforming growth factor-β1 on gene expression. *J. Biol. Chem.,* **271**, 16567.

38. Guasch, R. M., Jones, G. E., and Ridley, A. J. (1998) RhoE regulates actin cytoskeleton organization and cell migration. *Mol. Cell. Biol.,* **18**, 4761.

39. Self, A. J. and Hall, A. (1995) Purification of recombinant Rho/Rac/G25K from *Escherichia coli. Meth. Enzymol.,* **256**, 3.

40. Lin, R., Bagrodia, S., Cerione, R., and Manor, D. (1997) A novel Cdc42Hs mutant induces cellular transformation. *Curr. Biol., 7*, 794.

41. Cerione, R. A. and Zheng, Y. (1996) The Dbl family of oncogenes. *Curr. Opinion Cell. Biol., 8*, 216.

42. van Aelst, L. and d'Souza-Schory, C. (1997) Rho GTPases and signaling networks. *Genes Dev., 11*, 2295.

43. Eva, A. and Aaronson, S. A. (1985) Isolation of a new human oncogene from a diffuse B-cell lymphoma. *Nature, 316*, 273.

44. Hart, M. J., Eva, A., Zangrilli, D., Aaronson, S. A., Evans, T., Cerione, R. A., and Zheng, Y. (1994) Cellular transformation and guanine nucleotide exchange activity are catalyzed by a common domain on the *dbl* oncogene product. *J. Biol. Chem., 269*, 62.

45. Zheng, Y., Sangrilli, D., Cerione, R. A., and Eva, A. (1996) The pleckstrin homology domain mediates transformation by oncogenic Dbl through specific intracellular targeting. *J. Biol. Chem., 271*, 19017.

46. Zheng, J., Chen, R.-H., Corblan-Garcia, S., Cahill, S. M., Bar-Sagi, D., and Cowburn, D. (1997) The solution structure of the pleckstrin homology domain of human SOS1. *J. Biol. Chem., 272*, 30340.

47. Chen, R.-H., Corbalan-Garcia, S., and Bar-Sagi, D. (1997) The role of the PH domain in the signal-dependent membrane targeting of Sos. *EMBO J., 16*, 1351.

48. Olson, M. F., Sterpetti, P., Natata, K.-I., Toksoz, D., and Hall, A. (1997) Distinct roles for DH and PH domains in the Lbc oncogene. *Oncogene, 15*, 2827.

49. Michiels, F., Habets, G. G., Stam, J. C., Hordjick, P. L., van der Kammen, R. A., and Collard, J. G. (1995) A role for Rac in Tiam1-induced membrane ruffling and invasion. *Nature, 375*, 338.

50. Zheng, Y., Olson, M. F., Hall, A., Cerione, R. A., and Toksoz, D. (1995) Direct involvement of the small GTP-binding protein Rho in Lbc oncogene function. *J. Biol. Chem., 270*, 9031.

51. Alberts, A. S. and Treisman, R. (1998) Activation of RhoA and SAPK/JNK signalling pathways by the RhoA-specific exchange factor mNET1. *EMBO J., 17*, 4075.

52. Whitehead, I. P., Abe, K., Gorski, J. L., and Der, C. J. (1998) CDC42 and FGD1 cause distinct signaling and transforming activities. *Mol. Cell. Biol., 18*, 4689.

53. Fan, W.-T., Koch, C. A., de Hoog, C. L., Fam, N. P., and Moran, M. F. (1998) The exchange factor Ras-GRF2 activates Ras-dependent and Rac-dependent mitogen-activated protein kinase pathways. *Curr. Biol., 8*, 935.

54. Nimnual, A. S., Yatsula, B. A., and Bar-Sagi D (1998) Coupling of Ras and Rac guanosine triphosphatases through the Ras exchanger Sos. *Science, 279*, 560.

55. Debant, A., Serra-Pagès, C., Seipel, K., O'Brien, S., Tang, M., Park, S.-H., and Streuli, M. (1996) The multidomain protein Trio binds the LAR transmembrane tyrosine phosphatase, contains a protein kinase domain, and has separate rac-specific and rho-specific guanine nucleotide exchange factor domains. *Proc. Natl Acad. Sci., USA, 93*, 5466.

56. Habets, G. G. M., van der Kammen, R. A., Stam, J. C., Michiels, F., and Collard, J. G. (1995) Sequence of the human invasion-inducing Tiam1 gene, its conservation in evolution and its expression in tumor cell lines of different tissue origin. *Oncogene, 10*, 1371.

57. Zhou, K., Wang, Y., Gorski, J. L., Nomura, N., Collard, J., and Bokoch, G. M. (1998) Guanine nucleotide exchange factors regulate specificity of downstream signalling from Rac and Cdc42. *J. Biol. Chem., 273*, 16782.

58. Parker, P. (1995) PI 3-kinase puts GTP on the Rac. *Curr. Biol., 5*, 577.

59. Crespo, P., Schuebel, K. E., Ostrom, A. A., Gutkind, J. S., and Bustelo, X. R. (1997)

Phosphotyrosine-dependent activation of Rac-1 GDP/GTP exchange by the vav proto-oncogene product. *Nature,* **385**, 169.

60. Han, J., Luby-Phelps, K., Das, B., Shu, X., Xia, Y., Mosteller, R. D., Krishna, U. M., Falck, J. R., White, M. A., and Broek, D. (1998) Role of substrates and products of PI 3-kinase in regulating activation of Rac-related guanosine triphosphatases by Vav. *Science,* **279**, 558.

61. Michiels, F., Stam, J. C., Hordjick, P. L., van der Kammen, R. A., Ruuls-van Stalle, L., Feltkamp, C. A., and Collard, J. G. (1997) Regulated membrane localization of Tiam1, mediated by the NH$_2$-terminal pleckstrin homology domain, is required for Rac-dependent membrane ruffling and c-Jun NH$_2$-terminal kinase activation. *J. Cell Biol.,* **137**, 387.

62. Feig, L. A. (1994) Guanine nucleotide exchange factors: a family of positive regulators of Ras and related GTPases. *Curr. Opinion Cell Biol.,* **6**, 204.

63. Bourne, H. A., Sanders, D. A., and McCormick, F. (1991) The GTPase superfamily: conserved structure and molecular mechanism. *Nature,* **349**, 117.

64. Barbacid, M. (1987) *ras* genes. *Annu. Rev. Biochem.,* **56**, 779.

65. Lamarche, N. and Hall, A. (1994) GAPs for rho-related GTPases. *Trends Genet.,* **10**, 436.

66. Ridley, A. J., Self, A. J., Kasmi, F., Paterson, H. F., Hall, A., Marshall, C. J., and Ellis, C. (1993) rho Family GTPase Activating Proteins p190, bcr and rhoGAP Show Distinct Specificities in vitro and in vivo. *EMBO J.,* **12**, 5151.

67. Cicchetti, P., Ridley, A. J., Zheng, Y., Cerione, R. A., and Baltimore, D. (1995) 3BP-1, an SH3 domain binding protein, has GAP activity for Rac and inhibits growth factor-induced membrane ruffling in fibroblasts. *EMBO J.,* **14**, 3127.

68. Kozma, R., Ahmed, S., Best, A., and Lim, L. (1996) The GTPase-activating protein n-chimaerin cooperates with Rac1 and Cdc42Hs to induce the formation of lamellipodia and filopodia. *Mol. Cell. Biol.,* **16**, 5069.

69. Sharma, S. V. (1998) Rapid recruitment of p120RasGAP and its associated protein, p190RhoGAP, to the cytoskeleton during integrin mediated cell-substrate interaction. *Oncogene,* **17**, 271.

70. Takai, Y., Sasaki, T., Tanaka, K., and Nakanishi, H. (1995) Rho as a regulator of the cytoskeleton. *Trends Biochem. Sci.,* **20**, 227.

71. Nishiyama, T., Sasaki, T., Takaishi, K., Masaki, K., Hideaki, Y., Araki, K., Matsuura, Y., and Takai, Y. (1994) rac p21 is involved in insulin-induced membrane ruffling and rho p21 is involved in hepatocyte growth factor- and 12–0-tetradecanoylphorbol-13-acetate (TPA)-induced membrane ruffling in KB cells. *Mol. Cell. Biol.,* **14**, 2447.

72. Coso, O. A., Chiariello, M., Yu, J.-C., Teramoto, H., Crespo, P., Xu, N., Miki, T., and Gutkind, J. S. (1995) The small GTP-binding proteins Rac1 and Cdc42 regulate the activity of the JNK/SAPK signaling pathway. *Cell,* **81**, 1137.

73. Lelias, J. M., Adra, C. N., Wulf, G. M., Guillemot, J. C., Khagad, M., Caput, D., and Lim, B. (1993) cDNA cloning of a human mRNA preferentially expressed in hematopoietic cells and with homology to a GDP-dissociation inhibitor for the rho GTP-binding proteins. *Proc. Natl Acad. Sci., USA,* **90**, 1479.

74. Scherle, P., Behrens, T., and Staudt, L. M. (1993) Ly-GDI, a GDP-dissociation inhibitor of the RhoA GTP-binding protein, is expressed preferentially in lymphocytes. *Proc. Natl Acad. Sci., USA,* **90**, 7568.

75. Masuda, T., Tanaka, K., Nonaka, H., Yamochi, W., Maeda, A., and Takai, Y. (1994) Molecular cloning and characterization of yeast rho GDP dissociation inhibitor. *J. Biol. Chem.,* **269**, 19713.

76. Gosser, Y. Q., Nomanbhoy, T. K., Aghazadeh, B., Manor, D., Combs, C., Cerione, R. A.,

and Rosen, M. K. (1997) C-terminal binding domain of Rho GDP-dissociation inhibitor directs N-terminal inhibitory peptide to GTPases. *Nature*, **387**, 814.

77. Keep, N. H., Barnes, M., Barsukov, I., Badii, R., Lian, L. Y., Segal, A. W., Moody, P. C., and Roberts, G. C. (1997) A modulator of rho family G proteins, rhoGDI, binds these G proteins via an immunoglobulin-like domain and a flexible N-terminal arm. *Structure*, **5**, 623.

78. Gorvel, J. P., Chang, T. C., Boretto, J., Azuma, T., and Chavrier, P. (1998) Differential properties of D4/LyGDI versus RhoGDI: phosphorylation and rho GTPase selectivity. *FEBS Lett.*, **422**, 269.

79. Platko, J. V., Leonard, D. A., Adra, C. N., Shaw, R. J., Cerione, R. A., and Lim, B. (1995) A single residue can modify target-binding affinity and activity of the functional domain of the Rho-subfamily GDP dissociation inhibitors. *Proc. Natl Acad. Sci., USA*, **92**, 2974.

80. Zalcman, G., Closson, V., Camonis, J., Honore, N., Rousseau-Merck, M. F., Tavitian, A., and Olofsson, B. (1996) RhoGDI-3 is a new GDP dissociation inhibitor (GDI). *J. Biol. Chem.*, **271**, 30366.

81. Adra, C. N., Manor, D., Ko, J. L., Zhu, S., Horiuchi, T., Van Aelst, L., Cerione, R. A., and Lim, B. (1997) RhoGDIγ: a GDP-dissociation inhibitor for Rho proteins with preferential expression in brain and pancreas. *Proc. Natl Acad. Sci., USA*, **94**, 4279.

82. Tsukita S, Yonemura S, Tsukita S (1997) ERM proteins: head-to-tail regulation of actin-plasma membrane interaction. *Trends Biochem Sci* 22:53.

83. Takahashi, K., Sasaki, T., Mammoto, A., Takaishi, K., Kameyama, T., Tsukita, S., Tsukita, S., and Takai, Y. (1997) Direct interaction of the Rho GDP dissociation inhibitor with ezrin/radixin/moesin initiates the activation of the Rho small G protein. *J. Biol. Chem.*, **272**, 23371.

84. Takahashi, K., Sasaki, T., Mammoto, A., Hotta, I., Takaishi, K., Imamura, H., Nakano, K., Kodama, A., and Takai, Y. (1998) Interaction of radixin with Rho small G protein GDP/GTP exchange protein Dbl. *Oncogene*, **16**, 3279.

85. Guillemot, J. C., Kruskal, B. A., Adra, C. N., Zhu, S., Ko, J. L., Burch, P., Nocka, K., Seetoo, K., Sions, E., and Lim, B. (1996) Targeted disruption of guanosine diphosphate-dissociation inhibitor for Rho-related proteins, GDID4: normal hematopoietic differentiation but subtle defect in superoxide production by macrophages derived from in vitro embryonal stem cell differentiation. *Blood*, **88**, 2722.

86. Yin, L., Schwartzberg, P., Scharton-Kersten, T. M., Staudt, L., and Lenardo, M. (1997) Immune responses in mice deficient in Ly-GDI, a lymphoid-specific regulator of Rho GTPases. *Mol. Immunol.*, **34**, 481.

87. Danley, D. E., Chuang, T. H., and Bokoch, G. M. (1996) Defective Rho GTPases regulation by IL-1 beta-converting enzyme-mediated cleavage of D4 GDP dissociation inhibitor. *J. Immunol.*, **15**, 500.

88. Na, S., Chuang, T. H., Cunningham, A., Turi, T. G., Hanke, J. H., Bokochm G. M., and Danley, D. E. (1996) D4-GDI, a substrate of CPP32, is proteolyzed during Ras-induced apoptosis. *J. Biol. Chem.*, **271**, 11209.

89. Tolias, K. F., Cantley, L. C., and Carpenter, C. L. (1995) Rho family GTPases bind to phosphoinositide kinases. *J. Biol. Chem.*, **270**, 17656.

90. Ren, X.-D., Bokoch, G. M., Traynor-Kaplan, A., Jenkins, G. H., Anderson, R. A., and Schwartz, M. A. (1996) Physical association of the small GTPase Rho with a 68-kDa phosphatidylinositol 4-phosphate 5-kinase in Swiss 3T3 cells. *Mol. Biol. Cell*, **7**, 435.

91. Fanger, G. R., Lassignal Johnson, N., and Johnson, G. L. (1997) MEK kinases are regulated by EGF and selectively interact with Rac/Cdc42. *EMBO J.*, **16**, 4961.

92. Neudauer, C. L., Joberty, G., Tatsis, N., and Macara, I. G. (1998) Distinct cellular effects and interactions of the Rho-family GTPase TC10. *Curr. Biol.,* **8**, 1151.

93. Alberts, A. S., Bouquin, N., Johnston, L. H., and Treisman, R. (1998) Analysis of RhoA-binding proteins reveals an interaction domain conserved in heterotrimeric G protein beta subunits and the yeast response regulator protein Skn7. *J. Biol. Chem.,* **273**, 8616.

94. Fujisawa, K., Madaule, P., Ishizaki, T., Watanabe, G., Bito, H., Saito, Y., Hall, A., and Narumiya, S. (1998) Different regions of Rho determine Rho-selective binding of different classes of Rho target molecules. *J. Biol. Chem.,* **273**, 18943.

95. Sahai, E., Alberts, A. S., and Treisman, R. (1998) RhoA effector mutants reveal distinct effector pathways for cytoskeletal reorganization, SRF activation and transformation. *EMBO J.,* **17**, 1350.

96. Flynn, P., Mellor. H,, Palmer, R., Panayotou, G., and Parker, P. J. (1998) Multiple interactions of PRK1 with RhoA. Functional assignment of the HR1 repeat motif. *J. Biol. Chem.,* **273**, 2698.

97. Bae, C.-D., Min, D.-S., Fleming, I. N., and Exton, J. H. (1998) Determination of interaction sties on the small G protein RhoA for phospholipase D. *J. Biol. Chem.,* **273**, 11596.

98. Manser, E., Leung, T., Salihuddin, H., Zhao, Z.-S., and Lim, L. (1994) A brain serine/threonine kinase activated by Cdc42 and Rac1. *Nature,* **367**, 40.

99. Manser, E., Leung, T., Salihuddin, H., Tan, L., and Lim, L. (1993) A non-receptor tyrosine kinase that inhibits the GTPase activity of p21*cdc42*. *Nature,* **363**, 364.

100. Knaus, U. G. and Bokoch, G. M. (1998) The p21$^{Rac/Cdc42}$-activated kinases (PAKs). *Int. J. Biochem. Cell. Biol.,* **30**, 857.

101. Rudel, T. and Bokoch, G. M. (1997) Membrane and morphological changes in apoptotic cells regulated by caspase-mediated activation of PAK2. *Science,* **276**, 1571.

102. Yang, W. and Cerione, R. A. (1997) Cloning and characterization of a novel Cdc42-associated tyrosine kinase, ACK-2, from bovine brain. *J. Biol. Chem.,* **272**, 24819.

103. Satoh, T., Kato, J., Nishida, K., and Kaziro, Y. (1996) Tyrosine phosphorylation of ACK in response to temperature shift-down, hyperosmotic shock, and epidermal growth factor stimulation. *FEBS Lett.,* **386**, 230.

104. Lamarche, N., Tapon, N., Stowers, L., Burbelo, P. D., Aspenström, P., Bridges, T., Chant, J., and Hall, A. (1996) Rac and Cdc42 induce actin polymerization and G1 cell cycle progression independently of p65PAK and JNK/SAPK MAP kinase cascade. *Cell,* **87**, 519.

105. Westwick, J. K., Lambert, Q. T., Clark, G. J., Symons, M., van Aelst, L., Pestell, R. G., and Der, C. J. (1997) Rac regulation of transformation, gene expression, and actin organization by multiple, PAK-independent pathways. *Mol. Cell. Biol.,* **17**, 1324.

106. Knaus, U. G., Wang, Y., Reilly, A. M., Wornock, D., and Jackson, J. H. (1998) Structural requirement for PAK activation by Rac GTPases. *J. Biol. Chem.,* **273**, 21512.

107. Tapon, N. and Hall, A. (1997) Rho, Rac and Cdc42 GTPases regulate the organization of the actin cytoskeleton. *Curr. Opinion Cell Biol.,* **9**, 86.

108. Chrzanowska-Wodnicka, M. and Burridge, K. (1996) Rho-stimulated contractility drives the formation of stress fibres and focal adhesions. *J. Cell Biol.,* **133**, 1403.

109. Hirata, K., Kikuchi, A., Sasaki, T., Kuroda, S., Kaibuchi, K., Matsuura, Y., Seki, H., Saida, K., and Takai, Y. (1992) Involvement of rho p21 in the GTP-enhanced calcium ion sensitivity of smooth muscle contraction. *J. Biol. Chem.,* **267,** 8719.

110. Uehata, M., Ishiziki, T., Satoh, H., Ono, T., Kawahara, T., Morishita, T., Tamakawa, H., Yamagami, K., Inui, J., Maekawa, M., and Narumiya, S. (1997) Calcium sensitization of smooth muscle mediated by a Rho-associated protein kinase in hypertension. *Nature,* **389**, 990.

111. Burridge, K. and Chrzanowska-Wodnicka, M. (1996) Focal adhesions, contractility, and signaling. *Annu. Rev. Cell. Dev. Biol.,* **12,** 463.

112. Brock, J., Midwinter, K., Lewis, J., and Martin, P. (1996) Healing of incisional wounds in the embryonic chick wing bud: characterization of the actin purse-string and demonstration of a requirement for Rho activation. *J. Cell Biol.,* **135,** 1097.

113. Ridley, A. J., Comoglio, P. M., and Hall, A. (1995) Regulation of scatter factor/hepatocyte growth factor responses by Ras, Rac and Rho proteins in MDCK cells. *Mol. Cell. Biol.,* **15,** 1110.

114. Nusrat, A., Giry, M., Turner, J. R., Colgan, S. P., Parkos, C. A., Carnes, D., Lemichez, E., Bouquet, P., and Madara, J. L. (1995) Rho protein regulates tight junctions and peri-junctional actin organization in polarized epithelia. *Proc. Natl Acad. Sci., USA,* **92,** 10629.

115. Allen, W. E., Jones, G. E., Pollard, J. W., and Ridley, A. J. (1997) Rho, Rac and Cdc42 regulate actin organization and cell adhesion in macrophages. *J. Cell Sci.,* **110,** 707.

116. Jalink, K., van Corven, E. J., Hengeveld, T., Morii, N., Narumiya, S., and Moolenaar, W. H. (1994) Inhibition of lysophosphatidate- and thrombin-induced neurite retraction and neuronal cell rounding by ADP ribosylation of the small GTP-binding protein rho. *J. Cell Biol.,* **126,** 801.

117. Kozma, R., Sarner, S., Ahmed, S., and Lim, L. (1997) Rho family GTPases and neuronal growth cone remodelling: relationship between increased complexity induced by Cdc42Hs, Rac1, and acetylcholine and collapse induced by RhoA and lysophosphatidic acid. *Mol. Cell. Biol.,* **17,** 1201.

118. Norman, J. C., Price, L. S., Ridley, A. J., Hall, A., and Koffer, A. (1994) Actin filament organization in activated mast cells is regulated by heterotrimeric and small GTP-binding proteins. *J. Cell Biol.,* **126,** 1005.

119. Wójciak-Stothard, B., Entwistle, A., Garg, R., and Ridley, A. J. (1998) Regulation of TNF-a-induced reorganization of the actin cytoskeleton and cell-cell junctions by Rho, Rac and Cdc42 in human endothelial cells. *J. Cell Physiol.,* **176,** 150.

120. Adam, T., Giry, M., Boquet, P., and Sansonetti, P. (1996) Rho-dependent membrane folding causes *Shigella* entry into epithelial cells. *EMBO J.,* **15,** 3315.

121. Machesky, L. M. and Hall, A. (1997) Role of actin polymerization and adhesion to extracellular matrix in Rac- and Rho-induced cytoskeletal reorganization. *J. Cell Biol.,* **138,** 913.

122. Imamura, H., Tanaka, K., Hihara, T., Umikawa, M., Kamei, T., Takahashi, K., Sasaki, T., and Takai, Y. (1997) Bni1p and Bnr1p: downstream targets of the Rho family small G-proteins which interact with profilin and regulate actin cytoskeleton in *Saccharomyces cerevisiae. EMBO J.,* **16,** 2745.

123. Arellano, M., Duran, A., and Perez, P. (1997) Localisation of the *Schizosaccaromyces pombe* rho1p GTPase and its involvement in the organisation of the actin cytoskeleton. *J. Cell Sci.,* **110,** 2547.

124. Tanaka, K. and Takai, Y. (1998) Control of reorganization of the actin cytoskeleton by Rho family small GTP-binding proteins in yeast. *Curr. Opinion Cell Biol.,* **10,** 112.

125. Yamochi, W., Tanaka, K., Nonaka, H., Maeda, A., Musha, T., and Takai, Y. (1994) Growth site localization of Rho1 small GTP-binding protein and its involvement in bud formation in *Saccharomyces cerevisiae. J. Cell Biol.,* **125,** 1077.

126. Bussey, H. (1996) Cell shape determination: a pivotal role for Rho. *Science,* **272,** 224.

127. Barrett, K., Leptin, M., and Settleman, J. (1997) The Rho GTPase and a putative RhoGEF mediate a signaling pathway for the cell shape changes in *Drosophila* gastrulation. *Cell,* **91,** 905.

128. Häcker, U. and Perrimon, N. (1998) *DRhoGEF2* encodes a member of the Dbl family of oncogenes and controls cell shape changes during gastrulation in *Drosophila. Genes Dev.*, **12**, 274.

129. Ridley, A. J. (1996) Rho: theme and variations. *Curr. Biol.*, **6**, 1256.

130. Amano, M., Ito, M., Kimura, K., Rukata, Y., Chcihara, K., Nakano, T., Matsuura, Y., and Kaibuchi, K. (1996) Phosphorylation and activation of myosin by Rho-associated kinase (Rho-kinase*). J. Biol. Chem.*, **271**, 20246.

131. Kureishi, Y., Kobayashi, S., Amano, M., Kimura, K., Kanaide, H., Nakano, T., Kaibuchi, K., and Ito, M. (1997) Rho-associated kinase directly induces smooth muscle contraction through myosin light chain phosphorylation. *J. Biol. Chem.*, **272**, 12257.

132. Narumiya, S, Ishizaki, T., and Watanabe, N. (1997) Rho effectors and reorganization of actin cytoskeleton. *FEBS Lett.*, **410**, 68.

133. Kimura, K., Fukata, Y., Matsuoka, Y., Bennett, V., Matsuura, Y., Okawa, K., Iwamatsu, A., and Kaibuchi, K. (1998) Regulation of the association of adducin with actin filaments by Rho-associated kinase (Rho-kinase) and myosin phosphatase. *J. Biol. Chem.*, **273**, 5542.

134. Kosako, H., Amano, M., Yanagida, M., Tanabe, K., Nishi, Y., Kaibuchi, K., and Inagaki, M. (1997) Phosphorylation of glial fibrillary acidic protein at the same sites by cleavage furrow kinase and Rho-associated kinase. *J. Biol. Chem.*, **272**, 10333.

135. Tominaga, T., Ishizaki, T., Narumiya, S., and Barber, D. L. (1998) p160ROCK mediates RhoA activation of Na-H exchange. *EMBO J.*, **17**, 4712.

136. Matsui, T., Maeda, M., Doi, Y., Yonemura, S., Amano, M., Kaibuchi, K., Tsukita, S., and Tsukita, S. (1998) Rho-kinase phosphorylates COOH-terminal threonines of ezrin/radixin/moesin (ERM) proteins and regulates their head-to-tail association. *J. Cell Biol.*, **140**, 647.

137. Fujita, A., Saito, Y., Ishizaki, T., Maekawa, M., Fujisawa, K., Ushikubi, F., and Narumiya, S (1997) Integrin-dependent translocation of p160ROCK to cytoskeletal complex in thrombin-stimulated platelets. *Biochem. J.*, **328**, 769.

138. Watanabe, N., Madaule, P., Reid, T., Ishizaki, T., Watanabe, G. G., Kakizuka, A., Saito, Y., Nakao, K., Jockusch, B. M., and Narumiya, S. (1997) p140mDia, a mammalian homolog of *Drosophila* diaphanous, is a target protein for Rho small GTPase and is a ligand for profilin. *EMBO J.*, **16,** 3044.

139. Castrillon, D. H. and Wasserman, S. A. (1994) Diaphanous is required for cytokinesis in *Drosophila* and shares domains of similarity with the products of the limb deformity gene. *Development,* **120**, 3367.

140. Lynch, E. D., Lee, M. K., Morrow, J. E., Welcsh, P. L., León, P. E., and King, M.-C. (1997) Nonsyndromic deafness DFNA1 associated with mutation of a human homolog of the *Drosophila* gene *diaphanous. Science,* **278**, 1315.

141. Evangelista, M., Blundell, K., Longtine, M. S., Chow, C. J., Adames, J., Pringle, J. R., Peter, M., and Boone, C. (1997) Bni1p, a yeast formin linking Cdc42p and the actin cytoskeleton during polarized morphogenesis. *Science,* **276,** 118.

142. Stossel, T. P. (1993) On the crawling of mammalian cells. *Science,* **260**, 1086.

143. Barkalow, K., Witke, W., Kwiatkowski, D. J., and Hartwig, J. H. (1996) Coordinated regulation of platelet actin filament barbed ends by gelsolin and capping protein. *J. Cell Biol.,* **134**, 389.

144. Schafer, D. A., Jennings, P. B., and Cooper, J. A. (1996) Dynamics of capping protein and actin assembly in vitro: uncapping barbed ends by polyphosphoinositides. *J. Cell Biol.,* **135**, 169.

145. Hartwig JH, Kung S, Kovacsovics T, Janmey PA, Cantley LC, Stossel TP, Toker A (1996)

D3 phosphoinositides and outside-in integrin signaling by glycoprotein IIb-IIIa mediate platelet actin assembly and filopodial extension induced by phorbol 12-myristate 13-acetate. *J Biol Chem* 271, 32986.

146. Carpenter, C. L. and Cantley, L. C. (1996) Phosphoinositide kinases. *Curr. Opinion Cell Biol.,* **8**, 153.

147. Chong, L. D., Traynor-Kaplan, A., Bokoch, G., and Schwartz, M. A. (1994) The small GTP-binding protein rho regulates a phosphatidylinositol 4-phosphate 5-kinase in mammalian cells. *Cell,* **79**, 507.

148. Vincent, S. and Settleman, J. (1997) The PRK2 kinase is a potential effector target of both Rho and Rac GTPases and regulates actin cytoskeletal organization. *Mol. Cell. Biol.,* **17**, 2247.

149. Welch, M. D., Mallavarapu, A., Rosenblatt, J., and Mitchison, T. J. (1997) Actin dynamics in vivo. *Curr. Opinion Cell Biol.,* **9**, 54.

150. Mitchison, T. J. and Cramer, L. P. (1996) Actin-based cell motility and cell locomotion. *Cell,* **84**, 371.

151. Zigmond, S. H. (1998) The Arp2/3 complex gets to the point. *Curr. Biol.,* **8**, R654.

152. Arber, S., Barbayannis, F. A., Hanser, H., Schneider, C., Stanyon, C. A., Bernard, O., and Caroni, P. (1998) Regulation of actin dynamics through phosphorylation of cofilin by LIM-kinase. *Nature,* **393**, 805.

153. Yang, N., Higuchi, O., Ohashi, K., Nagata, K., Wada, A., Kangawa, K., Nishida, E., and Mizuno, K. (1998) Cofilin phosphorylation by LIK-kinase 1 and its role in Rac-mediated actin reorganization. *Nature,* **393**, 809.

154. Eaton, S., Auvinen, P., Luo, L., Jan, Y. N., and Simons, K. (1995) Cdc42 and Rac1 control different actin-dependent processes in the *Drosophila* wing disc epithelium. *J. Cell Biol.,* **131**, 151.

155. Takaishi, K., Sasaki, T., Kotanin, H., Nishioka, H., and Takai, Y. (1997) Regulation of cell-cell adhesion by Rac and Rho small G proteins in MDCK cells. *J. Cell Biol.,* **139**, 1047.

156. Bush, J., Franek, K., and Cardelli, J. (1993) Cloning and characterization of seven novel *Dictyostelium discoideum* rac-related gene belonging to the rho family of GTPases. *Gene,* **136**, 61.

157. Chen, W., Chen, S, Yap, S. F., and Lim, L. (1996) The *Caenorhabditis elegans* p21-activated kinase (CePAK) colocalizes with CeRac1 and Cdc42Ce at hypodermal cell boundaries during embryo elongation. *J. Biol. Chem.,* **271**, 26362.

158. Harden, N., Loh, H. Y., Chia, W., and Lim, L. (1995) A dominant inhibitory version of the small GTP-binding protein Rac disrupts cytoskeletal structures and inhibits developmental cell shape changes in *Drosophila. Development,* **121**, 903.

159. Nobes, C. D. and Hall, A. (1995) Rho, Rac, and Cdc42 GTPases regulate the assembly of multimolecular focal adhesion complexes associated with actin stress fibers, lamellipodia, and filopodia. *Cell,* **81**, 53.

160. Zigmond, S. H., Joyce, M., Borleis, J., Bokoch, G. M., and Devreotes, P. N. (1997) Regulation of actin polymerization in cell-free systems by GTP?S and Cdc42. *J. Cell Biol.,* **138**, 363.

161. Ma, L., Cantley, L. C., Janmey, P. A., and Kirschner, M. W. (1998) Corequirement of specific phosphoinositides and small GTP-binding protein Cdc42 in inducing actin assembly in *Xenopus* extracts. *J. Cell Biol.,* **140**, 1125.

162. Chant, J. (1996) Generation of cell polarity in yeast. *Curr. Opinion Cell Biol.,* **8**, 557.

163. Leberer, E., Thomas, D. Y., and Whiteway, M. (1997) Pheromone signalling and polarized morphogenesis in yeast. *Curr. Opinion Genet. Dev.,* **7**, 59.

164. Miller, P. J. and Johnson, D. I. (1994) Cdc42p GTPases is involved in controlling polarized cell growth in *Schizosaccharomyces pombe*. *Mol. Cell. Biol.*, **14**, 1075.

165. Welch, M. D., Holtzman, D. A., and Drubin, D. G. (1994) The yeast actin cytoskeleton. *Curr. Opinion Cell Biol.*, **6**, 110.

166. van Leeuwen, F. N., Kain, H. E. T., van der Kammen, R. A., Michiels, F., Krenenburg, O. W. and Collard, J. G. (1997) The guanine nucleotide exchange factor Tiam1 affects neuronal morphology: opposing roles for the small GTPases Rac and Rho. *J. Cell Biol.*, **139**, 797.

167. Lamoureux, P., Altun-Gultekin, S. F., Lin, C., Wagner, J. A., and Heidemann, S. R. (1997) Rac is required for growth cone function but not neurite assembly. *J. Cell Sci.*, **110**, 635.

168. Luo, L., Hensch, T. K., Ackerman, L., Barbel, S., Jan, L. Y., and Jan, Y. N. (1996) Differential effects of the Rac GTPase on Purkinje cell axons and dendritic trunks and spines. *Nature*, **379**, 837.

169. Threadgill, R., Bobb, K., and Ghosh, A. (1997) Regulation of dendritic growth and remodeling by Rho, Rac and Cdc42. *Neuron*, **19**, 625.

170. Albertinazzi, C., Gilardelli, D., Paris, S., Longhi, R., and de Curtis, I. (1998) Overexpression of a neural-specific Rho family GTPases, cRac1B, selectively induces enhanced neuritogenesis and neurite branching in primary neurons. *J. Cell Biol.*, **142**, 815.

171. Steven, R., Kubiseski, T. J., Zheng, H., Kulkarni, S., Mancillas, J., Ruiz Morales, A., Hogue, C. W., Pawson, T., and Culotti, J. (1998) UNC-73 activates the Rac GTPase and is required for cell and growth cone migrations in *C. elegans*. *Cell*, **92**, 785.

172. Sone, M., Hoshino, M., Suzuki, E., Kuroda, S., Kaibuchi, K., Nakagoshi, H., Saigo, K., Nabeshima, Y., and Hama, C. (1997) Still life, a protein in synaptic terminals of *Drosophila* homologous to GDP–GTP exchangers. *Science*, **275**, 543.

173. Luo, L., Liao, Y. J., Jan, L. Y., and Jan, Y. N. (1994) Distinct morphogenetic functions of similar small GTPases: *Drosophila* Drac1 is involved in axonal outgrowth and myoblast fusion. *Genes Dev.* **8**, 1787.

174. Sells, M. A., Knaus, U. G., Bagrodia, S., Ambrose, D. M., Bokoch, G. M., and Chernoff, J. (1997) Human p21-activated kinase (Pak1) regulates actin organization in mammalian cells. *Curr. Biol.*, **7**, 202.

175. Manser, E., Huang, H.-Y., Loo, T.-H., Chen, X.-Q., Dong, J.-M., Leung, T., and Lim, L (1997) Expression of constitutively active α-PAK reveals effects of the kinase on actin and focal complexes. *Mol. Cell. Biol.*, **17**, 1129.

176. Daniels, R. H., Hall, P. S., and Bokoch, G. M. (1998). Membrane targeting of p21-activated kinase 1 (PAK1) induces neurite outgrowth from PC12 cells. *EMBO J.* **17**, 754.

177. Dharmawardhane, S., Sanders, L. C., Martin, S. S., Daniels, R. H., and Bokoch, G. M. (1997) Localization of p21-activated kinase 1 (PAK1) to pinocytic vesicles and cortical actin structures in stimulated cells. *J. Cell Biol.*, **138**, 1265.

178. Obermeier, A., Ahmed, S., Manser, E., Yen, S.-C., Hall, C., and Lim, L. (1998) PAK promotes morphological changes by acting upstream of Rac. *EMBO J.*, **17**, 4328.

179. Lu, W., Katz, S., Gupta, R., and Mayer, B. J. (1997) Activation of Pak by membrane localization mediated by an SH3 domain from the adaptor protein Nck. *Curr. Biol.*, **7**, 85.

180. Manser, E., Loo, T.-H., Koh, C.-G., Zhao, Z.-S., Chen, X.-Q., Tan, L., Tan, I., Leung, T., and Lim, L. (1998) PAK kinases are directly coupled to the PIX family of nucleotide exchange factors. *Mol. Cell*, **1**, 183.

181. Chant, J. and Stowers, L. (1995) GTPase cascades choreographing cellular behavior: movement, morphogenesis, and more. *Cell*, **81**, 1.

182. Nikolic, M., Chou, M. M., Lu, W., Mayer, B. J., and Tsaim L.-H. (1998) The p35/Cdk5 kinase is a neuron-specific Rac effector that inhibits Pak1 activity. *Nature*, **395**, 194.

183. Cvrcková, F., De Virgilio, C., Manser, E., Pringle, J. R., and Nasmyth, K. (1995) Ste-20-like protein kinases are required for normal localization of cell growth and for cytokinesis in budding yeast. *Genes Dev.,* **9**, 1817.

184. Eby, J. J., Holly, S. P., van Drogen, F., Grishin, A. V., Peter, M., Drubin, D. G., and Blumer, K. J. (1998) Actin cytoskeleton organization regulated by the PAK family of protein kinases. *Curr. Biol.,* **8**, 967.

185. Machesky, L. M. (1998) Cytokinesis: IQGAPs find a function. *Curr. Biol.,* **8**, R202.

186. Bashour, A.-M., Fullerton, A. T., Hart, M. J., and Blook, G. S. (1997) IQGAP1, a Rac- and Cdc42-binding protein, directly binds and cross-links microfilaments. *J. Cell Biol.,* **137**, 1555.

187. Brill, S., Li, S., Lyman, C. W., Church, D. M., Wasmuth, J. J., Weissbach, L., Bernadrs, A., and Snijders, A. (1996) The Ras GTPase-activating-protein-related human protein IQGAP2 harbors a potential actin binding domain and interacts with calmodulin and Rho family GTPases. *Mol. Cell. Biol.,* **16**, 4869.

188. Kuroda, S., Fukata, M., Nakagawa, M., Fujii, K., Nakamura, T., Ookubo, T., Izawa, I., Nagase, T., Nomura, N., Tani, H., Shoui, I., Matsuura, Y., Yonehara, S., and Kaibuchi, K. (1998) Role of IQGAP1, a target of the small GTPases Cdc42 and Rac1, in regulation of E-cadherin-mediated cell adhesion. *Science,* **281**, 832.

189. Kirchhausen, T. and Rosen, F. S. (1996) Unravelling Wiskott–Aldrich syndrome. *Curr. Biol.,* **6**, 676.

190. Derry, J. M. J., Ochs, H. J., and Francke, U. (1994) Isolation of a novel gene mutated in Wiskott–Aldrich syndrome. *Cell,* **78**, 635.

191. Miki, H., Miura, K., and Takenawa, T. (1996) N-WASP, a novel actin-depolymerizing protein regulates the cortical cytoskeletal rearrangement in a PIP2-dependent manner downstream of tyrosine kinases. *EMBO J.,* **15**, 5326.

192. Symons, M., Derry, J. M., Karlak, B., Jiang, S., Lemahieu, V., Mccormick, F., Francke, U., and Abo, A. (1996) Wiskott–Aldrich syndrome protein, a novel effector for the GTPase CDC42Hs, is implicated in actin polymerization. *Cell,* **84**, 723.

193. Miki, H., Sasaki, T., Takai, Y., and Takenawa, T. (1998) Induction of filopodium formation by a WASP-related actin-depolymerizing protein N-WASP. *Nature,* **391**, 93.

194. Li, R. (1998) Bee1, a yeast protein with homology to Wiscott–Aldrich syndrome protein, is critical for the assembly of cortical actin cytoskeleton. *J. Cell Biol.,* **136**, 649.

195. Lechler, T. and Li, R. (1998) In vitro reconstitution of cortical actin assembly sites in budding yeast. *J. Cell Biol.,* **138**, 95.

196. Suetsugu, S., Miki, H., and Takenawa, T. (1998) The essential role of profilin in the assembly of actin for microspike formation. *EMBO J.,* **17**, 6516.

197. Ramesh, N., Antón, I. M., Hartwig, J. H., and Geha, R. S. (1997) WIP, a protein associated with Wiskott–Aldrich syndrome protein, induces actin polymerization and redistribution in lymphoid cells. *Proc. Natl Acad. Sci., USA,* **94**, 14671.

198. Naqvi, S. N., Sahn, R., Mitchell, D. A., Stevenson, B. J., and Munn, A. L. (1998) The WASp homologue Las17p functions with the WIP homologue End5p/verprolin and is essential for endocytosis in yeast. *Curr. Biol.,* **8**, 959.

199. Cory, G. O., MacCarthy-Morrogh, L., Banin, S., Gout, I., Brickell, P. M., Levinsky, R. J., Kinnon, C., and Lovering, R. C. (1997) Evidence that the Wiskott–Aldrich syndrome protein may be involved in lymphoid cell signaling pathways. *J. Immunol.,* **157**, 3791.

200. Tolias, K. F., Couvillon, A. D., Cantley, L. C., and Carpenter, C. L. (1998) Characterization of a Rac1- and RhoGDI-associated lipid kinase signaling complex. *Mol. Cell. Biol.,* **18**, 762.

201. Hartwig, J. H., Bokoch, G. M., Carpenter, C. L., Janmey, P. A., Taylor, L. A., Toker, A., and Stossel, T. P. (1995) Thrombin receptor ligation and activated Rac uncap actin filament barbed ends through phosphoinositide synthesis in permeabilized human platelets. *Cell*, **82**, 643.

202. Aspenström, P. (1997) A Cdc42 target protein with homology to the non-kinase domain of FER has a potential role in regulating the actin cytoskeleton. *Curr. Biol.*, **7**, 479.

203. van Aelst, L., Joneson, T., and Bar-Sagi, D. (1996) Identification of a novel Rac1-interacting protein involved in membrane ruffling. *EMBO J.*, **15**, 3778.

204. Salmon, E. D. (1989) Cytokinesis in animal cells. *Curr. Opinion Cell Biol.*, **1**, 541.

205. Fishkind, D. J. and Wang, Y.-I. (1993) Orientation and three-dimensional organization of actin filaments in dividing cultured cells. *J. Cell Biol.*, **123**, 837.

206. Kishi, K., Sasaki, T., Kuroda, S., Itoh, T., and Takai, Y. (1993) Regulation of cytoplasmic division of *Xenopus* embryo by *rho* p21 and its inhibitory GDP/GTP exchange protein (*rho* GDI). *J. Cell Biol.*, **120**, 1187.

207. Drechsel, D. N., Hyman, A. A., Hall, A., and Glotzer, M. (1996) A requirement for Rho and Cdc42 during cytokinesis in *Xenopus* embryos. *Curr. Biol.*, **7**, 12.

208. Maduale, P., Eda, M., Watanabe, N., Fujisawa, K., Matsuoka, T., Bito, H., Ishizaki, T., and Narumiya, S. (1998) Role of citron kinase as a target of the small GTPase Rho in cytokinesis. *Nature*, **394**, 491.

209. Dutartre, H., Davoust, J., Gorvel, J.-P., and Chavrier, P. (1996) Cytokinesis arrest and redistribution of actin-cytoskeleton regulatory components in cells expressing the Rho GTPase CDC42Hs. *J. Cell Sci.*, **109**, 367.

210. Eng, K., Naqvi, N. I., Wong, K. C. Y., and Balasubramanian, M. K. (1998) Rng2p, a protein required for cytokinesis in fission yeast, is a component of the actomyosin ring and the spindle pole body. *Curr. Biol.*, **8**, 611.

211. Larochelle, D. A., Vithalani, K.K., and De Lozanne, A. (1996) A novel member of the rho family of small GTP-binding proteins is specifically required for cytokinesis. *J. Cell Biol.*, **133**, 1321.

212. Faix, J. and Dittrich, W. (1996) DGAP1, a homologue of RasGTPase activating proteins that controls growth, cytokinesis, and development in *Dictyostelium discoideum*. *FEBS Lett.*, **394**, 251.

213. Lee, S., Escalante, R., and Firtel, R. A. (1997) A RasGAP is essential for cytokinesis and spatial patterning in *Dictyostelium. Development*, **124**, 983.

214. Adachi, H., Takahashi, Y., Hasebe, T., Shirouzu, M., Yokoyama, S., and Sutoh, K. (1997) *Dictyostelium* IQGAP-related protein specifically involved in the completion of cyto-kinesis. *J. Cell Biol.*, **137**, 891.

215. Gauthier-Rouvière, C., Vignal, E., Meriane, M., Roux, P., Montcourier, P., and Fort, P. (1998) RhoG GTPase controls a pathway that independently activates Rac1 and Cdc42Hs. *Mol. Biol. Cell*, **9**, 1379.

216. Murphy, A. M. and Montell, D. J. (1996) Cell type-specific roles for Cdc42, Rac and RhoL in *Drosophila* oogenesis. *J. Cell Biol.*, **133**, 617.

217. Schwartz, M. A. (1997) Integrins, oncogenes and anchorage independence. *J. Cell Biol.*, **139**, 575.

218. Barry, S. T., Flinn, H. M., Humphries, M., Critchley, D. R., and Ridley, A. J. (1997) Requirement for Rho in integrin signalling. *Cell Adh. Commun.*, **4**, 387.

219. Clark, E. A., King, W. G., Brugge, J. S., Symons, M., and Hynes, R. O. (1998) Integrin-mediated signals regulated by members of the Rho family of GTPases. *J. Cell Biol.*, **142**, 573.

220. Leng, L., Kashiwagi, H., Ren, X.-D., and Shattil, S. J. (1998) RhoA and the function of platelet integrin $\alpha_{IIb}\beta_3$. *Blood,* **91**, 4206.

221. Flinn, H. M. and Ridley, A. J. (1996) The small GTP-binding protein Rho stimulates tyrosine phosphorylation of focal adhesion kinase, p130 and paxillin. *J. Cell Sci.,* **109**, 1133.

222. Hotchin, N. A. and Hall, A. (1995) The assembly of integrin adhesion complexes requires both extracellular matrix and intracellular rho/rac GTPases. *J. Cell Biol.,* **131**, 1857.

223. Defilippi, P., Olivo, C., Tarone, G., Mancini, P., Torrisi, M. R., and Eva, A. (1997) Actin cytoskeleton polymerization in Dbl-transformed NIH3T3 fibroblasts is dependent on cell adhesion to specific extracellular matrix proteins. *Oncogene,* **14**, 1933.

224. Kozma, R., Ahmed, S., Best, A., and Lim, L. (1995) The Ras-related protein Cdc42Hs and bradykinin promote formation of peripheral actin microspikes and filopodia in Swiss 3T3 fibroblasts. *Mol. Cell. Biol.,* **15**, 1942.

225. Tominaga, T. and Barber, D. L. (1998) Na-H exchange acts downstream of RhoA to regulate integrin-induced cell adhesion and spreading. *Mol. Biol. Cell,* **9**, 2287.

226. Hooley, R., Yu, C.-Y., Symons, M., and Barber, D. L. (1996) Ga13 stimulates Na-H exchange through distinct Cdc42-dependent and Rho-dependent pathways. *J. Biol. Chem.,* **271**, 6152.

227. Price, L. S., Leng, J., Schwartz, M. A., and Bokoch, G. M. (1998). Activation of Rac and Cdc42 by integrins mediates cell spreading. *Mol. Biol. Cell,* **9**, 1863.

228. D'Souza-Schorey, C., Boettner, B., and van Aelst, L. (1998) Rac regulates integrin-mediated spreading and increased adhesion of T lymphocytes. *Mol. Cell. Biol.,* **18**, 3936.

229. Gumbiner, B. M. (1996) Cell adhesion: the molecular basis of tissue architecture and morphogenesis. *Cell,* **84**, 345.

230. Balda, M. S. and Matter, K. (1998) Tight junctions. *J. Cell Sci.,* **111**, 541.

231. Jou, T.-S. and Nelson, W. J. (1998) Effects of regulated expression of mutant RhoA and Rac1 small GTPases on the development of epithelial (MDCK) cell polarity. *J. Cell Biol.,* **142**, 85.

232. Jou, T.-S., Schneeberger, E.E., and Nelson, W. J. (1998) Structural and functional regulation of tight junctions by RhoA and Rac1 small GTPases. *J. Cell Biol.,* **142**, 101.

233. Braga, V. M. M., Machesky, L. M., Hall, A., and Hotchin, N. A. (1997) The small GTPases Rho and Rac are required for the establishment of cadherin-dependent cell–cell contacts. *J. Cell Biol.,* **137**, 1421.

234. Hordijk, P. L., ten Klooster, J. P., van der Kammen, R. A., Michiels, F., Oomen, L. C., and Collard, J. G. (1997) Inhibition of invasion of epithelial cells by Tiam1-Rac signaling. *Science,* **278**, 1464.

235. Potempa, S. and Ridley, A. J. (1998) Activation of both MAP kinase and phosphatidylinositide 3-kinase by Ras is required for HGF/SF-induced adherens junction disassembly. *Mol. Biol. Cell,* **9**, 2185.

236. Bretscher, A., Reczek, D., and Berryman, K. (1997) Ezrin: a protein requiring conformational activation to link microfilaments to the plasma membrane in the assembly of cell surface structures. *J. Cell Sci.,* **110**, 3011.

237. Kotani, H., Takaishi, K., Sasaki, T., and Takai, Y. (1997) Rho regulates association of both the ERM family and vinculin with the plasma membrane in MDCK cells. *Oncogene,* **14**, 1705.

238. Mackay, D. J. G., Esch, F., Furthmayr, H., and Hall, A. (1997) Rho- and Rac-dependent assembly of focal adhesion complexes and actin filaments in permeabilized fibroblasts: as essential role for ezrin/radixin/moesin proteins. *J. Cell Biol.,* **138**, 927.

239. Lauffenberger, D.A. and Horwitz, A. F. (1996) Cell migration: a physically integrated molecular process. *Cell,* **84**, 359.

240. Allen, W. E., Zicha, D., Ridley, A. J., and Jones, G. E. (1998) A role for Cdc42 in macrophage chemotaxis. *J. Cell Biol.,* **141**, 1147.

241. Takaishi, K., Kikuchi, A., Kuroda, S., Kotanin, K., Sasaki, T., and Takai, Y. (1993) Involvement of rho p21 and its inhibitory GDP–GTP exchange protein (rho GDI) in cell motility. *Mol. Cell Biol.,* **13**, 72.

242. Stowers, L., Yelon, D., Berg, L. J., and Chant, J. (1995) Regulation of the polarization of T cells toward antigen-presenting cells by Ras-related GTPase Cdc42. *Proc. Natl Acad. Sci., USA,* **92**, 5027.

243. Habets, G. G. M., Scholtes, E. H. M., Zuydgeest, D., van der Kammen, R. A., Stam, J. C., Berns, A., and Collard, J. G. (1994) Identification of an invasion-inducing gene, *Tiam*-1, that encodes a protein with homology to GDP–GTP exchangers for Rho-like proteins. *Cell,* **77**, 537.

244. Stam, J. C., Michiels, F., van der Kammen, R. A., Moolenaar, W. H., and Collard, J. G. (1998) Invasion of T-lymphoma cells: cooperation between Rho family GTPases and lysophospholipid receptor. *EMBO J.,* **17**, 4066.

245. Shaw, L. M., Rabinovitz, I., Wang, H. H. F., Toker, A., and Mercurio, A. M. (1997) Activation of phosphoinositide 3-OH kinase by the α6β4 integrin promotes carcinoma invasion. *Cell,* **91**, 949.

246. Keely, P. J., Westwick, J. K., Whitehead, I. P., Der, C. J., and Parise, L. V. (1997) Cdc42 and Rac1 induce integrin-mediated cell motility and invasiveness through PI(3)K. *Nature,* **390**, 632.

247. Sander, E. E., van Delft, S., ten Klooster, J. P., Reid, T., van der Kammen, R. A., Michiels, F., and Collard, J. G. (1998) Matrix-dependent Tiam1/Rac Signaling in epithelial Cells promotes either cell–cell adhesion or cell migration and is regulated by phosphatidyl-inositol 3-kinase. *J. Cell Biol.,* **143**, 1385.

248. Kheradmand, F., Werner, E., Tremble, P., Symons, M., and Werb, Z. (1998) Role of Rac1 and oxygen radicals in collagenase-1 expression induced by cell shape change. *Science,* **280**, 898.

249. Suwa, H., Ohshio, G., Imamura, T., Watanabe, G., Arii, S., Imamura, M., Narumiya, S., Hiai, H., and Fumumoto, M. (1998) Overexpression of the rhoC gene correlates with progression of ductal adenocarcinoma of the pancreas. *Br. J. Cancer,* **77**, 147.

250. Qui, R.-G., Abo, A., McCormick, F., and Symons, M. (1997) Cdc42 regulates anchorage-independent growth and is necessary for Ras transformation. *Mol. Cell. Biol.,* **17**, 3449.

251. Roux, P., Gauthier-Rouvière, C., Doucet-Brutin, S., and Fort, P. (1997) The small GTPases Cdc42Hs, Rac1 and RhoG delineate Raf-independent pathways that cooperate to transform NIH3T3 cells. *Curr. Biol.,* **7**, 629.

252. Symons, M. (1996) Rho family GTPases: the cytoskeleton and beyond. *Trends Biochem. Sci.,* **21**, 178.

253. Olson, M. F., Pasteris, N. G., Gorski, J. L., and Hall, A. (1996) Faciogenital dysplasia protein (FGD1) and Vav, two related proteins required for normal embryonic development, are upstream regulators of Rho GTPases. *Curr. Biol.,* **6**, 1628.

254. Westwick, J. K., Lee, R. J., Lambert, Q. T., Symons, M., Pestell, R. G., Der, C. J., and Whitehead, I. P. (1998) Transforming potential of Dbl family proteins correlates with transcription from the cyclin D1 promoter but not with activation of Jun NH2-terminal kinases, p38/Mpk2, serum response factor, or c-Jun. *J. Biol. Chem.,* **273**, 16739.

255. Olson, M. F., Ashworth, A., and Hall, A. (1995) An essential role for Rho, Rac and Cdc42 GTPases in cell cycle progression through G_1. *Science,* **269**, 1270.

256. Mólnar, A., Theodoras, A. M., Zon, L. I., and Kyriakis, J. M. (1997) Cdc42Hs, but not Rac1, inhibits serum-stimulated cell cycle progression at G1/S through a mechanism requiring p38/RK. *J. Biol. Chem.,* **272**, 13229.

257. Lorès, P., Morin, L., Luna, R., and Gacon, G. (1997) Enhanced apoptosis in the thymus of transgenic mice expressing constitutively activated forms of human Rac2GTPase. *Oncogene,* **15**, 601.

258. Fanger, G. R., Gerwins, P., Widmann, C., Jarpe, M. B., and Johnson, G. L. (1997) MEKKs, GCKs, MLKs, PAKs, TAKs, and Tpls: upstream regulators of the c-Jun amino-terminal kinases? *Curr. Opinion Genet. Dev.,* **7**, 67.

259. Kauffmann-Zeh, A., Rodriguez-Viciana, P., Ulrich, E., Gilbert, C., Coffer, P., Downward, J., and Evan, G. (1997) Suppression of c-Myc-induced apoptosis by Ras signalling through PI(3)K and PKB. *Nature,* **385**, 544.

260. Hill, C. S., Wynne, J., and Treisman, R. (1995) The rho family GTPases RhoA, Rac1, and Cdc42Hs regulate transcriptional activation by SRF. *Cell,* **81**, 1159.

261. Alberts, A. S., Geneste, O., and Treisman, R. (1998) Activation of SRF-regulated chromosomal templates by Rho-family GTPases requires a signal that also induces H4 hyperacetylation. *Cell,* **92**, 475.

262. Olson, M. F., Paterson, H. F., and Marshall, C. J. (1998) Signals from Ras and Rho GTPases interact to regulate expression of p21Waf1/Cip1. *Nature,* **394**, 295.

263. Hirai, A., Nakamura, S., Noguchi, Y., Yasuda, T., Kitagawa, M., Tatsuno, I., Oeda, T., Tahara, K., Terano, T., Narumiya, S., Kohn, L. D., and Saito, Y. (1997) Geranyl-geranylated rho small GTPase(s) are essential for the degradation of p27Kip1 and facilitate the progression from G1 to S phase in growth-stimulated rat FRTL-5 cells. *J. Biol. Chem.,* **272**, 13.

264. Tanaka, T., Tatsuno, I., Noguchi, Y., Uchida, D., Oeda, T., Narumiya, S., Yasuda, T., Higashi, H., Kitagawa, M., Nakayama, K., Saito, Y., and Hirai, A. (1998) Activation of cyclin-dependent kinase 2 (Cdk2) in growth-stimulated rat astrocytes. Geranyl-geranylated Rho small GTPase(s) are essential for the induction of cyclin E gene expression. *J. Biol. Chem.,* **273**, 26772.

265. Kyriakis, J. M. and Avruch, J. (1996) Protein kinase cascades activated by stress and inflammatory cytokines. *BioEssays,* **18**, 567.

266. Sulciner, D. J., Irani, K., Yu, Z. X., Ferrans, V. J., Goldschmidt-Clermont, P., and Finkel, T. (1996) rac1 regulates a cytokine-stimulated, redox-dependent pathway necessary for NF-kB activation. *Mol. Cell. Biol.,* **16**, 7115.

267. Perona, R., Montaner, S., Saniger, L., Sanchez, P. I., Bravo, R., and Lacal, J. C. (1997) Activation of the nuclear factor-κB by Rho, CDC42, and Rac-1 proteins. *Genes Dev.,* **11**, 463.

268. Hunter, T. and Pines, J. (1994) Cyclins and cancer II: cyclin D and CDK inhibitors come of age. *Cell,* **79**, 573.

269. Gjoerup, O., Lukas, J., Bartek, J., and Willumsen, B. M. (1998) Rac and Cdc42 are potent stimulators of E2F-dependent transcription capable of promoting retinoblastoma susceptibility gene product hyperphosphorylation. *J. Biol. Chem.,* **273**, 18812.

270. Ridley, A. J. (1995) Rac and Bcr regulate phagocytic phoxes. *Curr. Biol.,* **5**, 710.

271. Diekmann, D., Abo, A., Johnston, C., Segal, A. W., and Hall, A. (1994) Interaction of Rac with p67phox and regulation of phagocytic NADPH oxidase activity. *Science,* **265**, 531.

272. Dourseuil, O., Reibel, L., Bokoch, G. M., Camonis, J., and Gacon, G. (1996) The Rac target NADPH oxidase p67phox interacts preferentially with Rac2 rather than Rac1. *J. Biol. Chem.,* **271**, 83.

273. Diekmann, D., Nobes, C. D., Burbelo, P. D., Abo, A., and Hall, A. (1995) Rac GTPase interacts with GAPs and target proteins through multiple effector sites. *EMBO J.,* **14**, 5297.

274. Freeman, J. L., Abo, A., and Lambeth, J. D. (1996) Rac 'insert region' is a novel effector region that is implicated in the activation of NADPH oxidase, but not PAK65. *J. Biol. Chem.,* **271**, 19794.

275. Joseph, G. and Pick, E. (1996) 'Peptide walking' is a novel method for mapping functional domains in proteins. Its application to the Rac1-dependent activation of NADPH oxidase. *J. Biol. Chem.,* **270**, 29079.

276. Toporik, A., Gorzalczany, Y., Hirshberg, M., Pick, E., and Lotan, O. (1998) Mutational analysis of novel effector domains in Rac1 involved in the activation of nicotinamide adenine dinucleotide phosphate (reduced) oxidase. *Biochemistry,* **37**, 7147.

277. Ahmed, S., Prigmore, E., Govind, S., Veryard, C., Kozma, R., Wientjes, F. B., Segal, A. W., and Lim, L. (1998) Cryptic Rac-binding and p21(Cdc42Hs/Rac)-activated kinase phosphorylation sites of NADPH oxidase component p67[phox]. *J. Biol. Chem.,* **272**, 15693.

278. Allen, L. A. and Aderem, A. (1996) Mechanisms of phagocytosis. *Curr. Opinion Immunol.,* **8**, 36.

279. Cox, D., Chang, P., Zhang, Q., Reddy, P. G., Bokoch, G. M., and Greenberg, S. (1997) Requirements for both Rac1 and Cdc42 in membrane ruffling and phagocytosis in leukocytes. *J. Exp. Med.,* **186**, 1487.

280. Massol, P., Montcourrier, P., Guillemot, J.-C., and Chavrier, P. (1998) Fc receptor-mediated phagocytosis requires CDC42 and Rac1. *EMBO J.,* **17**, 6219.

281. Caron, E. and Hall, A. (1998) Identification of two distinct mechanisms of phagocytosis controlled by different Rho GTPases. *Science,* **282**, 1717.

282. Araki, N., Johnson, M. T., and Swanson, J. A. (1996) A role for phosphoinositide 3-kinase in the completion of macropinocytosis and phagocytosis by macrophages. *J. Cell Biol.,* **135**, 1249.

283. Jones, B. D., Paterson, H. F., Hall, A., and Falkow, S. (1993) *Salmonella typhimurium* induces membrane ruffling by a growth factor-receptor-independend mechanism. *Proc. Natl Acad. Sci., USA,* **90**, 10390.

284. Chen, L. M., Hobbie, S., and Galan, J. E. (1996) Requirement of Cdc42 for *Salmonella*-induced cytoskeletal and nuclear responses. *Science,* **274**, 2115.

285. Kelleher, J. F. and Titus, M. A. (1998) Intracellular motility: how can we all work together? *Curr. Biol.,* **9**, R394.

286. Price, L. S., Norman, J. C., Ridley, A. J., and Koffer, A. (1995) Rac and rho as regulators of secretion in mast cells. *Curr. Biol.* **5**, 68.

287. O'Sullivan, A. J., Brown, A. M., Freeman, H. N. M., and Gomperts, B. D. (1996) Purification and identification of FOAD-II, a cytosolic protein that regulates secretion in streptolysin-O-permeabilized mast cells, as a rac/rhoGDI complex. *Mol. Biol. Cell,* **7**, 397.

288. Brown, A. M., O'Sullivan, A. J., and Gomperts, B. D. (1998) Induction of exocytosis from permeabilized mast cells by the guanosine triphosphatases Rac and Cdc42. *Mol. Biol. Cell,* **9**, 1053.

289. Billadeau, D. D., Brumbaugh, K. M., Dick, C. J., Schoon, R. A., Bustelo, X. R., and Leibson, P. J. (1998) The Vav-Rac1 pathway in cytotoxic lymphocytes regulates the generation of cell-mediated killing. *J. Exp. Med.,* **188**, 549.

290. Norman, J. C., Price, L. S., Ridley, A. J., and Koffer, A. (1996) The small GTP-binding proteins, Rac and Rho, regulate cytoskeletal organisation and exocytosis in mast cells by parallel pathways. *Mol. Biol. Cell,* **7**, 1429.

291. Lamaze, C., Chuang, T.-H., Terlecky, L. J., Bokoch, G. M., and Schmid, S. L. (1996) Regulation of receptor-mediated endocytosis by Rho and Rac. *Nature,* **382**, 177.

292. Schmalzing, G., Richter, H.-P., Hansen, A., Schwartz, W., Just, I., and Aktories, K. (1995) Involvement of the GTP binding protein Rho in constitutive endocytosis in *Xenopus laevis* oocytes. *J. Cell Biol.,* **130**, 1319.

293. Lamaze, C., Fujimoto, L. M., Yin, H. L., and Schmid, S. L. (1997) The actin cytoskeleton is required for receptor-mediated endocytosis in mammalian cells. *J. Biol. Chem.,* **272**, 20332.

294. Frohman, M. A. and Morris, A. J. (1996) Rho is only ARF the story. *Curr. Biol.,* **6**, 945.

295. De Camilli, P., Emr, S. D., McPherson, P. S., and Novick, P. (1996) Phosphoinositides as regulators in membrane traffic. *Science,* **271**, 1533.

5 | Rab

RUTH N. COLLINS AND PATRICK BRENNWALD

1. Introduction

A hallmark of eukaryotic cells is their possession of a set of membrane-enclosed organelles. The traffic of proteins between different organelles of the secretory and endocytic pathways is mediated by membrane-bound containers, termed vesicles. Figure 1 illustrates the basic stages of transport. Vesicles form by budding off from a donor compartment. The vesicle is transported to the acceptor compartment where it docks onto the membrane. This is followed by the final stage of fusion which releases the vesicle contents into the acceptor compartment. A variety of experimental strategies have begun to identify components of the transport machinery, which appear to be conserved in all eukaryotic cells. Every transport event is thought to require representatives from a set of conserved protein families. One of these families is that of the Rab proteins, a group of small, monomeric GTP-binding proteins related to Ras. The Rab GTPases will be the focus of this chapter.

1.1 Rabs in yeast

The initial discovery of the role of Rab proteins in vesicle transport was made over 15 years ago, with the sequencing of the *SEC4* gene. *SEC4* is one of the original 23 *sec* genes that are required at different stages of the secretory pathway in the yeast *S. cerevisiae*. Yeast cells carrying the *sec4–8* temperature-sensitive lethal mutation are deficient in the final stage of secretion and accumulate secretory vesicles upon a shift to the restrictive temperature (1). Sequence analysis of *SEC4* demonstrated a similarity to both the Ras GTPase and another yeast gene, *YPT1 (2)*. *YPT1* was first identified based on its proximity to the *ACT1* gene, but its function in yeast was unknown (3). The homology to *SEC4* suggested that it might also function in vesicular transport, a prediction that was subsequently confirmed experimentally. *Ypt1–1* mutant cells were shown to be deficient in transport from the endoplasmic reticulum (ER) to the Golgi apparatus and, like *sec4* mutants, were found to accumulate vesicle intermediates. Both Sec4p and Ypt1p are essential elements in the yeast secretory pathway, each functioning and located at specific stages of transport: Ypt1p in ER-to-Golgi transport and Sec4p in Golgi-to-cell-surface transport. The

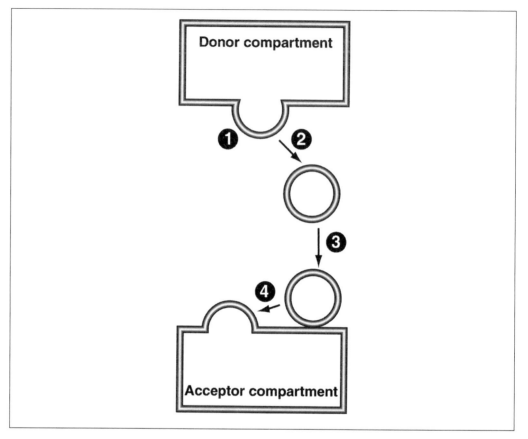

Fig. 1 This diagram illustrates the basic stages of membrane transport. (1) Membrane transport is initiated by the budding of a vesicle or tubule container from a donor compartment that contains selected cargo elements. (2) The transport vesicle 'pinches' off from the donor compartment and travels to the acceptor compartment. (3) The vesicle docks onto the target membrane of the acceptor compartment. (4) Fusion occurs, releasing the contents of the vesicle and its membrane into the acceptor compartment.

complete sequence of the *S. cerevisiae* genome reveals a total of 11 Rab GTPases (Fig. 2).

1.2 Rabs in mammals

Approximately 30 Rab GTPases have been identified to date in mammals. Mammalian Rabs were originally isolated from a <u>rat b</u>rain library, using PCR with oligonucleotides corresponding to GTP-binding domains conserved between *SEC4* and *YPT1*(4, 5). The acronym Rab is now commonly used for all members of this protein subfamily, regardless of their origin. Novel Rab sequences with 30–85% homology receive new numbers. Rab sequences sharing 85% or greater identity are given the same number and are distinguished by addition of a letter (e.g. Rab5A, Rab5B).

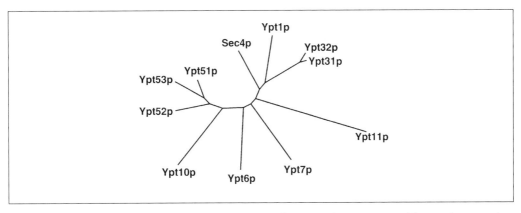

Fig. 2 Phylogenetic tree of yeast Rab proteins. Availability of the complete sequence of the yeast genome has enabled the identification of the complete set of Rab proteins present in this organism. Multiple sequence alignment and treefile creation were performed using Clustal W (version 1.7); the phylogenetic tree was plotted with Drawtree in PHYLIP package (version 3.572).

1.3   Rabs in other organisms

Three hundred and seventeen Rab sequences have been identified in other eukaryotic species. With the current level of genome sequencing efforts, this number will need to be continually revised. At present, this is the largest subfamily of the Ras superfamily. Tables 1 and 2 show the number of different Rab sequences presently identified in the most common divisions of taxonomy and for a few model organisms.

Table 1 Rab species distribution

Species	Number of Rab sequences
Viruses	None
Bacteria	None
Archaea	None
Eukaryota	317
Metazoa	137
Chordata	89
Arthropoda	13
Nematoda	27
Fungi	20
Viridiplantae (Plants)	128

Table 2 Number of Rab proteins in selected model organisms

Homo sapiens	30
Mus musculus	16
Drosophila melanogaster	12
Caenorhabditis elegans	26
Arabidopsis thaliana	26
Saccharomyces cerevisiae	11

1.4 Evolutionary conservation of Rab function

During evolution Rabs have been extremely well conserved. In a few cases, yeast and mammalian Rab proteins are functionally interchangeable. Rab1a can complement loss of Ypt1p in *S. cerevisiae* (6), with both proteins regulating transport between the ER and Golgi. Rab8, which shares high sequence homology with Sec4p, also regulates Golgi-to-cell-surface transport in mammalian cells (7), and can complement null mutations of Ypt2p in *S. pombe*, the functional homologue of *S. cerevisiae* Sec4p (8). Ypt51p, a homologue of mammalian Rab5 is also required at an early step in endocytic trafficking. Remarkably, Ypt51p expression in animal cells not only localizes to Rab5-positive early endosomes, but also stimulates endocytosis (9).

1.5 Functional specificity of Rab proteins

Rab proteins have additional regions of sequence homology with Sec4p and Ypt1p that are not present in other Ras-related GTPases. Figure 3 shows consensus alignments for conserved residues of the core GTPase domain. The Rab family is shown in comparison to the Ras and Rho families. Rab family members share many generic family residues that are different from residues in equivalent positions for the Ras and Rho families. Particularly after the DTAGQ motif, there are stretches of amino acids that are well conserved within each family but differ between families. Additional characteristics of Rab proteins are their N-terminal extensions (Table 3) and a double geranylgeranylation motif at the C terminus (see below). In Ras, there are only four residues from the N-terminal end to the first highly conserved residue (K5); however, the tertiary structure shows that this short tail sticks out of the major globular domain, and potentially could accommodate a larger N-terminal extension. Many Rab proteins contain extended N termini, which could form small, flexible

Table 3 Comparison of N-terminal extension of Rab proteins with other members of the Ras superfamily

Protein	N-terminal extension[a]
H-ras	MTEY–
Rap1A	MREY–
RhoA	MAAIRK–
Rac1	MQAI–
Cdc42	MQTL–
Rab1A	MSSMNPEYDYLF–
Rab8	MAKTYDYLF–
Rab3A	MASATDSRYGQKESSDQNFDYMF–
Rab5	MASRGATRPNGPNTGNKICQF–
Rab6	MSTGGDFGNPLRKF–
Sec4p	MSGLRTVSASSGNGKSYDSIM–

[a] The sequence of the N-terminal extension is given before the first strictly conserved residue

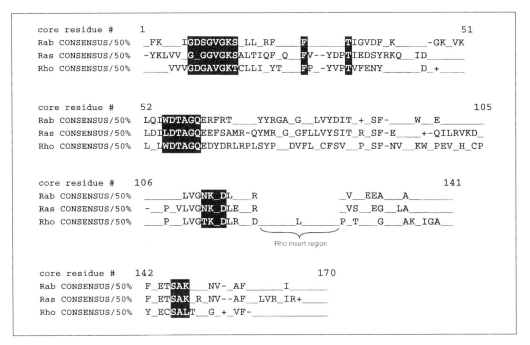

Fig. 3 Comparison of the core domain of Ras superfamily sequences between Rab, Ras, and Rho families. The core domain is aligned showing only those residues conserved at the 50% consensus level, i.e. 50% of sequences or more show this residue at the position indicated. Core residues are numbered according to the Rab core residue number and indicated above each line, these numbers are intended as a descriptive guide only. Positively and negatively charged residues are indicated with + and − respectively. The location of the Rho insert region is marked, this insert is not contained in the Ras or Rab families. For greater clarification, the G-protein conserved sequence elements are shown highlighted.

domains that may be involved in the interaction with specific target proteins (Table 3). Consistent with this idea, blocking Sec4p at its N terminus by fusion to GFP (green fluorescent protein) results in a protein that is unable to substitute for that produced from the wild-type gene, although it does retain partial function (N. B. Elkind, Yale University, personal communication).

What defines the unique identity of an individual Rab protein? Multiple sequence alignments, structural data, and molecular modelling, together with functional studies, have defined candidate regions for Rab functional specificity. These regions include the hypervariable C terminus that mediates membrane targetting specificity (discussed below), the N terminus, and regions of the switch 2 domain (helix α2/loop5 and helix α3/loop7). The important role of these regions for Rab function has been determined experimentally by functional analysis of hybrid molecules between Ypt1p and Sec4p (10, 11) and between Rab5 and Rab6 (12). In addition, taxonomic analysis identifies residues that define the individuality of a particular Rab protein member and the location of subclass-specific sequence motifs (13). The non-conserved stretches in the alignment shown in Fig. 3 also give an overview of these regions. For example, all Rab proteins from the Rab5 subclass share a motif,

LAPM in helix2, (residues 71–75 for Ypt51p); the Rab1 subclass of proteins share the sequence TIT(T/S) in this position (14). Sec4p and Ypt1p are part of the same Rab subclass and a four-amino-acid sequence in the effector domain (loop2) enables the Sec4p effector, Sec15p, to discriminate between them (15).

It may be simplistic to identify critical regions based on primary sequence alignments alone: this fact has been demonstrated elegantly by the recent structural determination of Rab3A complexed to its effector Rabphilin, where several non-contiguous regions of Rab3A combine to form a pocket for Rabphilin binding (16). The authors have termed this a Rab-CDR or complementarity determining region—analogous to the complementary surfaces presented by antibodies to their antigen. For a more extensive discussion of the structure of Ras superfamily members, the reader is referred to Chapter 9 and also to recent reviews (17, 18).

2. Post-translational modification

2.1 Prenylation

Rab proteins are present in two intracellular pools: a minor, cytosolic pool and a major, membrane-associated pool. The proteins are anchored to membranes by double geranylgeranylation on C-terminal cysteines. Deletion of the two cysteine residues at the C terminus of Sec4p results in a soluble, non-functional protein (19). This underscores the fact that localization of Sec4p to membranes is critical for its function. Rabs usually contain various cysteine motifs (-CXC, -CC, -CCX, -CCXX, -CCXXX) and geranylgeranylation requires upstream sequences in addition to the C-terminal cysteines; simple peptidomimetics to the C termini of Rab family members are not substrates for the prenylation machinery, and truncated Rab proteins that lack terminal cysteine acceptor residues inhibit prenylation of wild-type Rabs (20).

The Rab geranylgeranyl transferase (Rab GGTase II) is the enzyme that isoprenylates Rab proteins (21). This is a multi-subunit enzyme consisting of a catalytic component, a heterodimer of A and B subunits of molecular weights 60 and 36 kDa, respectively, and an accessory component—a Rab escort protein (REP) (22). REP-1 corresponds to the product of the mutated gene in human choroideraemia, an X-linked recessive syndrome of retinal degeneration (23, 24). REP forms a complex with newly synthesized Rab proteins bound to GDP and presents them to the catalytic component of the enzyme (25). After the transfer reaction is completed, REP removes the prenylated Rab from the catalytic machinery and delivers it the membrane (26). The REP homologue in yeast, Mrs6p, is the product of an essential gene and has similar biochemical properties to mammalian REP in geranylgeranylation of yeast Rab proteins (27). It should be mentioned that some Rabs, for example Rab8 and Rab23, contain other prenylation motifs at their C termini. Rab8 contains a CAAL motif (C, cysteine; A, aliphatic amino acid; L, leucine) and presumably is a candidate for a single round of geranylgeranylation mediated by a different enzyme, GGTase I. Other known GGTase I substrates include the Rho family and γ subunits of the heterotrimeric G proteins (28).

2.2 Phosphorylation

Rab1A and Rab4A contain a phosphorylation site for Cdc2 kinase (29). Phosphorylation of this site (Ser196) leads to the translocation of Rab4A into the cytosol of mitotic cells (30), due to its inability to be recruited onto membranes (31). Phosphorylation may also be a mechanism for controlling Rab function in differentiated cells. Human platelets phosphorylate Rab3B, Rab6, and Rab8 proteins in response to thrombin, an activator of platelet secretion (32). The abscisic acid responsive maize gene *RAB-17* codes for a protein having a cluster of eight serine residues followed by a putative casein kinase 2-type substrate consensus sequence. This protein is highly phosphorylated *in vivo* and is an *in vitro* substrate for casein kinase 2 (33). However, any functional significance of these latter examples remains to be determined and one should bear in mind that phosphorylation may not be a relevant post-translational modification for many Rab proteins.

3. Cellular localization

Like the yeast Rab proteins, mammalian Rabs are localized to specific intracellular compartments of both the exocytic and endocytic pathways (34, 35). Tables 4 and 5 list the Rab proteins in yeast and mammals, together with what is known about their subcellular distribution. This distribution is consistent with a function of Rab proteins in distinct intracellular transport processes. In every case examined, a Rab protein regulates a membrane transport step that is reflected by its characteristic localization.

Table 4 Localization of Rab proteins in yeast

Rab protein	Localization	Function	Yeast ORF	Functionally related to
Sec4p	Post-Golgi vesicles, PM	Golgi to cell surface	YFL005w	Rab8
Ypt1p	ER-derived vesicles, early Golgi	ER to Golgi	YFL038c	Rab1
Ypt31p	Golgi	Intra-Golgi (anterograde/retrograde?)	YER031c	ND
Ypt32p	Golgi	Intra-Golgi (anterograde/retrograde?)	YGL210w	ND
Vps21p alias Ypt51p	Endosomes	Early endocytic transport	YOR089c	Rab5
Ypt52p	Endosomes	Early endocytic transport	YKR014c	Rab5
Ypt53p	Endosomes	Early endocytic transport	YNL093w	Rab5
Ypt6p	ND	Intra-Golgi, Golgi to endosome	YLR262c	Rab6
Ypt7p	Vacuole	Late endocytic transport, vacuole fusion	YML001w	Rab7
Ypt10p	ND	ND	YBR264c	ND
Ypt11p	ND	ND	YNL304w	ND

ND, Not determined; PM, Plasma membrane.

Table 5 Localization of Rab proteins in mammals[a]

Rab protein	Localization	Notes
Rab1A, Rab1B	Intermediate compartment, CGN	Involved in ER to Golgi transport
Rab2	Intermediate compartment, CGN	Involved in ER to Golgi transport
Rab3A, Rab3B, Rab3C	Exocytic vesicles	Involved in Ca^{2+}-dependent exocytosis, Rab3B expressed in platelets
Rab3D	Regulated exocytic vesicles	Expressed in pancreas, adipocytes
Rab4A, Rab4B	Recycling endosomes	Involved in recycling from early endosomes to plasma membrane
Rab5A, Rab5B, Rab5C	Endocytic vesicles, PM, early endosomes (Rab5B localizes to PM)	Involved in PM to early endosome transport, fusion of early endosomes
Rab6	Golgi	Involved in intra-Golgi retrograde transport
Rab7	Late endosomes	Involved in early to late endosome transport
Rab8A	Post-Golgi vesicles, tight junction of epithelial cells	Involved in Golgi to basolateral plasma membrane transport in polarized cells
Rab8B	PM, vesicles	Highly expressed in spleen, testis, and brain
Rab9	Late endosomes, TGN	Involved in late endosome to TGN
Rab10	Golgi and post-Golgi vesicles	
Rab11A	TGN, recycling endosomes, post-Golgi vesicles	Involved in transport through recycling endosomes
Rab11B	ND	Expressed in brain, heart, and testis
Rab12	Golgi	
Rab13	Tight junctions in polarized epithelia	
Rab14	ND	
Rab15	ND	Specifically expressed in rat brain
Rab16	ND	Expressed predominantly in lung
Rab17	Basolateral PM in epithelial cell, apical endosomes	Involved in transport through apical recycling endosomes
Rab18	Apical endosomes in kidney tubules, also on basolateral domain of intestinal epithelial cells	Level of expression is high in the brain, moderate in the pituitary, and low in the liver
Rab19		Rab19 transcript detected at high levels in intestine, lung, and spleen, and at low levels in kidney
Rab20	Apical endosomes in kidney tubules and intestinal epithelial cells	
Rab21	ND	
Rab22	Endosomes, PM	
Rab23		Expressed predominantly in the brain
Rab24	ER, *cis* Golgi, late endosomes	
Rab25	ND	Expression confined to gastrointestinal mucosa, kidney, and lung
Rab26	ND	Expressed in pancreas, kidney, brain, submandibular gland, and lung
Rab27A	ND	Detected in melanocytes, platelets, and wide variety of cells, excluding brain
Rab27B	ND	Expressed mainly in testis, melanocytes, platelets
Rab28S, Rab28L	ND	Rab28S detected in most tissues (cortex, liver, kidney, skeletal muscle, adipose tissue, testis, and urothelium), whereas hRab28L is predominant in testis
Rab30	ND	
Rab33A	ND	Ubiquitously expressed
Rab33B	Golgi	Expression restricted to the brain and the immune system

ND, Not determined; CGN, *cis* Golgi network; PM, plasma membrane; TGN, trans-Golgi network
[a] See text for references

Differentiated cells that have specialized membrane-trafficking pathways express specialized Rab proteins (36). One example is provided by epithelial cells, which have distinct apical and basolateral plasma membranes, the establishment and maintenance of which require specialized membrane-trafficking routes. Several Rab proteins are specifically expressed in epithelial cells, including Rab17, Rab18, Rab 27, and Rab 29 (37–39). Rab17 is induced when mesenchymal cells differentiate into polarized epithelial cells during kidney development (40). This Rab protein is localized to both the basolateral plasma membrane and to apical endocytic structures, and regulates membrane trafficking through apical recycling endosomes (41). Specific Rab protein expression may correlate with cellular differentiation and the acquisition of distinctive patterns of intracellular traffic. For example, the level of Rab3D increases in 3T3L1 cells upon their differentiation into adipocytes, where it may act in the insulin-dependent exocytosis of vesicles containing glucose transporters (42). The expression pattern of Rab3D in the developing rat exocrine pancreas also correlates with the acquisition of regulated exocytosis (43). It is also clear that more than one Rab protein can be associated with the same organelle. For example, Rab4A and Rab5A are both localized to early endosomes (4, 44). The presence of multiple Rab proteins on a given compartment may reflect the different trafficking pathways intersecting at that organelle.

How do Rab proteins acquire their characteristic localization? *In vitro* assays demonstrate that Rab proteins themselves contain all the information for targeting to a specific subcellular compartment (45). Prenylation is a prerequisite for membrane association, although it does not, by itself, determine specificity. Chavrier *et al*. (1991) constructed chimeric forms of Rab2, Rab5, and Rab7 in order to determine the motifs involved in targeting (46). These Rab proteins are normally localized to the intermediate compartment, early endosome, and late endosome, respectively. Substituting 13 C-terminal amino acids of Rab5 with those of Rab7 did not affect localization; however, substituting 34 C-terminal residues directed the chimeric protein to late endosomes. Similarly, replacing 35 amino acids at the C terminus of Rab2 with those of Rab5 or Rab7 redirected the protein to early or late endosomes, respectively. These data indicate that the motif for specific membrane targetting of Rab proteins resides in the C-terminal region. The C-terminal region does not participate in nucleotide binding or hydrolysis and, with the exception of the C-terminal cysteines, displays a high degree of sequence variability between different Rab proteins of the same subclass; as a consequence, this region has been termed the hypervariable domain. Similar studies have extended this finding to other Rab proteins. For example, the C-terminal regions of Ypt1p and Sec4p constitute a localization domain for ER-derived vesicles and Golgi in the case of Ypt1p and exocytic vesicles in the case of Sec4p (10). This study also demonstrated that specific localization is not absolutely essential for Rab function, as a chimeric molecule of Ypt1p with the C-terminal region of Sec4p will complement a deletion of *YPT1* but not *SEC4*. Presumably, high expression levels can bypass the strict requirement for localization by ensuring a sufficient concentration of Rab protein at the correct site.

4. Biochemistry of point mutants

Similar to Ras, point mutations in highly conserved regions can stabilize Rab proteins in either the GDP-, GTP-, or nucleotide-free conformations. These regions are illustrated in Fig. 4. Dominant interfering Rab mutants can be generated by substitution of Asn with Ile in the G2 region, which results in low-affinity binding of both GDP and GTP. Substitution of Ser/Thr with Asn in the α and β phosphoryl-binding site (GXXXXGKS/T) is also a dominant-negative mutant with a high preference for GDP binding (47). Rab mutants that are stabilized in the GTP-bound form are obtained by replacement of an active-site Gln residue with Leu in the phosphoryl-binding site. Such a mutation blocks intrinsic GTPase activity, favouring formation of the activated GTP-bound form of the molecule, which for Ras results in transformation. The point mutation Gly to Val at codon 12 (Ras numbering) gives a very different result in Rab compared to the corresponding mutation in Ras and Rho proteins In Ras, glycines are found in positions 12 and 13, and replacement of G12 by any other amino acid except proline slows down the intrinsic GTP hydrolysis activity and the G12V mutant behaves *in vivo* as an activated protein (see Chapter 3). However, some Rab proteins, such as Rab3, have neither G12 nor G13 yet hydrolyse GTP with comparable efficiencies to Ras, indicating that these residues are not absolutely required for GTP hydrolysis. Corresponding mutations in other Rab proteins appear to mimic the GDP-bound conformation (48).

Recently, investigators have been making use of GTPases with point mutations engineered to change nucleotide specificity from guanine to xanthine. Such a point mutation was first identified for the bacterial elongation factor Tu (49) in the NXXD region conserved between all GTPases. Similar mutations (D to N) also change nucleotide specificity for other GTPases (50). The great power of such mutants is that

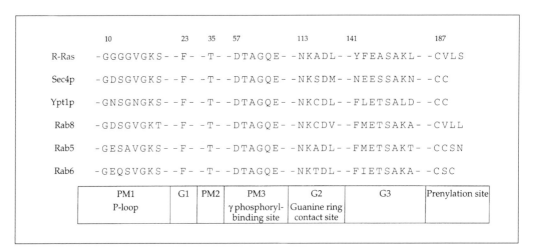

Fig. 4 Regions of Rab proteins involved in GTP-binding and hydrolysis and which are important for the generation of mutant proteins with predicted changes in these parameters. R-Ras is shown for comparison. The numbering is that of Ras. Compare to Fig. 3.

they provide a tool for the study of an individual Rab in crude extracts by assaying the xanthosine triphosphate (XTP) dependency (51, 52, 58).

5. Models for Rab protein function

A combination of genetic, biochemical, and morphological evidence indicates that each step of eukaryotic membrane trafficking requires the participation of specific Rab proteins. In recent years, identification of components of the membrane trafficking machinery has been the subject of intense investigation. Simply stated (refer to Fig. 1), membrane transport requires:

- the generation of a transport vesicle (or tubule) from a donor compartment;
- movement of the transport vesicle (or tubule) to the acceptor compartment; and
- its fusion with an acceptor compartment.

It is crucial that transport be regulated in order to deliver cargo molecules to the correct acceptor organelle while maintaining the integrity of distinct compartments. As well as Rab proteins, distinct members of three other families of proteins act at each individual stage of membrane trafficking. These are the Sec1-like family and the two membrane-protein families of v-SNAREs and t-SNAREs. Additionally, there are other factors, including GDI (see below), SNAPs, and NSF, that act ubiquitously at all stages of membrane trafficking. The SNARE proteins, the v-SNARE on the transport vesicle and the t-SNARE(s) on the acceptor membrane, participate in the final stage of membrane trafficking, namely the fusion of the two membranes. The Sec1-like family of molecules is thought to aid SNARE function directly, in a fashion that has not yet been determined (53, 54). Discussion of this topic is beyond the scope of this review, refer to (55) for clarification.

Figure 5 summarizes the current model for the cycle of Rab function. Cytosolic GDP-bound protein is complexed to GDP dissociation inhibitor (GDI). After delivery to the donor membrane compartment, GDI is released and a guanine nucleotide exchange factor (GEF) catalyses GDP/GTP exchange. After GTP hydrolysis on the acceptor membrane compartment, GDP-bound Rab is retrieved from the membrane by GDI. This cycle couples the events of membrane transport with the Rab protein GTP- and GDP-bound states.

Given the variety of ways that small GTPases can be utilized and our current knowledge of the membrane transport machinery, what could be plausible mechanisms for Rab function? Figure 6 illustrates three possibilities that are not mutually exclusive. Do Rab proteins function at the donor compartment, on the transport vesicle (or tubule), or when the vesicle reaches the target compartment?

Assuming only four classes of molecules (Rabs, v-SNAREs, t-SNAREs, and Sec1) are conserved as the core machinery of membrane transport, does this imply that Rab proteins interact directly with SNARES and Sec1-like family members? Such an idea is attractive because it would provide a functional mechanism for all Rab family members. Indeed, the yeast Rab, Ypt1p, has been proposed to function by directly

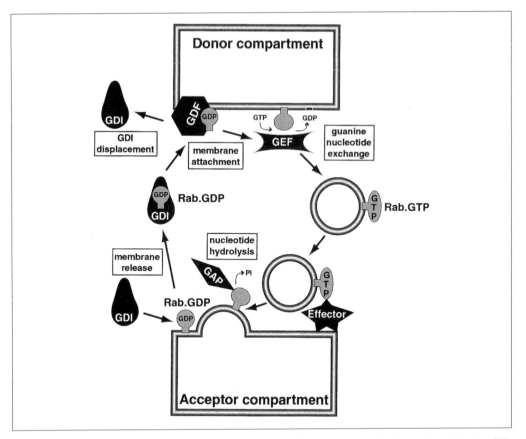

Fig. 5 The proposed functional cycle of Rab proteins. Rab protein is present on the transport vesicle in a GTP-bound form. Vesicle fusion at the target (acceptor) membrane is accompanied by GTP hydrolysis. The Rab-GDP is shuttled back to the donor compartment by GDI. At the donor compartment, GDI is displaced by a GDI displacement factor (GDF), following which the Rab protein is activated by a GEF for another round of transport.

activating the t-SNARE, Sed5p (56). However, these experiments did not address the specificity of interaction and remain controversial. All other biochemical attempts carried out so far to identify Rab proteins as components of the SNARE-complex machinery have failed. Although this may be for technical reasons, the most likely explanation is that these two classes of proteins do not interact directly and that there are other factors that contribute an additional layer of regulation between Rabs and SNAREs. Each organelle would evolve its own mechanism for this layer of regulation according to the individual functions of the organelle. This model would predict unique effector(s) for each Rab protein, with the Rabs and effector molecules evolving along with the physiology of the organelle. According to such a model, direct contacts between Rabs and SNAREs would only be observed in the most evolutionarily primitive membrane trafficking system. In higher eukaryotes, direct regulation of SNARE assembly is more likely to reside within one of the many other components that have been documented as SNARE interactors.

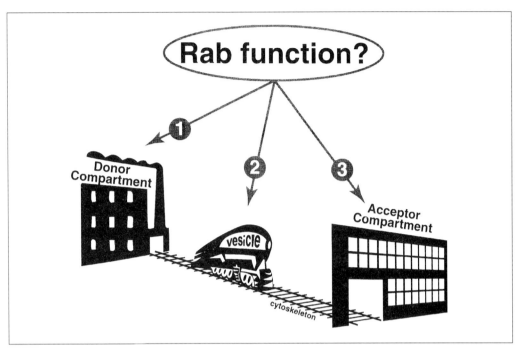

Fig. 6 Possibilities for Rab action: (1) Do Rab proteins act at the donor compartment as the transport vesicle is generated? (2) Do Rabs exert their function on the vesicle as it travels to the acceptor compartment? (3) Do Rabs act once the transport vesicle has reached the target membrane directly prior to fusion? Note that these options are not mutually exclusive.

By analogy with the Ras/Raf system (Chapter 3) (57), activated Rabs may recruit cytosolic factors to the membrane in their GTP-bound state. In this fashion, Rabs are acting as signalling molecules, assembling effector complexes at specific sites on the membrane. These assembled effector complexes may, in turn, signal downstream to SNARE proteins to directly promote fusion, or may set up a timed environment for SNARE pairs to explore the fidelity of their interactions before committing to fusion. SNAREs are known to interact promiscuously *in vitro* and the latter mechanism may be a way of proof-reading SNARE interactions. Such a mechanism has been proposed for the Rab5-regulated homotypic fusion of endocytic organelles (58). Rab proteins could also be the starting point for amplification of a secretory signal by regulating the level of GTP-bound Rab. Such a mechanism would implicate the nucleotide exchange event as a critical intervention point in the Rab protein cycle.

It is also important to remember that different categories of membrane transport may well have different requirements for Rab-protein function. Regulated secretion for instance, may require a Rab protein to be maintained in the active state, ready to respond rapidly to a secretory stimulus. In such a system, the Rab effector may negatively regulate the fusion event by acting as a 'clamp' that will only be released upon a secretory stimulus. Homotypic fusion between similar organelles may require a very transient activation of Rab protein—too much activation and subsequent effector recruitment will have a stimulatory effect on membrane fusion and will result in an enlarged, aberrant organelle (59).

Clearly, the assembly of SNARE complexes must be regulated in a directional manner, and an early hypothesis for Rab-protein function was that the specific subcellular localization must somehow be ensuring the specificity of vesicle delivery. An argument against this model was the finding that an artificial Rab protein can function in more than one step of the secretory pathway. A Sec4p–Ypt1p chimera was shown to complement a knockout of both *SEC4* and *YPT1*, two essential genes (10,11). The bifunctionality of this artificial construct means that Rab proteins do not by themselves function as specificity determinants; however, it is conceivable that the relevant domains of Ypt1p and Sec4p are both contained in the artificial fusion protein. Accumulating evidence exists that the effector molecules recruited by the GTP-bound Rab, rather than the Rab protein itself, provide the spatial regulation of vesicle delivery. This model has been termed the Velcro hypothesis (55), where Rabs direct vesicle docking through assembly of effector complexes (60).

How many rounds of GTP binding and hydrolysis are required for a single round of membrane docking and fusion? Several studies suggest that hydrolysis of Rab proteins is only required for their recycling. A hydrolysis-deficient mutant Sec4p^{L79} is able to function as the only copy of Sec4p in the cell (61). An analogous point mutation in Ypt1p has no observable growth phenotypes, and when overexpressed does not exhibit a dominant phenotype (62). *In vitro* assays of Rab5-mediated endosomal fusion demonstrate that GTP hydrolysis is not required for fusion (58). Together, these studies lead to the model that GTP hydrolysis is not essential for Rab activity, but for recycling of Rab between compartments. Recycling is not strictly required for Rab-protein function as Sec4p and Ypt1p with transmembrane tails instead of prenylation motifs are fully capable of participating in membrane trafficking, providing they are targetted to the correct compartment (63). Such Rab proteins need to be expressed at higher than normal levels, presumably they are 'single-use only' proteins, unable to be recycled and reused.

Rabs may have multiple functions during a single round of membrane transport and this would require them to undergo several cycles of GTP binding and hydrolysis. Strong connections between cytoskeletal elements and Rab proteins have been observed in several systems. Possibly, Rab proteins may help the transport intermediates to correctly target via the cytoskeleton and then also be involved in recruiting SNARE activation factors. One could easily imagine that such a system may involve at least two rounds of GTP binding and hydrolysis, which could even be regulated by distinct sets of exchange factors.

6. Genetic analysis of Rab function

6.1 Yeast

Yeast genetics, both classical and molecular, is a powerful tool for studying the cellular functions of Rab proteins. The biological importance of Rab proteins was first illustrated in yeast, with studies on the conditional lethal mutants *ypt1* and *sec4*, where accumulation of vesicles was observed at the non-permissive temperature. This suggested that Ypt1p and Sec4p function in vesicle docking or fusion, but not in budding. Genetic studies also offer a means of defining the temporal sequence of events along a pathway. In the case of Rab proteins, such studies have indicated that Rab proteins function upstream of the events mediated by SNARE proteins. The loss of Ypt1p can be efficiently bypassed when the VSNAres Sec22p and Bos1p are both overexpressed. Bos1p and Sec22p form a complex on vesicles, and in the absence of functional Ypt1p, this complex fails to form (64). Because Ypt1p does not directly interact with Bos1p or Sec22p, these data can be interpreted to mean that Ypt1p lies upstream, and controls events leading to the activation of vesicular SNAREs. In other words, constitutive activation of downstream components of the pathway (through overexpression), overcomes the loss of the upstream regulatory molecule (the Rab protein). This is a classic use of genetic suppressor analysis to identify and order components of a regulatory pathway, first illustrated for the yeast Ras/adenylate cyclase system (65, 66). A SNARE protein, Sec9p (a yeast homologue of the t-SNARE SNAP-25), was identified as a high-copy suppressor of an effector-domain mutant of Sec4p (67), providing further evidence for a pathway where Rab proteins act upstream of SNARE function in membrane fusion. Wickner and colleagues have analysed biochemically the relative roles of Rab and SNARE function in yeast vacuolar fusion. These studies again demonstrate that the Rab protein, Ypt7p, is required for the stage where the donor and acceptor membranes dock, and not for the membrane fusion process itself (68, 69).

6.2   Metazoan organisms

Model organisms such as *Caenorhabditis elegans* and *Drosophila* offer an experimental system, not just for studying Rab function in general but also for examining the roles of specialized Rabs in differentiated tissue. Genetic analysis of Rab function in multicellular organisms is a relatively unexplored subject. The major exception is the neuronal protein Rab3, which has been studied in *C. elegans* (70) and with a mouse knockout of Rab3A (71).

Metazoan organisms allow the use of electrophysiological techniques that offer outstanding time resolution and sensitivity. The kinetics of vesicle docking and fusion can be determined precisely using electrophysiological analysis. Such experiments are particularly powerful in combination with genetic studies where Rab proteins are mutated or absent. In a series of studies, Sudhof and colleagues have examined the electrophysiological parameters of neurons from mice deficient in Rab3A (71–73). Rab3A-deficient neurons generate the same numbers and size of vesicles as wild-type cells. However, in Rab3A-deficient neurons, greater numbers of vesicles are released in response to nerve stimulation (72). These data suggest that Rab3A functions to regulate the extent of vesicle fusion following a nerve impulse and not in vesicle docking. Such a mechanism would make the function of Rab3A synapse-specific (74), or perhaps there is partial redundancy between Rab3A and another Rab that functions in synaptic vesicle docking. Other organisms with less genetic redundancy may offer a simpler system to examine these issues. Analysis of *C. elegans rab-3* mutants suggests that Rab3 may regulate recruitment of vesicles to the active zone or docking of vesicles at release sites.

7. Cellular assays of Rab function

Rab-protein function has been studied with the full arsenal of techniques available to the modern cell biologist. Cellular assays of Rab function can be divided into three general categories:

- characterization of novel Rab proteins and identification of the stage of vesicular trafficking in which they act;
- assays to determine the mechanism of action of Rab proteins within a particular membrane trafficking pathway;
- identification of accessory factors that regulate Rab-protein function.

Originally, it was known that GTP-binding proteins were involved in membrane transport through use of the non-hydrolysable analogue of GTP, GTPγS. In systems that had been reconstituted *in vitro*, the addition of GTPγS was shown to affect membrane fusion reactions (75). However, GTPγS is a non-specific tool and will affect the function of all GTPases that may be operating in the system and is not specific for Rab proteins. Specific tools that have been developed to analyse the role of Rab proteins include Rab mutant constructs (see Section 4), small peptides corresponding to the effector domain, and antibodies to specific Rab proteins. The combination of all these studies has provided strong evidence for the involvement of a distinct Rab protein at each stage of the endocytic and exocytic pathways.

Below, we highlight some of the recent studies; for additional information the reader is referred to other reviews (60, 76–78).

8. Accessory factors

As for all Ras-related GTPases, the intrinsic rates of GDP release and GTP hydrolysis for Rab proteins are quite low, and are regulated by accessory proteins that modulate these properties. These include GAPs (GTPase activating proteins), GEFs (guanine nucleotide exchange factors), GDS (guanine nucleotide release factors), and molecules that bind specifically to either the GDP-bound conformation (GDI) or the GTP-bound conformation (effector molecules).

8.1 GDI

While the majority of Rabs are membrane associated, prenylated Rabs are also found in the cytosol bound to a GDP dissociation inhibitor (GDI). GDI shares sequence homology with the Rab escort protein (REP) involved in presenting and removing

Rab proteins from the prenylation machinery. GDI has several properties that under-score its role in coupling the guanine nucleotide cycle of Rab proteins with the cycle of membrane association/dissociation:

- it binds preferentially to the GDP-bound conformation of the molecule and slows the intrinsic loss of nucleotide dissociation;
- it requires the fully prenylated Rab protein for interaction and binds in such a way that the geranylgeranyl groups are shielded in a hydrophobic pocket; and
- it is a ubiquitous factor interacting with many different Rab proteins *in vitro* and *in vivo*; in *S. cerevisiae* a single gene encodes GDI function for all 11 Rab proteins.

These properties enable the Rab protein to exist in the aqueous environment of the cytoplasm as a soluble heterodimer with GDI, and facilitate recycling of the GDP-bound Rab back to the donor compartment. Consistent with this hypothesis, Rab proteins are complexed to GDI in the cytosol and depletion of GDI in yeast causes loss of the soluble pool of Rabs and a concomitant inhibition of transport in the secretory pathway (79). The amino-acid sequence of GDI is highly conserved from yeast to humans (>50% identity). GDI is composed of two main structural units, a multisheet domain I and a smaller α-helical domain II (80). The structural organ-ization of domain I is related to FAD-containing mono-oxygenases and oxidases, the functional significance of which is unclear at present. Sequence-conserved regions common to GDI and REP are the regions that have been shown by site-directed mutagenesis to play a critical role in the binding of Rab proteins. In mammals there are three characterized isoforms of GDI with >80% identity, although Southern blotting suggests the presence of up to five isoforms in mouse and rat. *In vitro* experi-ments have demonstrated that Rab/GDI heterodimers can be correctly targetted to their specific membranes (45). The lack of discrimination of GDI for binding to different Rab proteins further underscores the fact that Rab proteins themselves contain all the information for their intracellular localization. Membrane recruitment of Rabs is accompanied by GDI release, after which the Rab proteins are converted to a GTP-bound form. The process of Rab membrane recruitment is unclear, and the factors responsible for mediating these activities are unidentified. It is also unclear how this delivery system operates as a one-way process. Is GTP activation the crucial step that removes the Rab proteins from possible back-extraction by GDI? Inter-estingly, the GDI/Rab5 cytosolic complex has been shown to be phosphorylated, whereas a transient, membrane-associated form is not (81). The possibility that GDI may be affected by phosphorylation is intriguing and regulation of GDI function in general remains an open question.

8.2 Exchange factors

In contrast to what is known for many Ras superfamily members, there has been a relative paucity of knowledge about Rab-specific exchange factors. The several Rab GEFs that have been identified to date appear to have little sequence similarity to each other, although this is hardly surprising given that many different structures have evolved that are capable of performing nucleotide exchange on Ras superfamily members (82).

How do Rab GEFs function in membrane trafficking? GEFs might be needed exclusively for the activation of Rab proteins, an event that may occur several times for each round of transport. Alternatively, exchange factors could have additional functions, such as recruitment of Rab proteins to membranes, promoting assembly of activated protein complexes at the site of vesicle targetting, or could themselves input information from other signal transduction pathways. As has been found for GEFs of the Ras and Rho families, there is at least one example of different GEFs (Dss4p and Sec2p) which are able to perform nucleotide exchange on the same substrate (Sec4p). Such substrate overlap might indicate a scenario where one Rab needs to encounter two different effectors sequentially during a single round of transport. For each effector, the Rab protein goes through a separate GTP/GDP cycle, each activation event being catalysed by a different GEF. Isolation of the proteins that encode exchange factors for different Rabs and further characterization of their function will help to elucidate these issues.

8.2.1 Dss4

The yeast *DSS4* and mammalian *mss4* genes were identified based on their ability to suppress a *sec4* mutation in yeast (83, 84). Both Dss4p and Mss4 are small (17 kDa), zinc-binding proteins that can stimulate guanine nucleotide dissociation on a subset of Rab proteins (14, 48). Mss4 has the distinction of being the first Ras superfamily GEF for which a three-dimensional structure was obtained (85). The exact role that Mss4 and Dss4p play is unclear, as *DSS4* is not an essential gene and it may have redundant roles with other GEFs.

8.2.2 RabEx-5/Vps9p

RabEx-5 was purified from bovine brain as a protein of p60 bound to Rabaptin-5 (see below), and was shown to have GEF activity on Rab5, hence its name, <u>Rab</u>aptin-5-associated <u>ex</u>change factor for Rab5. The sequence of RabEx-5 shows homology to a yeast protein, Vps9p. *VPS* (vacuolar protein sorting) genes were originally identified through their involvement in vacuolar protein transport, *vps9* mutants mislocalize vacuolar enzymes and accumulate 40–50 nm vesicles. Vps9p is a GEF specific for Vps21p, the yeast counterpart of Rab5; it is also catalytically active on Rab5 but not Ypt7p (86).

8.2.3 Sec2p

SEC2 is one of 10 genes specifically involved in exocytosis in *S. cerevisiae*. Temperature-sensitive mutations in two of these genes, *sec2* and *sec15*, can be suppressed simply by giving cells one extra copy of the *SEC4* gene, implying an intimate relationship between the proteins involved (87, 88). This prediction has been verified experimentally, with Sec15p being shown to be a Sec4p effector (15), and Sec2p a guanine-nucleotide exchange factor for Sec4p (89). Interestingly, Sec2p shows homology to Rabin3, a protein of unknown function identified through its interaction with Rab3A/3D (90). The shared homology lies in a region of predicted coiled-coil sequence that is part of the minimal Sec2p catalytic domain (R. N. Collins, unpublished data).

8.2.4 Rab3A-GEP/aex-3

Rab3A guanine-nucleotide release factor was first identified as a biochemical activity that stimulates GDP release from the Rab3A protein (91). It was subsequently purified from bovine brain and shown to be a protein of molecular weight 178 kDa, with activity on Rab3A, Rab3C, and Rab3D, but not Rab3B (92). Rab3A-GEP shares sequence homology with the *C. elegans aex-3* gene, and the human protein MADD. Aex-3 is involved in synaptic vesicle release, consistent with its probable identity as a Rab3 exchange factor (93). The homology between Rab3A-GEP and MADD is extremely intriguing as MADD is a novel death-domain protein that interacts with the type 1 tumour necrosis factor (TNF) receptor and activates MAP kinase activity (94).

8.3 GAP proteins

The intrinsic GTPase activity of Rab proteins is slow and requires a GAP protein in order to cycle between the GDP- and GTP-bound states. Recent years have witnessed the identification of several factors with GAP activity for Rab family members, although the physiological role of these proteins and how they might act in membrane transport is less clear.

8.3.1 The GYP family

The first cloned gene of a GTPase-activating protein for a member of the Rab-protein family was *GYP6* (from <u>G</u>AP of <u>Y</u>pt6 protein) (95). Two other genes related to *GYP6* can be identified in the completed yeast genome: *GYP7* and *GYP1*, which also encode GAP proteins for Rab family members (96, 97). Between them, the three GYP proteins are active on all the Rab proteins in yeast (Fig. 7). For example, Gyp1p is active on Sec4p, Ypt1p, Ypt7p, and Ypt51p but not on Ypt6p and Ypt32p. The specificity towards Rabs of the GYP proteins is not determined by the effector regions, as Gyp6p is not active on the *S. pombe* and human homologues of Ypt6p (Ryh1 and Rab6), which have identical effector regions. However, a fusion protein containing the first 50 amino acids of Ypt6p and the remaining 151 amino acids from Rhy1p gives a substrate as responsive as the wild-type Ypt6p.

Deletion of the genes *GYP1, GYP6, GYP7* does not yield any conditional phenotype, as would be expected if (1) they are the only Rab GAPs, and (2) hydrolysis is crucial for Rab function. Other Rab GAP activities distinct from GYP have been measured in yeast lysates, although such proteins have not yet been identified (61, 98, 99). One role for GAPs in the Rab life cycle may be to ensure that newly synthesized Rab protein is in the GDP-bound state ready for interaction with the REP. Prenylation of a hydrolysis-deficient Rab mutant is inhibited in an *in vitro* system, but not when the substrate is present in intact cells, where presumably it can

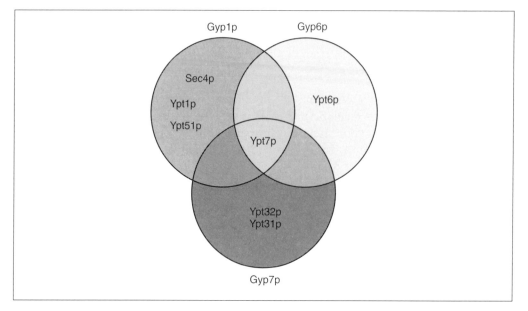

Fig. 7 Substrate specificity of GYP proteins. The three GYP proteins in yeast have substrate specificity profiles that can account for all yeast Rab proteins.

be acted upon by a GAP protein (100). Consistent with this finding is the fact that Gyp1p is active on unprenylated Rabs and can also greatly stimulate the GTPase activity of a hydrolysis-deficient mutant of Sec4p (97).

A more extensive sequence similarity between Gyp6p, Gyp7p, and many other proteins, including cell-cycle proteins Bub2p and Cdc16, has been described, using a searching algorithm designed to identify protein families (101, 102). The domain shared between these functionally apparently unrelated proteins has been termed the TBC domain (Fig. 8). The functional significance of these homologies is a fascinating question. Recently, Cdc16 was shown to have GAP activity on the small GTPase Spg1 that controls septum formation in fission yeast (103). Cdc16 together with the GYP proteins are the only examples of TBC domain proteins whose biochemical activity has been identified, but this raises the distinct possibility that all TBC domain proteins act as GAPs for members of the Ras superfamily. Interestingly, the GAP activity of Cdc16 is only observed in the presence of a third protein, Byr4. It is tantalizing to imagine that the GAP activity of TBC domain containing proteins may be regulated by independent subunits. One function of such subunits may be to dictate the substrate range amongst different Ras-related GTPases. Such a scenario would provide a mechanism for coordinating various small GTPase regulatory circuits.

8.3.2 Other Rab GAP proteins

Takai and colleagues have purified a GAP protein specific for the Rab3A family from rat brain (104). Rab3A-GAP has no sequence homology with the GYP proteins. The Rab3-GAP consists of two subunits termed p130 and p150, of which p130 contains

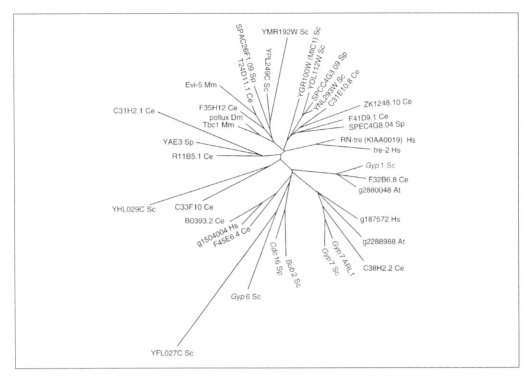

Fig. 8 Phylogenetic tree of proteins containing a TBC domain. The TBC-domain family contains a large number of sequences from different organisms of which only Gyp1, Gyp6, Gyp7, and Cdc16 have a defined function. Multiple sequence alignment and treefile creation were performed using Clustal W (version 1.7) and the phylogenetic tree was plotted with Drawtree in PHYLIP package (version 3.572).

the catalytic activity. Biochemical experiments have shown that the p150 subunit does not affect the activity of the catalytic subunit; at present, its physiological function is unknown (105).

Other GAP activities have been identified biochemically; however, these have not been isolated and characterized at a molecular level (106). For example, a Ypt1p-GAP activity associated with membranes is not affected by deletion of two genes that encode known Ypt1p GAPs, *GYP7* and *GYP1* (99), implying that cells contain an additional source of such activity.

8.4 Effector molecules

The formal definition of an effector molecule for Ras superfamily members is a factor that binds preferentially to the GTP-bound conformation of the GTPase and will compete for interaction with GAP proteins. In the past few years, several such effector molecules have been identified for Rab proteins (Fig. 9). Many of these molecules have been isolated through use of the two-hybrid system for detecting protein–protein interactions, often utilizing a hydrolysis-deficient Rab protein as

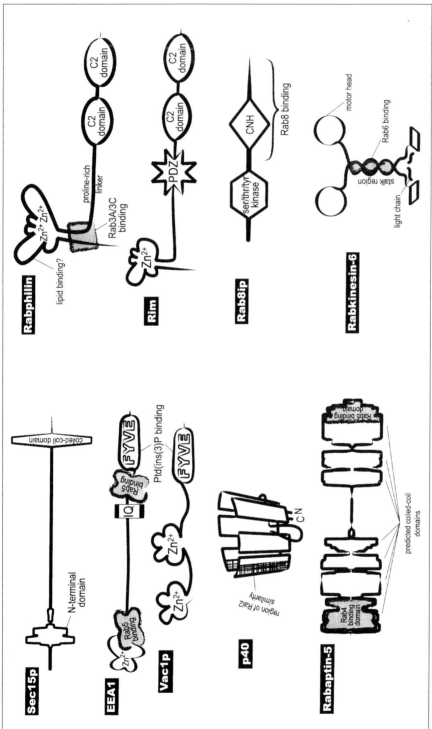

Fig. 9 Cartoon showing the major features of Rab effector proteins. Basic functional and structural domains are outlined (drawing not to scale). Minimal Rab interaction regions are also indicated if known. Both Rabphilin and Rim contain two C2 (protein kinase C conserved region 2) domains, a Zn^{2+}-binding domain, which includes the Rab3A/3C minimal binding region. In addition, Rim contains a PDZ domain, which mediates protein-protein interactions. Rab8ip contains a kinase domain, the specificity of which cannot be determined from sequence analysis, and a CNH domain found in NIK1-like kinases. The Rab8 interaction region comprises the C-terminal portion of the molecule (amino acids 430-821). Sec15p contains no commonly defined motifs, the protein has a coiled-coil stretch at the C terminus and a globular N terminus, presumably the site of Sec10p interaction. EEA1 and Vac1p contain FYVE fingers and zinc-binding domains; EEA1 is known to contain two binding sites for Rab5. The Rab6 binding site for Rabkinesin-6 lies in the predicted kinesin stalk region (amino acids 529-663). p40 consists primarily of the kelch repeats, predicted to form β-sheets that assemble into a barrel-like structure; the region of Ral2 similarity is indicated by shading. Rabaptin-5 contains several domains with high predicted propensity to form coiled-coil structures; sites of Rab4 and Rab5 interaction have been mapped to the N and C termini, respectively.

bait. One interesting characteristic of these effector molecules is that they are often part of a much larger multiprotein complex. This feature has led to the proposal that GTP-bound Rabs function by recruiting docking factors to the surface of transport vesicles to facilitate vesicle docking (54).

8.4.1 Rab3 effectors—Rabphilin-3A and Rim

Rabphilin-3A was the first effector to be identified for Rab proteins (107), it binds Rab3A-GTP and blocks the GAP-stimulated GTPase activity (108). Rab3A and Rab3C are the most abundant Rab proteins in neurons and are localized predominantly to synaptic vesicles. Rabphilin-3A consists of an N-terminal Rab3A interaction domain, a proline-rich linker region, and two tandem C2 domains (109). Similar to other C2 domains, the C2 domains of Rabphilin-3A bind to phosphatidylinositol 4,5-bisphosphate-containing vesicles in a Ca^{2+}-dependent manner (110, 111). The proline-rich linker region is a substrate for cAMP-dependent protein kinase and Ca^{2+}/calmodulin-dependent protein kinase II (112, 113). The N-terminal domain of Rabphilin binds specifically to Rab3A and Rab3C and possesses two conserved Zn^{2+}-binding motifs. The crystal structure of activated Rab3A bound to the effector domain of Rabphilin-3A has recently been solved (16). Rabphilin-3A contacts Rab3A in two distinct areas; the first interface involves the Rab3A switch 1 and switch 2 regions, which change conformation according to the nucleotide-binding state and the second interface consists of a pocket in Rab3A that interacts with a SGAWFF linear epitope of Rabphilin-3A.

How does Rabphilin-3A function in regulated exocytosis? Comparative studies of wild-type and Rab3A knockout mice have demonstrated interactions between Rab3A and Rabphilin-3A. Rabphilin-3A levels are decreased by 70% in Rab3A knockout mice despite normal mRNA levels, suggesting degradation of uncomplexed Rabphilin-3A protein *in vivo* (114). Furthermore, Rab3A may be required for transporting Rabphilin-3A from the neuronal cell body to the synapse, since Rabphilin-3A is retained in the cell body of Rab3A-defective neurons. Activated Rab3A reversibly recruits Rabphilin-3A to synaptic vesicles (115), although once recruited, Rabphilin-3A proceeds to interact with the synaptic vesicle in a manner independent of Rab3A (116). The electrostatic potential surface of the Rab3A/Rabphilin-3A complex has a highly asymmetrical distribution, with the Zn^{2+}-binding domain being positively charged. This domain would be expected to be close to the negatively charged membrane surface. The localization of Rabphilin-3A to synaptic vesicles by Rab3A may therefore promote interactions between Rabphilin-3A and components close to the site of the vesicle surface, such as cytoskeletal elements underlying the plasma membrane at the site of exocytosis. Rabphilin-3A binds to α-actinin, a factor that cross-links actin filaments during Ca^{2+}-dependent exocytosis (117). Binding of Rabphilin-3A to activated Rab3A inhibits the interaction with α-actinin. The Rabphilin-3A/Rab3A complex may play a role in actin filament reorganization before synaptic vesicle fusion occurs. One possibility is that activated Rab3A prevents actin filament reorganization by sequestering all free Rabphilin-3A molecules. Upon GTP hydrolysis, Rabphilin-3A would be released and would be available to reorganize actin filaments. Rab3A may thus act as a timer to prepare synaptic vesicles for docking or fusion by cortical cytoskeletal reorganization.

Noc2 is a soluble, cytoplasmic protein of molecular weight 38 kDa, with high sequence similarity to the Rab-binding domain of Rabphilin-3A. In contrast to Rabphilin-3A, it lacks the two C2 domains. Noc2 is expressed predominantly in endocrine tissue and hormone-secreting cell lines (118) and interacts with the LIM domain-containing protein, zyxin, a component of the cytoskeleton. Noc2 may therefore be involved in regulating exocytosis by interaction with the cytoskeleton, perhaps in a manner similar to Rabphilin-3A.

The SGAWFF epitope of Rabphilin-3A is also observed in another Rab3A effector, Rim (119) and, like Rabphilin-3A, Rim interacts specifically with Rab3A and Rab3C. Rim localizes to the active zone of the presynaptic plasma membrane instead of to synaptic vesicles. It has two Zn^{2+}-binding motifs similar to Rabphilin-3A, C2 domains, and a PDZ domain, which may be the domain responsible for the restricted plasma membrane localization of Rim. Because Rim localizes to the area where synaptic vesicles dock, it has been suggested to play a negative regulatory role in synaptic vesicle transmission, serving as a clamp for Ca^{2+}-dependent exocytosis (119).

8.4.2 Rabaptin-5

Rabaptin-5 was originally identified as a specific Rab5 effector that played a positive role in endosome fusion, as depletion of Rabaptin-5 inhibited *in vitro* fusion assays (120). Rabaptin-5 is a 100 kDa protein, which homodimerizes through a coiled-coil domain and is found in a complex of approximate molecular weight 500 kDa in the cytosol. The exact role of Rabaptin-5 is extremely intriguing, as recently it has also been identified as being capable of interaction with Rab4, Rab3A, Rabphilin-3A, RabEx-5, and the tumour suppressor gene product tuberin (121–124). In addition, a homologue of Rabaptin-5 has been identified, Rabaptin-5β (125). Rabaptin-5β does not dimerize with Rabaptin-5 and forms distinct cytosolic complexes that may share subunits with the Rabaptin-5 complexes, raising the potential complexity of the system.

The interaction of Rabaptin-5 with Rab5, Rab4, and Rab3A raises the question of the physical and functional relationship between the different membrane trafficking systems regulated by these molecules. Specialized cell types may have specific requirements for linking different stages of membrane traffic. A classical example is during the process of synaptic vesicle transmission. After fusion of synaptic vesicles with presynaptic membrane and secretion of the vesicle contents into the synaptic cleft, the vesicular membrane is immediately retrieved by endocytosis for re-use, without being routed through intermediate endosome-like compartments (126). Rabaptin-5 may act as a master regulator to link directly Rab3A-mediated exocytosis and Rab5-mediated endocytosis in neuronal cells. Interestingly, specific and selective cleavage of Rabaptin-5 is observed in cellular models of apoptosis (127), where a coordinated response of various membrane transport steps to the apoptotic stimulus may be required.

8.4.3   EEA1/Vac1p

EEA1 was first identified as a 162 kDa autoantigen associated with subacute cutaneous systemic lupus erythematosus (128). It is a coiled-coil protein localized to early endosomes and cytosol. At its C terminus, it contains a cysteine-rich zinc-binding motif, which is shared with Vps27p, Fab1p, Vac1p, (yeast proteins implicated in endocytic membrane trafficking), Hrs (a mammalian endosomal protein), and Hrs-2 (a mammalian protein involved in Ca^{2+}-regulated exocytosis). This motif has been termed a FYVE finger. Fusion between early endosomes is known to require phosphatidylinositol-3 hydroxykinase (PI3K) activity and EEA1 was identified as a protein rapidly released from cell membranes in response to wortmannin, a potent inhibitor of mammalian PI3 kinases (129). The FYVE finger is part of a Zn^{2+}-binding domain which binds phosphatidylinositol 3-phosphate (PI3P) with high specificity (130–132). EEA1 was also shown to be required for endosome fusion (133), and this probably accounts for the sensitivity of endocytosis to inhibitors of PI3K.

Using the two-hybrid system and direct *in vitro* assays, Simonsen *et al.* identified EEA1 as a specific Rab5 effector (134). A yeast homologue of EEA1, Vac1p, is also an effector for Vps21p (135, 136). EEA1 contains two binding sites for Rab5, one at the N terminus and one at the C terminus adjacent to the FYVE finger. The C-terminal Rab5-binding site contains a region of homology to the Rab5-binding site of Rabaptin-5. EEA1 association with the endosomal membrane requires both Rab5-GTP and PI3K activity. If PI3K is inhibited, excess Rab5-GTP stabilizes the membrane association of EEA1 and restores endosome fusion. Interestingly, the sequences of the EEA1 and Rabphilin-3A Zn^{2+}-binding domains are related, suggesting that the Rabphilin-3A Zn^{2+}-binding domain may bind to a distinct lipid in addition to Rab3A-GTP (although Gaullier and co-workers have demonstrated that this lipid is not PI3P (131)).

8.4.4   Rabkinesin

Rabkinesin-6, a kinesin protein, was originally identified as an effector of Rab6 (137). Rab6 is a Golgi-associated Rab protein that regulates intra-Golgi traffic (138). The kinesin superfamily comprises a large and structurally diverse group of microtubule-based motor proteins (139, 140). The microtubule cytoskeleton is intimately associated with the Golgi and plays an important role in Golgi structure and function (141). Rabkinesin-6 localizes to the Golgi apparatus and has an active role in the dynamics of this organelle, suggesting that it serves to coordinate membrane dynamics and intra-Golgi traffic. Interestingly, a multicopy suppressor of a deletion of Ypt6p, the yeast homologue of Rab6, is the protein Imh1p, which bears homology to the C-terminal portion of the cytoskeletal proteins integrin and myosin, although it is not known whether this represents a direct link (142).

8.4.5   Sec15p

Genetic analysis of exocytosis in yeast has identified a number of gene products required for this stage of secretion (143). Seven of these proteins (Sec6p, Sec8p,

Sec15p, Sec3p, Sec5p, Sec10p, Exo70p) function together in a complex required for exocytosis, termed the Exocyst (144). The Exocyst localizes to, and helps to define, sites on the plasma membrane where the exocytic vesicles are targetted (145). Sec15p interacts with a hydrolysis-deficient mutant of Sec4p by two-hybrid analysis and can be co-immunoprecipitated with anti-Sec15p antisera. Mutations in *SEC4* or *SEC2*, which encodes the exchange factor for Sec4p, result in a failure of vesicles to accumulate at the correct pole of the cell (89), suggesting that activated Rab protein is responsible for vesicle targetting. The direct connection of activated Sec4p to Sec15p provides a mechanism for the linking of the activated Sec4p to vesicle docking. In rat brain, mammalian homologues of Sec6p and Sec8p are present together in an analogous high molecular weight complex composed of eight proteins (146). A mammalian homologue of Sec15p has been identified in this complex and it will be interesting to see whether this, too, is an effector for a Rab protein. In addition, several other protein complexes have been identified which may play roles in vesicle docking (147–149). Although these are distinct entities that function in particular membrane trafficking steps, it is tempting to speculate that these complexes may all link mechanistically to an activated Rab GTPase.

8.4.6 p40

In 1997 Diaz and co-workers identified a Rab9 effector termed p40 (150). p40 positively mediates the Rab9-dependent transport of mannose 6-phosphate receptors from endosomes to the trans-Golgi network. Addition of recombinant protein stimulates, and anti-p40 antibodies inhibit, an *in vitro* transport assay. The sequence of p40 reveals it to be an exciting novel class of Rab effector, as it is comprised almost entirely of six kelch repeats. Kelch repeats are sequences of *c*. 50 amino acids, and are found in a wide variety of proteins of completely unrelated function (151). They are predicted to form four-stranded, antiparallel sheets that assemble into propeller-like barrel structures, where active sites are created by the loops located at the top of the barrel structure. In addition, p40 shares a 50-amino-acid stretch of 44% identity with the *S. pombe* protein, Ral2p. Ral2 shows genetic interactions with Ras1 and is involved in the activation of Ras1p (152). Perhaps p40 serves to connect Rabs with other Ras superfamily members. Further characterization of p40 is eagerly awaited.

8.4.7 Rab8ip

Rab8ip was isolated from a mouse MPC-11 myeloma cDNA library as an interacting partner for Rab8 (153). It is the murine homologue of the GC kinase, a serine/threonine protein kinase recently identified in human lymphoid tissue that is activated in the stress response (154). Rab8 plays a role in vesicular transport from the trans-Golgi network to the basolateral plasma membrane in polarized epithelial cells (MDCK), and to the dendritic surface in hippocampal neurons. Endogenous Rab8ip in MDCK cells is present both in the cytosol and as a peripheral membrane protein, with the same localization as Rab8. It is not known whether Rab8ip is directly required for Rab8-mediated vesicle transport. One possibility is suggested by the homology between Rab8ip and PAK, an effector of the Rho family GTPase

Cdc42, which itself is localized to the Golgi complex and has been implicated in directing Golgi-derived membrane traffic. Although Rab8ip does not possess the Cdc42 binding sequence present in PAK, perhaps it could compete for the same set of substrates. In this manner, Rab8ip would be part of the machinery that directs the polarity of Rab8-mediated membrane traffic (155). Rab8 is not a substrate for the kinase activity of Rab8ip. Determination of the cellular targets of the Rab8ip kinase should prove revealing.

9. Outstanding questions

9.1 Identification of novel Rab-protein regulators

Many studies are directed toward the identification of novel components that interact with Rab proteins and function as part of the Rab-protein cycle. One immediate question is how do Rabs become incorporated into transport vesicles? Several studies have shown that Rab–GDI complexes possess adequate information to deliver Rab GTPases to their correct membrane-bound compartments. After membrane delivery, Rabs are converted to their active, GTP-bound conformations (156, 157). The factors responsible for this recruitment of Rab proteins to the membrane and subsequent displacement of GDI are presently unknown.

9.2 Roles of lipid modification

To date, little attention has been paid to the sensitivity of regulators and accessory factors for Ras-related GTPases towards lipid modification of their substrate. Many biochemical studies have been performed using unprenylated substrate proteins purified from bacterial sources. Such work may need to be re-evaluated in light of emerging evidence that prenylation may affect activity to a significant extent, even if is not required for detection of activity. For example, both the Rab3-GAP and GEP are active on the lipid-modified forms of their substrate and relatively inactive towards the lipid unmodified form (92, 104). Although the Rho GEF activity of p115 was originally identified using a recombinant substrate (158, 159), it is 100-fold more active on prenylated forms of its substrate (P. C. Sternweis, University of Texas Southwestern Medical Center, personal communication). For a protein such as GDI, the role of prenylation is clearly understood; however, for many other factors the effects of prenylation may be insufficiently appreciated. The potential effects of protein lipid modification may have practical considerations. Complexes involving Ras-related proteins may be more stable with the fully modified form of the substrate; using the fully modified form to isolate novel proteins by the two-hybrid or biochemical pull-down methods may be more informative than with the unprenylated protein. These results may also have functional implications; for example, a GEF that interacts strongly with the prenylated form of its substrate might be capable of recruiting Rabs and other Ras-related proteins directly onto membranes.

9.3 How do Rabs communicate with the SNARE machinery?

Although it is clear that the Rab family of proteins does not contribute to the core docking interaction mediated by SNAREs, the mechanisms by which Rabs communicate with the SNARE machinery remain unknown. In the yeast endocytic pathway, Vps21p function has been mechanistically linked to Vps45p via an adaptor protein, Vac1p (135, 136). Vps45p is a member of the Sec1-like protein family, which interacts with SNAREs and affects SNARE function. These data provide the outline of a mechanistic link between Rabs and SNAREs. It is possible that Rab proteins may act to present the SNARE proteins in a more reactive state (64), or may act as a proof-reading mechanism to impart additional specificity to SNARE interactions (52). Further work in yeast and other systems is needed to evaluate these possibilities.

9.4 Cross-talk of Rab proteins with other signalling cascades?

Until recently, Ras was though to be part of a simple signalling cascade. It is now appreciated that Ras is a component of a very complex signalling circuitry. The Ras signalling network involves cross-talk, feedback loops, branch-points, parallel pathways, and multi-component signalling complexes. Amongst the components of these pathways are GTPases from not just other branches of the Ras superfamily but also heterotrimeric G proteins (see, for example, 158–160). It is probable that Rab family members will also be included in such signalling networks. Mechanistic clues for such an involvement have already emerged. The catalytic domain of p120 Ras-GAP will stimulate the GTPase activity of Rab5, but not other Rab proteins, suggesting a direct mechanistic link between Ras signalling and vesicular transport (161). Rosa et al. have identified an unusually large human protein (molecular weight 532 kDa) that stimulates guanine nucleotide exchange on Arf1 and Rab proteins (162). This protein contains multiple structural domains, including two regions related to the Ran exchange factor RCC1, seven WD40 repeats, three putative SH3 binding sites, a putative leucine zipper and a carboxy-terminal HECT domain. p532 has been hypothesized to be involved in membrane transport as it can form a complex with clathrin heavy chain (163), but its primary sequence clearly hints at regulatory input from other cellular processes.

Heterotrimeric G proteins are known to regulate Golgi structure (164), and signalling via MEK1 is required for Golgi fragmentation during mitosis (165). In light of this, it is interesting that a number of other signalling molecules localize to the Golgi compartment. For example, the Ras-related proteins Rap1A and Rap1B reside on the Golgi complex (166), as does MKK4 (167). Mammalian Cdc42 is a brefeldin A-sensitive component of the Golgi apparatus (168) and it is tantalizing to imagine that Cdc42 may act on targets in the Golgi that direct polarized membrane traffic.

Other signalling systems may also act on Rab proteins. In mossy fibre synapses of the hippocampal CA3 region, long-term potentiation is induced by cAMP and acts through Rab3A (73). The mechanism by which this occurs is unknown, although phosphorylation of Rabphilin-3A may play a role (169).

9.5   Links between Rab proteins and the cytoskeleton

Vesicular transport is a dynamic process that requires coordinated interactions between membranes and the cytoskeleton. The identification of Rabkinesin-6 (discussed above) demonstrates that Rab proteins may provide a connection between these processes. Membrane trafficking has also been linked to cytoskeletal processes regulated by Rho GTPase family members. Activated RhoD causes rearrangements of the actin cytoskeleton and cell surface, and regulates Rab5-mediated early endosome motility and distribution (170). Another Rho family member, Rho3p has been identified in *S. cerevisiae* as having a critical role in cell polarity and bud growth. *SEC4* was isolated as a multi-copy suppressor of a *rho3Δ*, suggesting a possible connection between Rho3p and the Sec4p-mediated regulation of exocytosis (171). Understanding the connections between membrane traffic and the cytoskeleton, and clarifying the possible roles of Rabs in this process, is an important puzzle.

9.6   Rabs and lipid metabolism

Many signal transduction processes are known to be regulated by lipids, and several lipids have been identified as being essential for membrane trafficking (172, 173). It is likely that further studies will illuminate roles for specific lipids in many other stages of transport. One possible use of such lipids may be to regulate various aspects of Rab function and to coordinate membrane trafficking with other cellular events, such as cytoskeletal rearrangements. For example, insulin stimulates the guanine nucleotide exchange on Rab4 via a phosphatidylinositol 3-kinase (PI3K)-dependent signalling pathway (174), and in endosomes, the Rab5 effector, EEA1, provides a molecular link between PI3K and Rab5 (134).

9.7   Genetic studies of Rab function in metazoan organisms

Most genetic studies of Rab-protein function to date have utilized the simple unicellular eukaryote *S. cerevisiae*. The identification of cell- and tissue-specific Rabs in higher eukaryotes opens up new questions as to the role these proteins play in the differentiated systems of metazoan organisms. In this light it is interesting to note that X-linked, non-specific mental retardation can be caused by mutations in the human GDI gene (175), and several Rab genes have been implicated in other genetic diseases. Future work to examine individual Rab functions in metazoan organisms with model genetic systems should prove illuminating.

10.   Conclusions

Clearly, a major question for the future is to understand how the unique functions of each Rab will contribute to the functions of the organelle with which it associates. Tremendous progress has been made in identifying factors that regulate Rab function, as well as details of how individual family members act. Rab-protein function may share a unifying theme, being involved in the process of membrane trafficking prior to SNARE-mediated fusion. However, despite the conservation of protein sequence, a universal mechanism for Rab proteins has not yet been determined. Understanding both the specifics and generalities of Rab-protein function remains an important challenge.

Acknowledgements

We would like to thank our colleagues for scientific advice and for sharing unpublished information. We apologize to investigators whose work was not cited due to limited space. Many thanks to Gary Whittaker for critical comments on the manuscript and Nena Winand for stimulating discussions.

References

1. Novick, P., Field, C., and Schekman, R. (1980) Identification of 23 complementation groups required for post-translational events in the yeast secretory pathway. *Cell*, **21**, 205.
2. Salminen, A. and Novick, P. J. (1987) A ras-like protein is required for a post-Golgi event in yeast secretion. *Cell*, **49**, 527.
3. Gallwitz, D., Donath, C., and Sander, C. (1983) A yeast gene encoding a protein homologous to the *c-has/bas* proto-oncogene product. *Nature*, **306**, 704.
4. Chavrier, P., Vingron, M., Sander, C., Simons, K., and Zerial, M. (1990) Molecular cloning of YPT1/SEC4-related cDNAs from an epithelial cell line. *Mol. Cell. Biol.*, **10**, 6578.
5. Chavrier, P., Simons, K., and Zerial, M. (1992) The complexity of the Rab and Rho GTP-binding protein subfamilies revealed by a PCR cloning approach. *Gene*, **112**, 261.
6. Haubruck, H., Prange, R., Vorgias, C., and Gallwitz, D. (1989) The ras-related mouse ypt1 protein can functionally replace the YPT1 gene product in yeast. *EMBO J.*, **8**, 1427.
7. Huber, L. A., Pimplikar, S., Parton, R. G., Virta, H., Zerial, M., and Simons, K. (1993) Rab8, a small GTPase involved in vesicular traffic between the TGN and the basolateral plasma membrane. *J. Cell Biol.*, **123**, 35.
8. Craighead, M. W., Bowden, S., Watson, R., and Armstrong, J. (1993) Function of the ypt2 gene in the exocytic pathway of *Schizosaccharomyces pombe*. *Mol. Biol. Cell*, **4**, 1069.
9. Singer-Kruger, B., Stenmark, H., and Zerial, M. (1995) Yeast Ypt51p and mammalian Rab5: counterparts with similar function in the early endocytic pathway. *J. Cell Sci.*, **108**, 3509.
10. Brennwald, P. and Novick, P. (1993) Interactions of three domains distinguishing the Ras-related GTP-binding proteins Ypt1 and Sec4. *Nature*, **362**, 560.
11. Dunn, B., Stearns, T., and Botstein, D. (1993) Specificity domains distinguish the Ras-related GTPases Ypt1 and Sec4. *Nature*, **362**, 563.
12. Stenmark, H., Valencia, A., Martinez, O., Ullrich, O., Goud, B., and Zerial, M. (1994) Distinct structural elements of rab5 define its functional specificity. *EMBO J.*, **13**, 575.
13. Moore, I., Schell, J., and Palme, K. (1995) Subclass-specific sequence motifs identified in Rab GTPases. *Trends Biochem. Sci.*, **20**, 10.
14. Burton, J. L., Slepnev, V., and De Camilli, P. V. (1997) An evolutionarily conserved domain in a subfamily of Rabs is crucial for the interaction with the guanyl nucleotide exchange factor Mss4. *J. Biol. Chem.*, **272**, 3663.

15. Guo, W., Roth, D., Walch-Solimena, C., and Novick, P. (1999) The exocyst is an effector for Sec4p, targeting secretory vesicles to sites of exocytosis. *EMBO J.*, **18**, 1071.

16. Ostermeier, C. and Brunger, A. T. (1999) Structural basis of Rab effector specificity: crystal structure of the small G Protein Rab3A complexed with the effector domain of Rabphilin-3A. *Cell*, **96**, 363.

17. Kjeldgaard, M., Nyborg, J., and Clark, B. F. (1996) The GTP binding motif: variations on a theme. *FASEB J.*, **10**, 1347.

18. Sprang, S. R. (1997) G protein mechanisms: insights from structural analysis. *Annu. Rev. Biochem.*, **66**, 639.

19. Walworth, N. C., Goud, B., Kabcenell, A. K., and Novick, P. J. (1989) Mutational analysis of SEC4 suggests a cyclical mechanism for the regulation of vesicular traffic. *Cell*, **8**, 1685.

20. Khosravi-Far, R., Clark, G. J., Abe, K., Cox, A. D., McLain, T., Lutz, R. J., Sinensky, M., and Der, C. J. (1992) Ras (CXXX) and Rab (CC/CXC) prenylation signal sequences are unique and functionally distinct. *J. Biol. Chem.*, **267**, 24363.

21. Seabra, M. C. (1998) Membrane association and targeting of prenylated Ras-like GTPases. *Cell Signal.*, **10**, 167.

22. Desnoyers, L., Anant, J. S., and Seabra, M. C. (1996) Geranylgeranylation of Rab proteins. *Biochem. Soc. Trans.*, **24**, 699.

23. Cremers, F. P., van de Pol, D. J., van Kerkhoff, L. P., Wieringa, B., and Ropers, H. H. (1990) Cloning of a gene that is rearranged in patients with choroideraemia. *Nature*, **347**, 674.

24. Andres, D. A., Seabra, M. C., Brown, M. S., Armstrong, S. A., Smeland, T. E., Cremers, F. P., and Goldstein, J. L. (1993) cDNA cloning of component A of Rab geranylgeranyl transferase and demonstration of its role as a Rab escort protein. *Cell*, **73**, 1091.

25. Seabra, M. C., Goldstein, J. L., Sudhof, T. C., and Brown, M. S. (1992) Rab geranylgeranyl transferase. A multisubunit enzyme that prenylates GTP-binding proteins terminating in Cys–X–Cys or Cys–Cys. *J. Biol. Chem.*, **267**, 14497.

26. Wilson, A. L., Erdman, R. A., and Maltese, W. A. (1996) Association of Rab1B with GDP-dissociation inhibitor (GDI) is required for recycling but not initial membrane targeting of the Rab protein. *J. Biol. Chem.*, **271**, 10932.

27. Jiang, Y., Rossi, G., and Ferro-Novick, S. (1995) Characterization of yeast type-II geranyl-geranyltransferase. *Meth. Enzymol.*, **257**, 21.

28. Casey, P. J. and Seabra, M. C. (1996) Protein prenyltransferases. *J. Biol. Chem.*, **271**, 5289.

29. Bailly, E., McCaffrey, M., Touchot, N., Zahraoui, A., Goud, B., and Bornens, M. (1991) Phosphorylation of two small GTP-binding proteins of the Rab family by p34cdc2. *Nature*, **350**, 715.

30. van der Sluijs, P., Hull, M., Huber, L. A., Male, P., Goud, B., and Mellman, I. (1992) Reversible phosphorylation–dephosphorylation determines the localization of rab4 during the cell cycle. *EMBO J.*, **11**, 4379.

31. Ayad, N., Hull, M., and Mellman, I. (1997) Mitotic phosphorylation of rab4 prevents binding to a specific receptor on endosome membranes. *EMBO J.*, **16**, 4497.

32. Karniguian, A., Zahraoui, A., and Tavitian, A. (1993) Identification of small GTP-binding rab proteins in human platelets: thrombin-induced phosphorylation of rab3B, rab6, and rab8 proteins. *Proc. Natl Acad. Sci., USA*, **90**, 7647.

33. Plana, M., Itarte, E., Eritja, R., Goday, A., Pages, M., and Martinez, M. C. (1991) Phosphorylation of maize RAB-17 protein by casein kinase 2. *J. Biol. Chem.*, **266**, 22510.

34. Chavrier, P., Parton, R. G., Hauri, H. P., Simons, K., and Zerial, M. (1990) Localization of low molecular weight GTP binding proteins to exocytic and endocytic compartments. *Cell*, **62**, 317.

35. Simons, K. and Zerial, M. (1993) Rab proteins and the road maps for intracellular transport. *Neuron*, **11,** 789.

36. Lutcke, A., Parton, R. G., Murphy, C., Olkkonen, V. M., Dupree, P., Valencia, A., Simons, K., and Zerial, M. (1994) Cloning and subcellular localization of novel rab proteins reveals polarized and cell type-specific expression. *J. Cell Sci.*, **107,** 3437.

37. McMurtrie, E. B., Barbosa, M. D., Zerial, M., and Kingsmore, S. F. (1997) Rab17 and rab18, small GTPases with specificity for polarized epithelial cells: genetic mapping in the mouse. *Genomics*, **45,** 623.

38. Massmann, S., Schurmann, A., and Joost, H. G. (1997) Cloning of two splicing variants of the novel Ras-related GTPase Rab29 which is predominantly expressed in kidney. *Biochim. Biophys. Acta*, **1352,** 48.

39. Seabra, M. C., Ho, Y. K., and Anant, J. S. (1995) Deficient geranylgeranylation of Ram/Rab27 in choroideremia. *J. Biol. Chem.*, **270,** 24420.

40. Lutcke, A., Jansson, S., Parton, R. G., Chavrier, P., Valencia, A., Huber, L. A., Lehtonen, E., and Zerial, M. (1993) Rab17, a novel small GTPase, is specific for epithelial cells and is induced during cell polarization. *J. Cell Biol.*, **121,** 553.

41. Zacchi, P., Stenmark, H., Parton, R. G., Orioli, D., Lim, F., Giner, A., Mellman, I., Zerial, M., and Murphy, C. (1998) Rab17 regulates membrane trafficking through apical recycling endosomes in polarized epithelial cells. *J. Cell Biol.*, **140,** 1039.

42. Baldini, G., Hohl, T., Lin, H. Y., and Lodish, H. F. (1992) Cloning of a Rab3 isotype predominantly expressed in adipocytes. *Proc. Natl Acad. Sci., USA*, **89,** 5049.

43. Valentijn, J. A., Gumkowski, F. D., and Jamieson, J. D. (1996) The expression pattern of rab3D in the developing rat exocrine pancreas coincides with the acquisition of regulated exocytosis. *Eur. J. Cell Biol.*, **71,** 129.

44. Van Der Sluijs, P., Hull, M., Zahraoui, A., Tavitian, A., Goud, B., and Mellman, I. (1991) The small GTP-binding protein rab4 is associated with early endosomes. *Proc. Natl. Acad. Sci. USA*, **88,** 6313.

45. Pfeffer, S. R., Dirac-Svejstrup, A. B., and Soldati, T. (1995) Rab GDP dissociation inhibitor: putting rab GTPases in the right place. *J. Biol. Chem.*, **270,** 17057.

46. Chavrier, P., Gorvel, J. P., Stelzer, E., Simons, K., Gruenberg, J., and Zerial, M. (1991) Hypervariable C-terminal domain of rab proteins acts as a targeting signal. *Nature*, **353,** 769.

47. Feig, L. A. and Cooper, G. M. (1988) Inhibition of NIH 3T3 cell proliferation by a mutant ras protein with preferential affinity for GDP. *Mol. Cell. Biol.*, **8,** 3235.

48. Collins, R. N., Brennwald, P., Garrett, M., Lauring, A., and Novick, P. (1997) Interactions of nucleotide release factor Dss4p with Sec4p in the post-Golgi secretory pathway of yeast. *J. Biol. Chem.*, **272,** 18281.

49. Hwang, Y. W. and Miller, D. L. (1987) A mutation that alters the nucleotide specificity of elongation factor Tu, a GTP regulatory protein. *J. Biol. Chem.*, **262,** 13081.

50. Zhong, J. M., Chen-Hwang, M. C., and Hwang, Y. W. (1995) Switching nucleotide specificity of Ha-Ras p21 by a single amino acid substitution at aspartate 119. *J. Biol. Chem.*, **270,** 10002.

51. Jones, S., Litt, R. J., Richardson, C. J., and Segev, N. (1995) Requirement of nucleotide exchange factor for Ypt1 GTPase mediated protein transport. *J. Cell. Biol.*, **130,** 1051.

52. Barbieri, M. A., Hoffenberg, S., Roberts, R., Mukhopadhyay, A., Pomrehn, A., Dickey, B. F., and Stahl, P. D. (1998) Evidence for a symmetrical requirement for Rab5-GTP in in vitro endosome-endosome fusion. *J. Biol. Chem.*, **273,** 25850.

53. Wu, M. N., Littleton, J. T., Bhat, M. A., Prokop, A., and Bellen, H. J. (1998) ROP, the

Drosophila Sec1 homolog, interacts with syntaxin and regulates neurotransmitter release in a dosage-dependent manner. *EMBO J.*, **17**, 127.

54. Halachmi, N. and Lev, Z. (1996) The Sec1 family: a novel family of proteins involved in synaptic transmission and general secretion. *J. Neurochem.*, **66**, 889.

55. Pfeffer, S. R. (1996) Transport vesicle docking: SNAREs and associates. *Annu. Rev. Cell Dev. Biol.*, **12**, 441.

56. Lupashin, V. V. and Waters, M. G. (1997) t-SNARE activation through transient interaction with a rab-like guanosine triphosphatase. *Science*, **276**, 1255.

57. Leevers, S. J., Paterson, H. F., and Marshall, C. J. (1994) Requirement for Ras in Raf activation is overcome by targeting Raf to the plasma membrane. *Nature*, **369**, 411.

58. Rybin, V., Ullrich, O., Rubino, M., Alexandrov, K., Simon, I., Seabra, C., Goody, R., and Zerial, M. (1996) GTPase activity of Rab5 acts as a timer for endocytic membrane fusion. *Nature*, **383**, 266.

59. Stenmark, H., Parton, R. G., Steele-Mortimer, O., Lutcke, A., Gruenberg, J., and Zerial, M. (1994) Inhibition of rab5 GTPase activity stimulates membrane fusion in endocytosis. *EMBO J.*, **13**, 1287.

60. Schimmoller, F., Simon, I., and Pfeffer, S. R. (1998) Rab GTPases, directors of vesicle docking. *J. Biol. Chem.*, **273**, 22161.

61. Walworth, N. C., Brennwald, P., Kabcenell, A. K., Garrett, M., and Novick, P. (1992) Hydrolysis of GTP by Sec4 protein plays an important role in vesicular transport and is stimulated by a GTPase-activating protein in *Saccharomyces cerevisiae*. *Mol. Cell. Biol.*, **12**, 2017.

62. Richardson, C. J., Jones, S., Litt, R. J., and Segev, N. (1998) GTP hydrolysis is not important for Ypt1 GTPase function in vesicular transport. *Mol. Cell. Biol.*, **18**, 827.

63. Ossig, R., Laufer, W., Schmitt, H. D., and Gallwitz, D. (1995) Functionality and specific membrane localization of transport GTPases carrying C-terminal membrane anchors of synaptobrevin-like proteins. *EMBO J.*, **14**, 3645.

64. Lian, J. P., Stone, S., Jiang, Y., Lyons, P., and Ferro-Novick, S. (1994) Ypt1p implicated in v-SNARE activation. *Nature*, **372**, 698.

65. Toda, T., Uno, I., Ishikawa, T., Powers, S., Kataoka, T., Broek, D., Cameron, S., Broach, J., Matsumoto, K., and Wigler, M. (1985) In yeast, *RAS* proteins are controlling elements of the adenylate cyclase. *Cell*, **40**, 27.

66. Cannon, J. F., Gibbs, J. B., and Tatchell, K. (1986) Suppressors of the ras2 mutation of *Saccharomyces cerevisiae*. *Genetics*, **113**, 247.

67. Brennwald, P., Kearns, B., Champion, K., Keranen, S., Bankaitis, V., and Novick, P. (1994) Sec9 is a SNAP-25-like component of a yeast SNARE complex that may be the effector of Sec4 function in exocytosis. *Cell*, **79**, 245.

68. Sato, K. and Wickner, W. (1998) Functional reconstitution of ypt7p GTPase and a purified vacuole SNARE complex. *Science*, **281**, 700.

69. Mayer, A. and Wickner, W. (1997) Docking of yeast vacuoles is catalyzed by the Ras-like GTPase Ypt7p after symmetric priming by Sec18p (NSF). *J. Cell Biol.*, **136**, 307.

70. Nonet, M. L., Staunton, J. E., Kilgard, M. P., Fergestad, T., Hartwieg, E., Horvitz, H. R., Jorgensen, E. M., and Meyer, B. J. (1997) *Caenorhabditis elegans* rab-3 mutant synapses exhibit impaired function and are partially depleted of vesicles. *J. Neurosci.*, **17**, 8061.

71. Geppert, M., Bolshakov, V. Y., Siegelbaum, S. A., Takei, K., De Camilli, P., Hammer, R. E., and Sudhof, T. C. (1994) The role of Rab3A in neurotransmitter release. *Nature*, **369**, 493.

72. Geppert, M., Goda, Y., Stevens, C. F., and Sudhof, T. C. (1997) The small GTP-binding protein Rab3A regulates a late step in synaptic vesicle fusion. *Nature*, **387**, 810.

73. Lonart, G., Janz, R., Johnson, K. M., and Sudhof, T. C. (1998) Mechanism of action of rab3A in mossy fiber LTP. *Neuron*, **21,** 1141.

74. Bean, A. J. and Scheller, R. H. (1997) Better late than never: a role for rabs late in exocytosis. *Neuron*, **19,** 751.

75. Ferro-Novick, S. and Novick, P. (1993) The role of GTP-binding proteins in transport along the exocytic pathway. *Annu. Rev. Cell Biol.*, **9,** 575.

76. Martinez, O. and Goud, B. (1998) Rab proteins. *Biochim. Biophys. Acta*, **1404,** 101.

77. Novick, P. and Zerial, M. (1997) The diversity of Rab proteins in vesicle transport. *Curr. Opinion Cell Biol.*, **9,** 496.

78. Olkkonen, V. M. and Stenmark, H. (1997) Role of Rab GTPases in membrane traffic. *Int. Rev. Cytol.*, **176,** 1.

79. Garrett, M. D., Zahner, J. E., Cheney, C. M., and Novick, P. J. (1994) GDI1 encodes a GDP dissociation inhibitor that plays an essential role in the yeast secretory pathway. *EMBO J.*, **13,** 1718.

80. Schalk, I., Zeng, K., Wu, S. K., Stura, E. A., Matteson, J., Huang, M., Tandon, A., Wilson, I. A., and Balch, W. E. (1996) Structure and mutational analysis of Rab GDP-dissociation inhibitor. *Nature*, **381,** 42.

81. Steele, M. O., Gruenberg, J., and Clague, M. J. (1993) Phosphorylation of GDI and membrane cycling of rab proteins. *FEBS Lett.*, **329,** 313.

82. Pai, E. F. (1998) The alpha and beta of turning on a molecular switch. *Nature, Struct. Biol.*, **5,** 259.

83. Moya, M., Roberts, D., and Novick, P. (1993) DSS4–1 is a dominant suppressor of sec4–8 that encodes a nucleotide exchange protein that aids Sec4p function. *Nature*, **361,** 460.

84. Burton, J., Roberts, D., Montaldi, M., Novick, P., and De Camilli, P. (1993) A mammalian guanine-nucleotide-releasing protein enhances function of yeast secretory protein Sec4. *Nature*, **361,** 464.

85. Yu, H. and Schreiber, S. L. (1995) Structure of guanine-nucleotide-exchange factor human Mss4 and identification of its Rab-interacting surface. *Nature*, **376,** 788.

86. Hama, H., Tall, G. G., and Horazdovsky, B. F. (1999) Vps9p is a guanine nucleotide exchange factor involved in vesicle-mediated vacuolar protein transport. *J. Biol. Chem.*, **274,** 15284.

87. Salminen, A. and Novick, P. J. (1989) The Sec15 protein responds to the function of the GTP binding protein, Sec4, to control vesicular traffic in yeast. *J. Cell Biol.*, **109,** 1023.

88. Nair, J., Muller, H., Peterson, M., and Novick, P. (1990) Sec2 protein contains a coiled-coil domain essential for vesicular transport and a dispensable carboxy terminal domain. *J. Cell Biol.*, **110,** 1897.

89. Walch-Solimena, C., Collins, R. N., and Novick, P. J. (1997) Sec2p mediates nucleotide exchange on Sec4p and is involved in polarized delivery of post-Golgi vesicles. *J. Cell Biol.*, **137,** 1495.

90. Brondyk, W. H., McKiernan, C. J., Fortner, K. A., Stabila, P., Holz, R. W., and Macara, I. G. (1995) Interaction cloning of Rabin3, a novel protein that associates with the Ras-like GTPase Rab3A. *Mol. Cell. Biol.*, **15,** 1137.

91. Miyazaki, A., Sasaki, T., Araki, K., Ueno, N., Imazumi, K., Nagano, F., Takahashi, K., and Takai, Y. (1994) Comparison of kinetic properties between MSS4 and Rab3A GRF GDP/GTP exchange proteins. *FEBS Lett.*, **350,** 333.

92. Wada, M., Nakanishi, H., Satoh, A., Hirano, H., Obaishi, H., Matsuura, Y., and Takai, Y. (1997) Isolation and characterization of a GDP/GTP exchange protein specific for the Rab3 subfamily small G proteins. *J. Biol. Chem.*, **272,** 3875.

93. Iwasaki, K., Staunton, J., Saifee, O., Nonet, M., and Thomas, J. H. (1997) aex-3 encodes a novel regulator of presynaptic activity in *C. elegans. Neuron*, **18,** 613.

94. Schievella, A. R., Chen, J. H., Graham, J. R., and Lin, L. L. (1997) MADD, a novel death domain protein that interacts with the type 1 tumor necrosis factor receptor and activates mitogen-activated protein kinase. *J. Biol. Chem.*, **272,** 12069.

95. Strom, M., Vollmer, P., Tan, T. J., and Gallwitz, D. (1993) A yeast GTPase-activating protein that interacts specifically with a member of the Ypt/Rab family. *Nature*, **361,** 736.

96. Vollmer, P. and Gallwitz, D. (1995) High expression cloning, purification, and assay of Ypt-GTPase-activating proteins. *Meth. Enzymol.*, **257,** 118.

97. Du, L. L., Collins, R. N., and Novick, P. J. (1998) Identification of a Sec4p GTPase-activating protein (GAP) as a novel member of a Rab GAP family. *J. Biol. Chem.*, **273,** 3253.

98. Tan, T. J., Vollmer, P., and Gallwitz, D. (1991) Identification and partial purification of GTPase-activating proteins from yeast and mammalian cells that preferentially act on Ypt1/Rab1 proteins. *FEBS Lett.*, **291,** 322.

99. Jones, S., Richardson, C. J., Litt, R. J., and Segev, N. (1998) Identification of regulators for Ypt1 GTPase nucleotide cycling. *Mol. Biol. Cell*, **9,** 2819.

100. Wilson, A. L., Sheridan, K. M., Erdman, R. A., and Maltese, W. A. (1996) Prenylation of a Rab1B mutant with altered GTPase activity is impaired in cell-free systems but not in intact mammalian cells. *Biochem. J.*, **318,** 1007.

101. Richardson, P. M. and Zon, L. I. (1995) Molecular cloning of a cDNA with a novel domain present in the tre-2 oncogene and the yeast cell cycle regulators BUB2 and cdc16. *Oncogene*, **11,** 1139.

102. Neuwald, A. F. (1997) A shared domain between a spindle assembly checkpoint protein and Ypt/Rab-specific GTPase-activators. *Trends Biochem. Sci.*, **22,** 243.

103. Furge, K. A., Wong, K., Armstrong, J., Balasubramanian, M., and Albright, C. F. (1998) Byr4 and Cdc16 form a two-component GTPase-activating protein for the Spg1 GTPase that controls septation in fission yeast. *Curr. Biol.*, **8,** 947.

104. Fukui, K., Sasaki, T., Imazumi, K., Matsuura, Y., Nakanishi, H., and Takai, Y. (1997) Isolation and characterization of a GTPase activating protein specific for the Rab3 subfamily of small G proteins. *J. Biol. Chem.*, **272,** 4655.

105. Nagano, F., Sasaki, T., Fukui, K., Asakura, T., Imazumi, K., and Takai, Y. (1998) Molecular cloning and characterization of the noncatalytic subunit of the Rab3 subfamily-specific GTPase-activating protein. *J. Biol. Chem.*, **273,** 24781.

106. Jena, B. P., Brennwald, P., Garrett, M. D., Novick, P., and Jamieson, J. D. (1992) Distinct and specific GAP activities in rat pancreas act on the yeast GTP-binding proteins Ypt1 and Sec4. *FEBS Lett.*, **309,** 5.

107. Shirataki, H., Kaibuchi, K., Sakoda, T., Kishida, S., Yamaguchi, T., Wada, K., Miyazaki, M., and Takai, Y. (1993) Rabphilin-3A, a putative target protein for smg p25A/rab3A p25 small GTP-binding protein related to synaptotagmin. *Mol. Cell. Biol.*, **13,** 2061.

108. Kishida, S., Shirataki, H., Sasaki, T., Kato, M., Kaibuchi, K., and Takai, Y. (1993) Rab3A GTPase-activating protein-inhibiting activity of Rabphilin-3A, a putative Rab3A target protein. *J. Biol. Chem.*, **268,** 22259.

109. Yamaguchi, T., Shirataki, H., Kishida, S., Miyazaki, M., Nishikawa, J., Wada, K., Numata, S., Kaibuchi, K., and Takai, Y. (1993) Two functionally different domains of rabphilin-3A, Rab3A p25/smg p25A-binding and phospholipid- and Ca(2$^+$)-binding domains. *J. Biol. Chem.*, **268,** 27164.

110. Nalefski, E. A. and Falke, J. J. (1996) The C2 domain calcium-binding motif: structural and functional diversity. *Protein Sci.*, **5,** 2375.

111. Chung, S. H., Song, W. J., Kim, K., Bednarski, J. J., Chen, J., Prestwich, G. D., and Holz, R. W. (1998) The C2 domains of Rabphilin3A specifically bind phosphatidylinositol 4,5-bisphosphate containing vesicles in a Ca^{2+}-dependent manner. *In vitro* characteristics and possible significance. *J. Biol. Chem.,* **273,** 10240.

112. Kato, M., Sasaki, T., Imazumi, K., Takahashi, K., Araki, K., Shirataki, H., Matsuura, Y., Ishida, A., Fujisawa, H., and Takai, Y. (1994) Phosphorylation of Rabphilin-3A by calmodulin-dependent protein kinase II. *Biochem. Biophys. Res. Comm.,* **205,** 1776.

113. Fykse, E. M. (1998) Depolarization of cerebellar granule cells increases phosphorylation of rabphilin-3A. *J. Neurochem.,* **71,** 1661.

114. Li, C., Takei, K., Geppert, M., Daniell, L., Stenius, K., Chapman, E. R., Jahn, R., De Camilli, P., and Sudhof, T. C. (1994) Synaptic targeting of rabphilin-3A, a synaptic vesicle Ca^{2+}/phospholipid-binding protein, depends on rab3A/3C. *Neuron,* **13,** 885.

115. Stahl, B., Chou, J. H., Li, C., Sudhof, T. C., and Jahn, R. (1996) Rab3 reversibly recruits rabphilin to synaptic vesicles by a mechanism analogous to raf recruitment by ras. *EMBO J.,* **15,** 1799.

116. Shirataki, H., Yamamoto, T., Hagi, S., Miura, H., Oishi, H., Jin-no, Y., Senbonmatsu, T., and Takai, Y. (1994) Rabphilin-3A is associated with synaptic vesicles through a vesicle protein in a manner independent of Rab3A. *J. Biol. Chem.,* **269,** 32717.

117. Kato, M., Sasaki, T., Ohya, T., Nakanishi, H., Nishioka, H., Imamura, M., and Takai, Y. (1996) Physical and functional interaction of rabphilin-3A with alpha-actinin. *J. Biol. Chem.,* **271,** 31775.

118. Kotake, K., Ozaki, N., Mizuta, M., Sekiya, S., Inagaki, N., and Seino, S. (1997) Noc2, a putative zinc finger protein involved in exocytosis in endocrine cells. *J. Biol. Chem.,* **272,** 29407.

119. Wang, Y., Okamoto, M., Schmitz, F., Hofmann, K., and Sudhof, T. C. (1997) Rim is a putative Rab3 effector in regulating synaptic-vesicle fusion. *Nature,* **388,** 593.

120. Stenmark, H., Vitale, G., Ullrich, O., and Zerial, M. (1995) Rabaptin-5 is a direct effector of the small GTPase Rab5 in endocytic membrane fusion. *Cell,* **83,** 423.

121. Horiuchi, H., Lippe, R., McBride, H. M., Rubino, M., Woodman, P., Stenmark, H., Rybin, V., Wilm, M., Ashman, K., Mann, M., and Zerial, M. (1997) A novel Rab5 GDP/GTP exchange factor complexed to Rabaptin-5 links nucleotide exchange to effector recruitment and function. *Cell,* **90,** 1149.

122. Ohya, T., Sasaki, T., Kato, M., and Takai, Y. (1998) Involvement of Rabphilin3 in endocytosis through interaction with Rabaptin5. *J. Biol. Chem.,* **273,** 613.

123. Vitale, G., Rybin, V., Christoforidis, S., Thornqvist, P., McCaffrey, M., Stenmark, H., and Zerial, M. (1998) Distinct Rab-binding domains mediate the interaction of Rabaptin-5 with GTP-bound Rab4 and Rab5. *EMBO J.,* **17,** 1941.

124. Xiao, G. H., Shoarinejad, F., Jin, F., Golemis, E. A., and Yeung, R. S. (1997) The tuberous sclerosis 2 gene product, tuberin, functions as a Rab5 GTPase activating protein (GAP) in modulating endocytosis. *J. Biol. Chem.,* **272,** 6097.

125. Gournier, H., Stenmark, H., Rybin, V., Lippe, R., and Zerial, M. (1998) Two distinct effectors of the small GTPase Rab5 cooperate in endocytic membrane fusion. *EMBO J.,* **17,** 1930.

126. Murthy, V. N. and Stevens, C. F. (1998) Synaptic vesicles retain their identity through the endocytic cycle. *Nature,* **392,** 497.

127. Cosulich, S. C., Horiuchi, H., Zerial, M., Clarke, P. R., and Woodman, P. G. (1997) Cleavage of rabaptin-5 blocks endosome fusion during apoptosis. *EMBO J.,* **16,** 6182.

128. Mu, F. T., Callaghan, J. M., Steele-Mortimer, O., Stenmark, H., Parton, R. G., Campbell, P. L., McCluskey, J., Yeo, J. P., Tock, E. P., and Toh, B. H. (1995) EEA1, an early endosome-associated protein. EEA1 is a conserved alpha-helical peripheral membrane protein flanked by cysteine 'fingers' and contains a calmodulin-binding IQ motif. *J. Biol. Chem.*, **270**, 13503.

129. Patki, V., Virbasius, J., Lane, W. S., Toh, B.H., Shpetner, H. S., and Corvera, S. (1997) Identification of an early endosomal protein regulated by phosphatidylinositol 3-kinase. *Proc. Natl Acad. Sci., USA*, **94**, 7326.

130. Burd, C. G. and Emr, S. D. (1998) Phosphatidylinositol(3)-phosphate signaling mediated by specific binding to RING FYVE domains. *Mol. Cell*, **2**, 157.

131. Gaullier, J. M., Simonsen, A., D'Arrigo, A., Bremnes, B., Stenmark, H., and Aasland, R. (1998) FYVE fingers bind PtdIns(3)P. *Nature*, **394**, 432.

132. Wiedemann, C. and Cockcroft, S. (1998) Vesicular transport. Sticky fingers grab a lipid. *Nature*, **394**, 426.

133. Mills, I. G., Jones, A. T., and Clague, M. J. (1998) Involvement of the endosomal autoantigen EEA1 in homotypic fusion of early endosomes. *Curr. Biol.*, **8**, 881.

134. Simonsen, A., Lippe, R., Christoforidis, S., Gaullier, J. M., Brech, A., Callaghan, J., Toh, B. H., Murphy, C., Zerial, M., and Stenmark, H. (1998) EEA1 links PI(3)K function to Rab5 regulation of endosome fusion. *Nature*, **394**, 494.

135. Tall, G., Hama, H., DeWald, D. B., and Horazdovsky, B. F. (1999) The phosphatidylinositol 3-phosphate binding protein, Vac1p, interacts with a Rab GTPase and a Sec1p homologue to facilitate vesicle-mediated vacuolar protein Sorting. *Mol. Biol. Cell*, **10**, 1873.

136. Peterson, M. R., Burd, C. G., and Emr, S. D. (1999) Vac1p coordinates Rab and phosphatidylinositol 3-kinase signaling in Vps45p-dependent vesicle docking/fusion at the endosome. *Curr. Biol.*, **9**, 159.

137. Echard, A., Jollivet, F., Martinez, O., Lacapere, J. J., Rousselet, A., Janoueix-Lerosey, I., and Goud, B. (1998) Interaction of a Golgi-associated kinesin-like protein with Rab6. *Science*, **279**, 580.

138. Mayer, T., Touchot, N., and Elazar, Z. (1996) Transport between cis and medial Golgi cisternae requires the function of the Ras-related protein Rab6. *J. Biol. Chem.*, **271**, 16097.

139. Hirokawa, N. (1998) Kinesin and dynein superfamily proteins and the mechanism of organelle transport. *Science*, **279**, 519.

140. Vale, R. D. and Fletterick, R. J. (1997) The design plan of kinesin motors. *Annu. Rev. Cell Dev. Biol.*, **13**, 745.

141. Burkhardt, J. K. (1998) The role of microtubule-based motor proteins in maintaining the structure and function of the Golgi complex. *Biochim. Biophys. Acta*, **1404**, 113.

142. Li, B. and Warner, J. R. (1996) Mutation of the Rab6 homologue of *Saccharomyces cerevisiae*, YPT6, inhibits both early Golgi function and ribosome biosynthesis. *J. Biol. Chem.*, **271**, 16813.

143. Novick, P., Garrett, M. D., Brennwald, P., Lauring, A., Finger, F. P., Collins, R., and TerBush, D. R. (1995) Control of exocytosis in yeast. *Cold Spring Harb. Symp. Quant. Biol.*, **60**, 171.

144. TerBush, D. R., Maurice, T., Roth, D., and Novick, P. (1996) The Exocyst is a multiprotein complex required for exocytosis in Saccharomyces cerevisiae. *EMBO J.*, **15**, 6483.

145. Finger, F. P., Hughes, T. E., and Novick, P. (1998) Sec3p is a spatial landmark for polarized secretion in budding yeast. *Cell*, **92**, 559.

146. Grindstaff, K. K., Yeaman, C., Anandasabapathy, N., Hsu, S. C., Rodriguez-Boulan, E.,

Scheller, R. H., and Nelson, W. J. (1998) Sec6/8 complex is recruited to cell–cell contacts and specifies transport vesicle delivery to the basal–lateral membrane in epithelial cells. *Cell*, **93**, 731.

147. Lupashin, V. V., Hamamoto, S., and Schekman, R. W. (1996) Biochemical requirements for the targeting and fusion of ER-derived transport vesicles with purified yeast Golgi membranes. *J. Cell Biol.*, **132**, 277.

148. Sonnichsen, B., Lowe, M., Levine, T., Jamsa, E., Dirac-Svejstrup, B., and Warren, G. (1998) A role for giantin in docking COPI vesicles to Golgi membranes. *J. Cell Biol.*, **140**, 1013.

149. Sacher, M., Jiang, Y., Barrowman, J., Scarpa, A., Burston, J., Zhang, L., Schieltz, D., Yates, J.R. 3rd, Abeliovich, H., and Ferro-Novick, S. (1998) TRAPP, a highly conserved novel complex on the cis-Golgi that mediates vesicle docking and fusion. *EMBO J.*, **17**, 2494.

150. Diaz, E., Schimmoller, F., and Pfeffer, S. R. (1997) A novel Rab9 effector required for endosome-to-TGN transport. *J. Cell Biol.*, **138**, 283.

151. Bork, P. and Doolittle, R. F. (1994) Drosophila kelch motif is derived from a common enzyme fold. *J. Mol. Biol.*, **236**, 1277.

152. Fukui, Y., Miyake, S., Satoh, M., and Yamamoto, M. (1989) Characterization of the *Schizosaccharomyces pombe* ral2 gene implicated in activation of the ras1 gene product. *Mol. Cell. Biol.*, **9**, 5617.

153. Ren, M., Zeng, J., De Lemos-Chiarandini, C., Rosenfeld, M., Adesnik, M., and Sabatini, D. D. (1996) In its active form, the GTP-binding protein rab8 interacts with a stress-activated protein kinase. *Proc. Natl Acad. Sci. USA*, **93**, 5151.

154. Katz, P., Whalen, G., and Kehrl, J. H. (1994) Differential expression of a novel protein kinase in human B lymphocytes. Preferential localization in the germinal center. *J. Biol. Chem.*, **269**, 16802.

155. Peranen, J., Auvinen, P., Virta, H., Wepf, R., and Simons, K. (1996) Rab8 promotes polarized membrane transport through reorganization of actin and microtubules in fibroblasts. *J. Cell Biol.*, **135**, 153.

156. Ullrich, O., Horiuchi, H., Bucci, C., and Zerial, M. (1994) Membrane association of Rab5 mediated by GDP-dissociation inhibitor and accompanied by GDP/GTP exchange. *Nature*, **368**, 157.

157. Soldati, T., Shapiro, A. D., Svejstrup, A. B., and Pfeffer, S. R. (1994) Membrane targeting of the small GTPase Rab9 is accompanied by nucleotide exchange. *Nature*, **369**, 76.

158. Hart, M. J., Jiang, X., Kozasa, T., Roscoe, W., Singer, W. D., Gilman, A. G., Sternweis, P. C., and Bollag, G. (1998) Direct stimulation of the guanine nucleotide exchange activity of p115 RhoGEF by Gα13. *Science*, **280**, 2112.

159. Kozasa, T., Jiang, X., Hart, M. J., Sternweis, P. M., Singer, W. D., Gilman, A. G., Bollag, G., and Sternweis, P. C. (1998) p115 RhoGEF, a GTPase activating protein for Gα12 and Gα13. *Science*, **280**, 2109.

160. Vojtek, A. B. and Der, C. J. (1998) Increasing complexity of the Ras signaling pathway. *J. Biol. Chem.*, **273**, 19925.

161. Liu, K. and Li, G. (1998) Catalytic domain of the p120 Ras GAP binds to Rab5 and stimulates its GTPase activity. *J. Biol. Chem.*, **273**, 10087.

162. Rosa, J. L., Casaroli-Marano, R. P., Buckler, A. J., Vilaro, S., and Barbacid, M. (1996) p619, a giant protein related to the chromosome condensation regulator RCC1, stimulates guanine nucleotide exchange on ARF1 and Rab proteins. *EMBO J.*, **15**, 4262.

163. Rosa, J. L. and Barbacid, M. (1997) A giant protein that stimulates guanine nucleotide exchange on ARF1 and Rab proteins forms a cytosolic ternary complex with clathrin and Hsp70. *Oncogene*, **15**, 1.

164. Jamora, C., Takizawa, P. A., Zaarour, R. F., Denesvre, C., Faulkner, D. J., and Malhotra, V. (1997) Regulation of Golgi structure through heterotrimeric G proteins. *Cell*, **91**, 617.

165. Acharya, U., Mallabiabarrena, A., Acharya, J. K., and Malhotra, V. (1998) Signaling via mitogen-activated protein kinase kinase (MEK1) is required for Golgi fragmentation during mitosis. *Cell*, **92**, 183.

166. Pizon, V., Desjardins, M., Bucci, C., Parton, R. G., and Zerial, M. (1994) Association of Rap1a and Rap1b proteins with late endocytic/phagocytic compartments and Rap2a with the Golgi complex. *J. Cell Sci.*, **107**, 1661.

167. Gerwins, P., Blank, J. L., and Johnson, G. L. (1997) Cloning of a novel mitogen-activated protein kinase kinase kinase, MEKK4, that selectively regulates the c-Jun amino terminal kinase pathway. *J. Biol. Chem.*, **272**, 8288.

168. Erickson, J. W., Zhang, C., Kahn, R. A., Evans, T., and Cerione, R. A. (1996) Mammalian Cdc42 is a brefeldin A-sensitive component of the Golgi apparatus. *J. Biol. Chem.*, **271**, 26850.

169. Lonart, G. and Sudhof, T. C. (1998) Region-specific phosphorylation of rabphilin in mossy fiber nerve terminals of the hippocampus. *J. Neurosci.*, **18**, 634.

170. Murphy, C., Saffrich, R., Grummt, M., Gournier, H., Rybin, V., Rubino, M., Auvinen, P., Lutcke, A., Parton, R.G., and Zerial, M. (1996) Endosome dynamics regulated by a Rho protein. *Nature*, **384**, 427.

171. Imai, J., Toh-e, A., and Matsui, Y. (1996) Genetic analysis of the *Saccharomyces cerevisiae* RHO3 gene, encoding a rho-type small GTPase, provides evidence for a role in bud formation. *Genetics*, **142**, 359.

172. De Camilli, P., Emr, S. D., McPherson, P. S., and Novick, P. (1996) Phosphoinositides as regulators in membrane traffic. *Science*, **271**, 1533.

173. Fang, M., Rivas, M. P., and Bankaitis, V. A. (1998) The contribution of lipids and lipid metabolism to cellular functions of the Golgi complex. *Biochim. Biophys. Acta*, **1404**, 85.

174. Shibata, H., Omata, W., and Kojima, I. (1997) Insulin stimulates guanine nucleotide exchange on Rab4 via a wortmannin-sensitive signaling pathway in rat adipocytes. *J. Biol. Chem.*, **272**, 14542.

175. D'Adamo, P., Menegon, A., Lo Nigro, C., Grasso, M., Gulisano, M., Tamanini, F., Bienvenu, T., Gedeon, A. K., Oostra, B., Wu, S. K., Tandon, A., Valtorta, F., Balch, W. E., Chelly, J., and Toniolo, D. (1998) Mutations in GDI1 are responsible for X-linked non-specific mental retardation. *Nature Genet.*, **19**, 134

6 | Arf

MICHAEL G. ROTH

1. Introduction

The six mammalian ADP-ribosylation factors form a distinct family within the Ras GTPase superfamily (1). The amino-acid sequences of these 20 kDa guanine nucleotide-binding proteins are at least 70% identical in plants, protozoa, and animals, attesting to their importance in all eukaryotic cells. Two genes encode redundant Arf proteins in *S. cerevisiae* that are essential for vegetative growth and secretory function (2). A third gene, *ARF3*, encoding a protein 52% identical to Arf1p, cannot compensate for the loss of *ARF1* and *ARF2* (1). In mammalian cells, the major function of Arfs is to regulate a variety of membrane transport processes, although Arf proteins also activate phospholipase D enzymes (3, 4) and one member of the family, Arf6, may be involved in regulating the actin cytoskeleton (5). Sar1, a small GTPase that is 37% identical to Arf1, functions in the assembly of the COPII coated vesicles that are responsible for membrane transport from the endoplasmic reticulum to the Golgi complex (6). The molecular details of Sar1 function are better understood than those of Arf proteins (7) and Sar1 has therefore been used as a model for understanding Arfs, particularly Arf1. However, this is likely to be misleading, since the regulation of Arf1 is clearly quite different from that of Sar1, and Arf1 appears to have a greater number of activities than Sar1.

2. The Arf family

Arf1 was discovered as a cytosolic factor that could enhance the ADP-ribosylation of $G_{\alpha s}$ by cholera toxin (8, 9), and this activity continues to serve as a functional test for members of the Arf family. When overexpressed, all six mammalian Arfs can complement the deletion of the two *ARF* genes in *S. cerevisiae*, which provides a second criterion for inclusion of GTPases in this family. However, this complementation does not necessarily indicate that all Arfs normally function in membrane transport (10). Since there is significant conservation of structure among Arfs, when overexpressed they may participate in activities that normally they would not. In addition to the Arf proteins, there is a growing family of Arf-related proteins, the Arls, that have approximately 50% sequence identity with Arf1 but which cannot substitute for Arf proteins in various assays (11). By this definition, Arf3 in *S. cerevisiae* is an Arl and not an Arf. The six mammalian Arfs can be subdivided into three classes based on their amino-acid sequence similarities and their *in vitro* activities (1, 12, 13). Class I Arfs, which include Arf1–3, are the best understood members of the family and appear to be functionally redundant. These proteins control the formation of vesicle coats in the secretory and endocytic pathways. The biological role of Arfs 4 and 5 (class II) is unknown, while Arf6, the only class III Arf, acts at the plasma membrane to regulate both the actin cytoskeleton and an endosomal compartment (14, 15). The literature on Arf6 is currently somewhat contradictory.

3. Biochemical properties of Arf proteins

Like other small GTP-binding proteins, the Arfs act as molecular switches. They are biologically active when bound to GTP and inactive when bound to GDP. The intrinsic rate of hydrolysis of GTP by Arfs is extremely slow and they require interaction with a GTPase activating protein (GAP) for GTP hydrolysis to occur. The affinity of Arfs for GDP is high so that at physiological concentrations of magnesium the rate of spontaneous exchange of GDP for GTP is slow. A number of guanine nucleotide exchange proteins (GEFs) specific for Arfs have been identified that accelerate the release of GDP from Arf by more than four orders of magnitude (16). Arf proteins are modified post-translationally by the removal of the amino-terminal methionine and the addition of myristic acid to the new amino-terminus. This modification is required for ARFs to bind tightly to membranes. Inactive Arf bound to GDP has a low affinity for membranes and has the amino-terminal myristate tucked into a hydrophobic pocket on the surface of the protein (see Chapter 9). The exchange of GTP for GDP on Arf takes place only on membranes and requires that the amino-terminal myristate be inserted into the lipid bilayer before exchange can occur. In this way, active Arfs are restricted to membrane surfaces.

Arf proteins have three known biochemical activities. First, Arf stimulates the ADP-ribosylation of $G_{\alpha s}$ by cholera toxin *in vitro*, presumably by altering the conformation of the heterotrimeric G protein in some way. Arf–GDP has been shown to bind to G-protein α subunits *in vitro*, and it has been proposed that Arf may compete with G_{α} for binding to βγ (17, 18). The biological significance of this interaction is unknown. The second activity of Arf is to stimulate a phospholipase D, PLD1. PLD1 removes choline from phosphatidylcholine (PC) to produce phosphatidic acid (PA), an important lipid precursor and potent second messenger in several signalling cascades. PA stimulates the activity of a phosphatidylinositide 5-kinase (PI5K) that phosphorylates phosphatidylinositol 4-phosphate (PI4P) to produce phosphatidyl-inositol 4,5-bisphosphate (PI4,5P$_2$). PA can also be dephosphorylated by phosphatidate hydrolase (PAH) to produce diacylglycerol (DAG). The formation of PI4,5P$_2$ and DAG is important for the production of vesicles regulated by Arf (19, 20). The third activity of Arf is to interact with a novel protein, Arfaptin-1 (identified through a yeast two-hybrid screen), in a GTP-dependent manner (21). Arfaptin-1 has no homologue in the genome of *S. cerevisiae*. Arfaptin inhibits Arf in both the *in vitro* ADP-ribosylation assay and the PLD assay (22). Furthermore, the binding of Arfaptin to Golgi membranes requires prior activation of Arf, suggesting that

Arfaptin may be a downstream effector for Arf. If this turns out to be the case, it will indicate that membrane transport processes in mammalian cells differ in some biochemical respects from the analogous processes in yeast.

4. Cellular activities of Arf proteins

Both forward and reverse genetics have been used to investigate the role of Arf in intact cells. Yeast lacking *ARF1* or both *ARF1* and *ARF2* grow poorly and have gross defects in the morphology of endosomes and Golgi (23, 24). Secretion is defective, but not totally inhibited by the loss of Arf activity. Genes capable of suppressing or increasing the growth defect caused by the loss of Arf in yeast have been identified (see below). In mammalian cells, dominant-negative mutants of Arf proteins have been expressed to determine which cellular processes are affected. Expression of Arf1(Q71L), a mutant deficient in GTPase activity, traps coat proteins on membranes and inhibits both secretion and endocytosis (25, 26). Expression of Arf1(T31N), a mutant unable to bind guanine nucleotides, causes the Golgi complex to disassemble (26). Similar mutants of Arf6 do not affect the secretory pathway but cause actin spikes to form at the plasma membrane and inhibit movement of proteins through an endosomal compartment (14, 27). The relationship of this endosomal compartment to those that carry well-studied endocytic receptors is currently uncertain.

To investigate the mechanisms by which Arf proteins might cause the effects observed in living cells, a number of *in vitro* assays have been developed that measure the effects of Arf on various aspects of membrane transport. Taken together, these assays suggest that class I Arfs (Arf1, 2, and 3) regulate the formation of transport vesicles at many steps in constitutive trafficking pathways and, in some cell types, the fusion of regulated secretory vesicles (4) (Fig. 1). They have also been reported to regulate, in some way, the binding of spectrin to the Golgi complex (28). The number of different coated vesicles reported to be regulated by class I Arfs is remarkable. Arf1 regulates the formation of COPI (coat protein I-coated) vesicles (29, 30) for retrograde transport between the Golgi and the ER (31–33), Arf1 or 3 are required for the formation of clathrin/AP1 (adapter protein 1 complex-associated) vesicles at the trans-Golgi network (TGN) (34) and on immature secretory vesicles (35), and Arfs 1, 2, and 3 control the formation of vesicles containing the AP3 adaptor on endosomes (36, 37). When bound to GTPγS, Arf1 inhibits endosome fusion reactions *in vitro* (38) and causes clathrin/AP2 to coat endosomes (39) and lysosomes (40). Arf1 also stimulates the formation of transport vesicles from the TGN for which the coat proteins have not been identified (41, 42).

The variety of functions ascribed to class I Arfs is in marked contrast to the activity of other members of the Ras superfamily that function in membrane trafficking. Sar1 and Rab proteins appear to function only at single steps of membrane transport pathways (43, 44). It is unlikely, therefore, that Rabs or Sar1 will be good models for understanding the regulation of Arfs. Recent data suggest that Arfs can be involved in signal transduction events (19, 20, 45–47 and the regulation of Arfs may therefore have more in common with the regulation of Ras and Rho proteins.

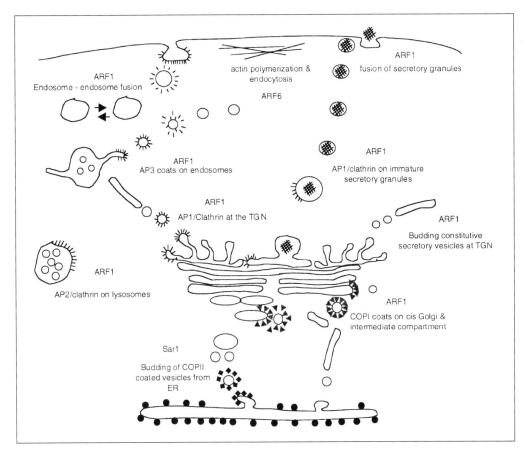

Fig. 1 Locations and activities of Arfs in a mammalian cell.

4.1 Arf function in yeast

Although yeast have two genes for Arf proteins, *Arf2p* is expressed at only 10% the level of *Arf1p* and cannot support normal growth when the gene for Arf1p is disrupted (*arf1Δ*). *arf1Δ* yeast grow slowly and have defects in both anterograde and retrograde secretory pathways and in the delivery of material from endosomes to the vacuole (2, 48, 49). In addition, the morphology of organelles on both the exocytic and endocytic pathways is grossly disturbed.

Chen and Graham have employed the growth defect in *arf1Δ* yeast to isolate mutations that prevent growth when combined with *arf1Δ* (50). One of the mutants identified by this approach expressed a defective clathrin heavy chain. Interestingly, defects in clathrin alone result in poor growth and a loss of the ability to correctly retain proteins in the Golgi complex (51, 52). Thus, one role of Arfs in yeast might be to control the formation of vesicles that recycle proteins from endosomes to the Golgi.

As an alternative approach, Kahn and colleagues expressed *arf1-3*, a temperature-sensitive allele of *ARF1*, on an *arf1Δ* background and isolated a high-copy suppressor, *SAT1,* that allowed normal growth at the non-permissive temperature (53). Sat1p contained a zinc-finger motif also found in Gsc1p, a known GAP protein for Arf1p. Gsc1p was also able to suppress the loss of Arf function for growth, but did not correct a defect in the secretion of invertase. Although Sat1p and Gsc1p might stabilize Arf1-3p so that a critical level of activity remained at the high, non-permissive temperature, a more interesting possibility is that Gsc1p functions as an effector of Arf1p. A similar situation has been found with the RGS proteins that serve as both GAPs and effectors for heterotrimeric G proteins (see Chapters 1 and 2) (54, 55).

In mammalian cells, Arf stimulates the activity of PLD1 but not a related enzyme, PLD2. In *S. cerevisiae*, there is a single phospholipase D, Spo14p, equally related in sequence to both PLD1 and PLD2. Yeast Arfs do not regulate Spo14p (56) and, furthermore, Spo14p is not required for vegetative growth or secretion. This observation has been interpreted to indicate that, in animal cells, PLD1 cannot be an essential effector of Arf for membrane trafficking (53, 56). This assumes, of course, that all essential aspects of membrane transport are conserved between *S. cerevisiae* and mammals.

4.2 Studies of Arf function in mammalian cells using reverse genetics

Based upon sequence relationships with Ras proteins, mutants of Arf were constructed that were either incapable of binding nucleotide (Arf(T31N)) or incapable of hydrolysing GTP (Arf(Q71L) or Arf(Q71I)). When expressed in animal cells, cDNAs encoding these proteins inhibited secretion (25, 26, 57). The phenotype of cells expressing Arf(T31N) was similar to that observed when cells were treated with the drug brefeldin A (26). In the presence of the mutant or the drug, the Golgi complex fused with the ER, and coat proteins regulated by Arf were found in the cytosol. In contrast, Arf(Q71L) caused coat proteins to accumulate on membranes and although membrane proteins could still exit the ER, they could not enter the Golgi (26). COPI proteins regulated by Arf are known to coat the intermediate compartment between the ER and Golgi (58) and this is the likely site of inhibition of protein transport. It is important to note that none of these studies showed the accumulation of coated vesicles that would be predicted by the early models of Arf function that postulated that GTP hydrolysis on Arf is required for vesicle uncoating, but not for budding (59, 60). The accumulation of horseradish peroxidase by fluid-phase endocytosis was inhibited by expression of Arf(Q71L), although whether this was a direct or indirect effect was not determined (25).

Similar experiments have been performed with mutants of the class III Arf6 (15, 61). Arf6 is located at the plasma membrane and on endosome-like structures but not on the Golgi. Even high levels of expression did not cause Arf6 to bind to Golgi membranes or affect Golgi coat proteins. There is disagreement as to whether wild-type Arf6 induces changes in cellular morphology. However, when overexpressed, the GTPase-defective Arf6 mutant (Arf6(Q67L)) was preferentially located at the plasma membrane and caused a depletion of endosomes, a redistribution of trans-ferrin receptors to the plasma membrane, and induced extensive folding of the plasma membrane. The mutant likely to be the nucleotide-free form, Arf6(T27N), accumulated on heavily coated endosomal structures (61) and caused transferrin receptors to be trapped within the cell (15). The nature of these endosomes is uncertain. In subsequent experiments carried out by one laboratory, this com-partment was shown to contain transferrin receptors and seemed likely to be the recycling endosome (27). Another laboratory, however, did not observe transferrin receptors in the compartment that is enlarged in cells expressing Arf6(T27N), but did find glycolipid-linked proteins in those membranes. This compartment could also be enlarged by the drug cytochalasin D, which prevents actin polymerization. It is currently not clear whether Arf6 acts primarily to control a novel endocytic pathway, or if this effect is secondary to an effect of Arf6 on the actin cytoskeleton. However, it is certain that the functions of Arf6 are completely different to those of the class I Arfs.

4.3 Studies using inhibitors in intact mammalian cells

A number of studies have used reagents that fix Arf in ether the GTP- or GDP-bound state. The most important of the reagents that can be used in living cells has been brefeldin A, a fungal metabolite that inhibits the interaction between Arf and GEFs that reside on the Golgi membrane (62–65). Within seconds of entering cells, brefeldin A prevents the binding to the Golgi of membrane coats that are regulated by Arf. As a consequence, coat proteins that are released from membranes cannot return. Subsequently, Golgi membranes fuse with the endoplasmic reticulum and the trans-Golgi network compartment fuses with the endocytic compartment (66, 67). This process of specific membrane fusion is intriguing but is hard to reconcile with models in which that small, discrete coated vesicles carry membrane between organelles. If vesicle formation is blocked, those models predict that membrane trafficking will stop and the organelles will remain in place. In spite of the fact that the effects of brefeldin A on membrane are more complicated than is predicted by the known effect of the drug on Arf, this reagent is often called a specific inhibitor of Arf. Membrane traffic that is sensitive to the drug is thought to require Arf.

Aluminium fluoride (AlF_4), which probably stabilizes the complex between Arf and Arf GAP (68), has been used to block coat protein release (69). This reagent was originally thought to act only on heterotrimeric G proteins and there are several reports in which the effects of AlF_4 on membrane activities known to be regulated by Arf have been interpreted as supporting a role for G proteins in these pathways. Although G proteins are found on membranes that also contain Arf, currently there are no strong data to indicate that G proteins control membrane traffic.

Non-hydrolysable analogues of GTP trap Arf in the activated state and inhibit membrane trafficking in a manner similar to mutant Arfs that are defective in the hydrolysis of GTP. Since cells contain many GTP-binding proteins, it is difficult to interpret the use of these analogues in intact cells, and they have been more useful in experiments *in vitro*. Peptides having the sequence of the Arf1 amino terminus have also been used to interfere with the interaction between Arf and its effector proteins (70) and were the basis for early claims that Arf is required for anterograde transport between the ER and the Golgi complex. Antibodies specific for Arf, or for the coat proteins thought to be controlled by Arf, have also been used to inhibit membrane traffic between the ER and the Golgi (71, 72). These reagents cannot cross the plasma membrane and must be microinjected into cells or used with permeabilized cells. However, our ignorance of the number of cellular targets for each of these reagents limits the interpretation of the studies that use them.

4.4 *In vitro* assays with broken cells or isolated membranes

One of the most useful approaches for understanding the roles of Arf proteins and the mechanisms by which they act has been to reconstitute Arf activities *in vitro*. These techniques have been used extensively to study the requirements for binding of coat proteins to membranes (29, 34, 39, 73, 74) and for producing coated vesicles (75–78). Based upon the earliest of such studies, a simple model for the function of Arf was proposed in which the Arf GEF protein controls the location interaction of Arf with membranes and its subsequent activation. In this model, coat proteins then bind stoichiometrically to the activated Arf, to form a vesicle that buds with Arf–GTP in the coat. Hydrolysis of GTP on Arf, by a recruited GAP, causes uncoating of the vesicle (60). However, the use of Arf–GTPγS in these assays precludes detailed investigation of the role of either the GEF or the GAP reaction in regulating Arf activity. In fact, recent evidence suggests that both of these reactions are more complicated than was originally thought and this currently popular model of Arf function is under revision (see Section 7, below).

5. Arf GEF proteins

Two families of Arf GEF proteins have been identified that share a common domain, the Sec7 domain, sufficient for catalysing guanine nucleotide exchange on Arf proteins (Fig. 2) (79, 80). One family contains the large proteins Gea1, Gea2, and Sec7 in yeast and p200 in mammalian cells (80–82). These proteins are sensitive to brefeldin A and are candidates for controlling class I Arfs on the Golgi, TGN, and perhaps endosomes. Gea1 and Gea2 are 50% identical and functionally redundant proteins involved in COPI traffic between the ER and Golgi (31, 80). Sec7 is more distantly related to the Gea proteins, having approximately 35% identical amino acids in the catalytic domain. Sec7 does not complement loss of the Gea proteins and appears to function in some way for post-Golgi membrane traffic. Mammalian p200 is more closely related to Sec7 than to either of the yeast Gea proteins and its site of action in the cell is unknown. The brefeldin A-sensitive exchange proteins can catalyse guanine nucleotide exchange on all classes of Arf proteins *in vitro*, although

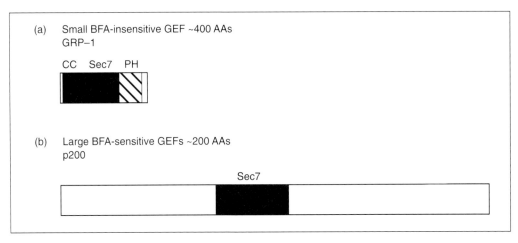

Fig. 2 Two families of guanine nucleotide exchange proteins for Arfs: (a) a representative of the small Arf GEFs, GRP-1; and (b) one of the large GEFs, mammalian p200. The figures are drawn to scale and identified domains are labelled.

with varying efficiency (82). Since the distribution of Arf6 within cells is not changed by brefeldin A, it is likely that the large Arf GEFs do not regulate Arf6 *in vivo*.

The second family of Arf GEFs currently contains three smaller proteins, ARNO1, GRP-1, and cytohesin-1, that are insensitive to brefeldin A (79, 83, 84). These proteins each contain a pleckstrin homology (PH) domain. PH domains bind phosphorylated phosphatidylinositides with varying affinities. ARNO binds to $PI4,5P_2$ and GRP1 binds $PI3,4,5P_3$ with very high affinity (85). The catalytic Sec7 domain of GRP-1 stimulates guanine nucleotide exchange on all Arfs and even the Arf-related (Arl) proteins *in vitro*, whereas the intact protein has a more restricted activity and does not react with Arls (84). This suggests that portions of the GEFs outside their catalytic domain may contribute to the target specificity. There is disagreement about the activities of the smaller GEFs. ARNO1 has been reported to be located at the plasma membrane, where it could be the GEF for Arf6 (86). Overexpression of ARNO causes the disassembly of actin stress fibres and ARNO is required for the formation of lamellipodia in cells treated with phorbol esters (87). Another group did not observe GEF activity of any of the small GEFs on Arf6 but found that overexpression of ARNO3, the human homologue of GRP-1, produced the same effects as brefeldin A (88). Other workers reported that GRP-1 stimulates guanine nucleotide exchange on Arf1 and Arf5, but not on Arf6 (84, 85). Much more work is required to determine where in the cell each GEF functions and which Arf each activates.

The structural basis for Sec7-induced catalysis of guanine nucleotide exchange on Arf has been elucidated by recent reports of the structure of the Sec7 domains of cytohesin-1 (89) and ARNO1 (90, 91) and of the complex of the Sec7 domain of Gea2 with Arf–GTP (see Chapter 9) (92). As their sequence conservation would suggest, Sec7 domains have a common structure, composed of a platform of ten α-helices. The GEF binds to the switch 1 and switch 2 regions of Arf and inserts residues into the nucleotide-binding pocket to displace the magnesium ion and repel the β-phosphate of the GDP ligand. Surprisingly, only the conformation of Arf–GTP, and not Arf–GDP, is compatible with binding of the Sec7 domain. Therefore, considerable conformational change must occur on Arf before GDP can be released (92). A surprising aspect of all the known Arf GEFs is that they are soluble proteins and their binding to membranes appears to depend upon Arf. *In vitro*, Arf GEFs cannot bind myristoylated Arf unless lipids are present (93, 94). Put together, these observations suggest that Arf must bind to membranes before the exchange reaction can occur. In fact, myristoylated Arf bound to GTP has been shown to bind to lipid vesicles (95). Thus, the regulation of Arf is very different from that of its cousin, Sar1, which controls the assembly of the COPII coat protein on the ER. The GEF for Sar1, Sec12, is a resident ER transmembrane protein and appears to determine where the COPII coat will form by activating guanine nucleotide exchange on Sar1, causing it to bind to membranes (7). It is currently unclear how the location of Arf membrane recruitment is specified prior to formation of the vesicle coats. A rather speculative description of how this might occur, based upon recent biochemical data and the three-dimensional structure of a Sec7 domain complexed with Arf-GTP, is described in Section 7.

6. Arf GAP proteins

Two types of Arf GTPase-activating proteins (Arf GAPs) have been identified (Fig. 3) (46, 47, 96, 97). Both families of Arf GAPs contain a small domain that is sufficient to catalyse GTP hydrolysis on Arf. The GAP domain contains a zinc-finger motif that has four cysteines required for catalysis. The first Arf GAP to be cloned, rat p47ArfGAP (ArfGAP1), binds to Golgi membranes in a manner sensitive to brefeldin A (96). When overexpressed, ArfGAP1 creates the same phenotype as Arf(T31N). Both the GAP and the mutant Arf cause the Golgi complex to fuse with the ER, similar to the effect of brefeldin A. By this criterion, ArfGAP1 is the protein responsible for regulating the formation of COPI vesicles from the Golgi. ArfGAP1 is 27% identical and 45% similar to Gcs1p, which has been shown to have GAP activity for Arf1p and Arf2p in *S. cerevisiae* (97). Gcs1p and ArfGAP1 are of similar size. Sequences with similar size and degrees of similarity are found in a number of other species, including *C. elegans* and *Drosophila*. ArfGAP1 has more limited sequence homology to several other genes in *S. cerevisiae*, of which two, *GLO3* and *SAT2*, may function as Arf GAPs, at least at high copy number (53).

The amino-terminal zinc-finger domain is less than one-third of ArfGAP1. Although this domain is sufficient to catalyse GTP hydrolysis on Arf in solution, it cannot produce the dominant-negative effects observed when ArfGAP1 is overexpressed *in vivo* (98). Some feature in the 125 amino acids following the GAP domain appears to be required for targetting the protein to the Golgi. It is likely that the small Arf GAPs will have biological functions in addition to inactivating Arf, since *GSC1* acts as a multi-copy suppressor of *arf1-3* in *S. cerevisiae* (53). ArfGAP1 may play a role linking the collection of cargo with the budding of COPI vesicles. In mammalian cells, ArfGAP1 can be coprecipitated with a transmembrane receptor, ERD2, that binds

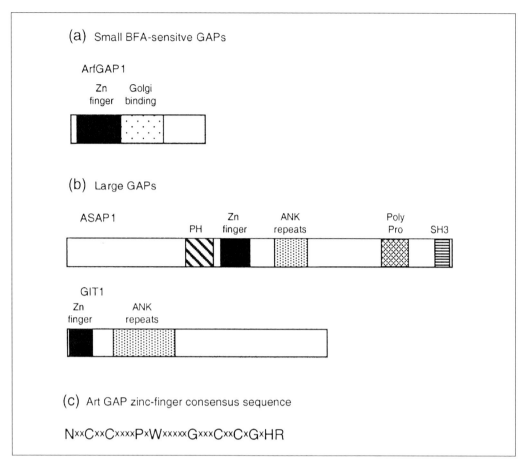

Fig. 3 Arf GAPs. (a, b) Linear representations of the small p47ArfGAP1 and two larger GAPs, ASAP1 and GIT1 are drawn to scale. The black box denotes the GAP catalytic domain. (c) The consensus sequence of the catalytic zinc-finger domain of Arf GAPs is shown in single-letter code. Non-conserved positions are indicated by x. N, Asparagine; C, cysteine; P, proline; W, tryptophan; G, glycine; H, histidine; R, arginine.

Golgi proteins having a carboxyl-terminal KDEL ER-retrieval signal. When overexpressed, ERD2 produces the same phenotype as the overexpression of ArfGAP1 (99). Overexpression of a protein bearing a KDEL sequence recruits more ArfGAP1 to Golgi membranes (100).

Class I Arfs control the formation of clathrin/AP1 coats on the TGN. Using an *in vitro* assay, Kornfeld and colleagues have shown that Arf acts to form transient, high-affinity binding sites for AP1. GTP hydrolysis on Arf can occur prior to the binding of AP1 to these sites and Arf is not found in vesicles coated by clathrin/AP1 (78). This suggests that Arf GAP activity occurs during vesicle assembly, rather than during vesicle uncoating as was previously thought (Section 4.4).

ArfGAP1 is stimulated by diacylglycerols (DAGs) that have monounsaturated acyl chains, and not at all by DAGs that have unsaturated acyl chains or polyunsaturated chains (101). DAG with monounsaturated acyl chains is produced mainly through the sequential action of phospholipase D, converting PC to PA, followed by PA hydrolase converting PA to DAG. Thus, it is possible that Arf recruits its own GAP to membranes through stimulation of PLD. It is currently not known whether there is a connection between the stimulation of Arf GAPs by DAG and the increase in Arf GAPs on membranes caused by ligand binding to ERD2.

The second family of Arf GAP proteins was identified originally through the characterization of a GAP activity in cytosol that was stimulated by $PI4,5P_2$ (102). Two groups have cloned cDNAs encoding large proteins that have the Arf GAP activity (46) (47). ASAP1 is a multidomain protein of 1146 amino acids that contains an amino-terminal PH domain followed by the zinc-finger GAP domain, ankyrin repeats, a proline-rich region containing SH3 binding sites, and a SH3 domain. ASAP1 was independently identified as a protein that binds to the Src kinase SH3 domain and is phosphorylated on tyrosine residues in cells in which Src is activated. A number of sequences in the database show significant similarity to ASAP1 and have the same PH–zinc finger–ankyrin repeat organization (Fig. 3). ASAP1 has GAP activity for Arf1 and Arf5 and but shows little activity on Arf6. This GAP activity is greatly stimulated by $PI4,5P_2$, much less by PA, and not at all by $PI3,4,5P_3$. Unlike ArfGAP1, ASAP1 was not observed to bind to Golgi membranes and its over-expression did not inhibit secretion or disturb Golgi morphology. Thus, ASAP1 is likely to regulate Arf function outside the secretory pathway, perhaps on endosomes.

A search for proteins that interact with G-protein-coupled receptor kinases identified a 770 amino acid protein, GIT1, that lacks an amino-terminal PH domain but contains the Arf GAP zinc-finger domain followed by ankyrin repeats. GIT1 has GAP activity for Arf1 and overexpression of the protein interferes with the internalization of the β_2-adrenergic receptor. This effect requires an intact zinc-finger domain (47). The discovery of ASAP1 and GIT1 suggest that Arf proteins might act in signal transduction events, perhaps to coordinate changes in membrane traffic.

7. A revised model for guanine nucleotide exchange on Arf

In its inactive, GDP-bound form, Arf adopts a quite distinct conformation from that observed for other members of the Ras superfamily (103, 104). However, in the 'active', GTP-bound form, Arf adopts a conformation similar to that observed for other small GTPases. This extensive conformational change requires that the amino terminus of Arf, along with its myristate, leaves a surface binding pocket, and that a loop between strands β2 and β3 invades the space that is vacated. The recent crystal structure for the complex between the Sec7 domain of the Gea2 exchange protein bound to Arf (92) shows that in its 'inactive' conformation Arf cannot fit into the binding pocket of the GEF. Furthermore, binding studies show that the GEF can bind to myristoylated Arf–GDP only when the latter is on membranes (94) and that myristoylated Arf–GDP can bind lipid vesicles and undergo a slow guanine nucleotide exchange reaction in the absence of a GEF (95). It appears, therefore, that Arf–GDP must first bind to membranes and insert the myristate into the bilayer, and then undergo a conformational change before the GEF can bind. This 'open' conformation must occur on an Arf still loaded with GDP and it may not, therefore, be exactly the same conformation as observed for Arf–GTP. Thus, the GEF is not responsible for determining the initial binding of Arf to membranes. Furthermore, the recruitment of Arf GEFs to the Golgi is sensitive to brefeldin A, indicating that this depends upon their interaction with Arf. Therefore, the GEF cannot by itself determine where Arf will be active nor can it specify where the coated vesicles regulated by Arf will form. Obviously, a more complicated mechanism must define the budding site for coated vesicles.

A hint as to what this mechanism might be comes from investigations of the action of specific lipids on the GTPase cycle of Arf. In the absence of phospholipids, ARNO can stimulate nucleotide exchange on a truncated Arf that lacks 17 amino-terminal residues. This reaction is unaffected by the presence of PI4,5P_2, indicating that this lipid is not required for the catalytic activity of ARNO (93). However, ARNO cannot stimulate exchange on myristoylated, full-length Arf in solution unless phospholipids are present (93). PI4,5P_2 stimulates this activity but, importantly, it also stimulates the activity of the isolated Sec7 domain from ARNO, which does not itself bind to lipids. Therefore, one action of PI4,5P_2 is to make Arf more accessible to the exchange protein. Arf interacts with membranes in two ways, by binding to acidic lipids through a patch of basic residues on one surface (105) and by inserting residues of its amino terminus and the attached myristate into the hydrophobic core of the bilayer. It is likely that the initial binding of Arf to membranes is through an ionic interaction with PI4,5P_2, or perhaps some other acidic lipids, followed by insertion of the amino-terminal myristate into the lipid bilayer. Since phospholipid vesicles (95) or trypsin-treated Golgi membranes (64) can stimulate a slow exchange of GTP for GDP on Arf, it is possible that the localized activation of Arf is controlled by enzymes that generate specific lipids, in particular PI4,5P_2. These lipids would facilitate Arf binding to membranes or the transition from the 'inactive' to the 'open' conformation of Arf. Thus, at steady state there might be a population of Arf on membranes cycling slowly between its active and inactive forms. This concentration might be too low to initiate the formation of a coated vesicle and would require the activity of other proteins to increase the local concentration of active Arf by generating the lipids that bind ARF and its GEF (Fig. 4, steps 1–4).

PI4,5P_2 and activated Arf stimulate PLD to make PA which, in turn, stimulates the PI4P 5-kinase to make more PI4,5P_2. Thus, the site where a coated vesicle forms might be determined by bringing active Arf, PLD, and the PI4P 5-kinase together to generate a positive feedback loop that would increase the exchange reaction on Arf (Fig. 4, steps 4–6) (106–108). The production of PI4,5P_2 requires ATP and, in agreement with this model, ATP stimulates the binding of Arf-GTPγS to Golgi membranes (65) and increases the production of coated vesicles *in vitro*, even though nucleotide exchange on Arf does not require ATP (Bi and Roth, unpublished results).

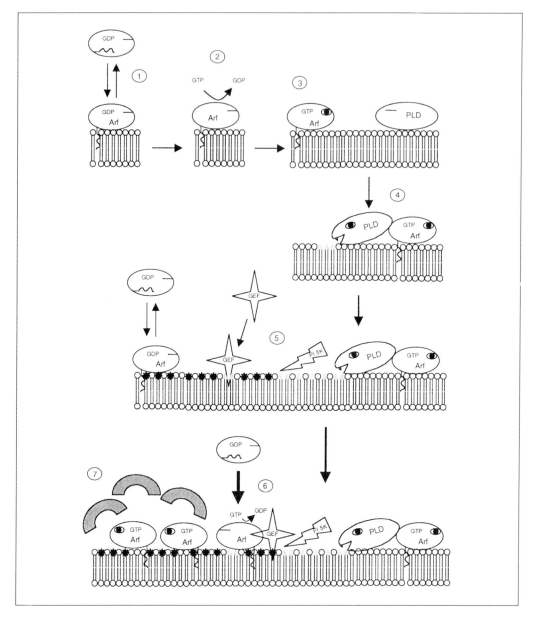

Fig. 4 Positive feedback regulation of guanine nucleotide exchange on Arf. (1) Arf-GDP binds with low affinity and reversibly to membranes. (2) Insertion of the N-terminal myristate into the bilayer can result in a low basal rate of exchange of GTP for GDP on Arf. (3) The concentration of Arf-GTP on membranes due to the basal rate of exchange is too low to cause vesicle coat proteins to bind to membranes. (4) Arf-GTP can activate a phospholipase D, PLD1, to produce phosphatidic acid, PA. (5) PI5K is activated by PA and produces PI4,5P$_2$. Both lipids attract the Arf GEF. (6)Arf GEF catalyses the exchange of GTP for GDP on Arfs, activating them. This creates a positive feedback loop from step (6) to step (4). (7) Cytosolic proteins bind to membranes in response to the presence of Arf, acidic lipids, and perhaps other proteins.

8. Negative feedback regulation through Arf GAP

It is likely that such a positive feedback cycle would be under the regulation of other enzymes that would prevent the concentration of PI4,5P$_2$ from becoming too high. One such enzyme that has been reported to be located on Golgi membranes is a PI 5-phosphatase, OCRL (109), while another is PA hydrolase. By producing DAG from PA, PA hydrolase would both lower PA concentrations and recruit an Arf GAP to membranes (Fig. 5) (101). Arf GAP would stimulate hydrolysis of Arf-bound GTP, terminating its activities. There is an active PA hydrolase on Golgi membranes, and Golgi membranes from cells overexpressing PLD contain elevated DAG levels and bind more ArfGAP1 (110). Thus, the GTPase cycle on Arf appears to be modulated by the rise and fall in concentration of certain lipids, and the place where coated vesicles will form and the extent to which they form may be regulated by whatever controls the location of the lipid-modifying enzymes.

9. Sar1

Sar1 is a small GTPase that is 37% identical to class I Arfs and 41% identical to Arf6. However, because Sar1 functions in vesicle formation, it is often considered to be similar to Arf. Currently, the only known function of Sar1 is to control the assembly of the COPII coat at the ER. The GTPase cycle of Sar1 has been studied extensively in yeast, where proteins important for the function of Sar1p were first identified in yeast mutants that were defective in secretion (*sec* mutants). Sar1p appears to be regulated by a mechanism similar to the one first proposed for Arf (Fig. 6). A transmembrane protein of the ER, Sec12p, catalyses the exchange of GTP for GDP on Sar1p, allowing the latter to bind to the ER membrane and form the binding site for COPII coat proteins. The Sec23p–Sec24p complex binds Sar1p–GTP. Sec23p contains GAP activity for Sar1p, but the exact step in vesicle budding at which hydrolysis occurs is unknown. However, without hydrolysis, disassembly of COPII coats does not occur. An attractive model, proposed by Schekman and Orci, suggests that the Sec23p–Sec24p complex binds to cargo and vesicle targeting molecules and that this multi-molecular assembly is bound by the remaining COPII components, Sec13p and Sec31p, which cross-link the coat and perhaps stimulate the GAP activity of Sec23p (7). Evidence directly supporting this hypothesis has been reported recently (111, 112). The stoichiometry of the binding of Sar1p to the other COPII proteins is not known precisely. The mechanism controlling the extent of polymerization of the COPII coat (or for that matter, any vesicle coat) and its fission from the donor membrane are also not understood precisely. COPII vesicles formed from ER membranes in the presence of non-hydrolyzable GTPγS appear to be resistant to uncoating. Thus hydrolysis of GTP on Sar1p controls uncoating rather than coat growth. Under conditions where there is a high ratio of COPII proteins to liposomes that contain some acidic lipids, smaller COPII vesicles will bud off the liposomes (111). This reaction occurs on ice without addition of ATP and requires only Sar1p bound to GTPγS, and the four COPII proteins, suggesting that the growth

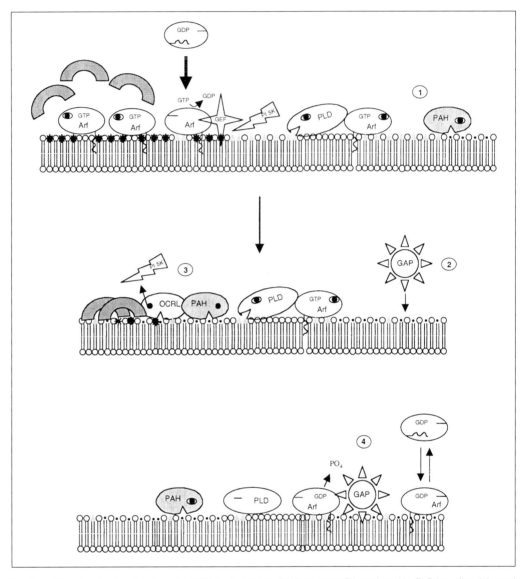

Fig. 5 Negative feedback regulation of Arf. (1) PA hydrolase (PAH) converts PA produced by PLD into diacylglycerol (DAG). (2) Arf GAP binds to membranes containing DAG. (3) Reduction of PA by PAH decreases the activity of PI 5-kinase. Diffusion of OCRL, a PI 5-phosphatase, reduces the local concentration of PI4,5P$_2$. (4) Arf GAP stimulates hydrolysis of GTP on Arf, returning Arf to the basal state of low-affinity binding to membranes.

of the coat, membrane curvature, and membrane fission can be self-contained events. However, just as polypeptides can fold into native proteins *in vitro* in the absence of accessory factors, but do not do so in the cell, it is likely that the mechanisms that determine the location and rate of coated vesicle formation require additional layers of regulation *in vivo*. In fact *SEC16*, an essential gene, encodes a protein not required

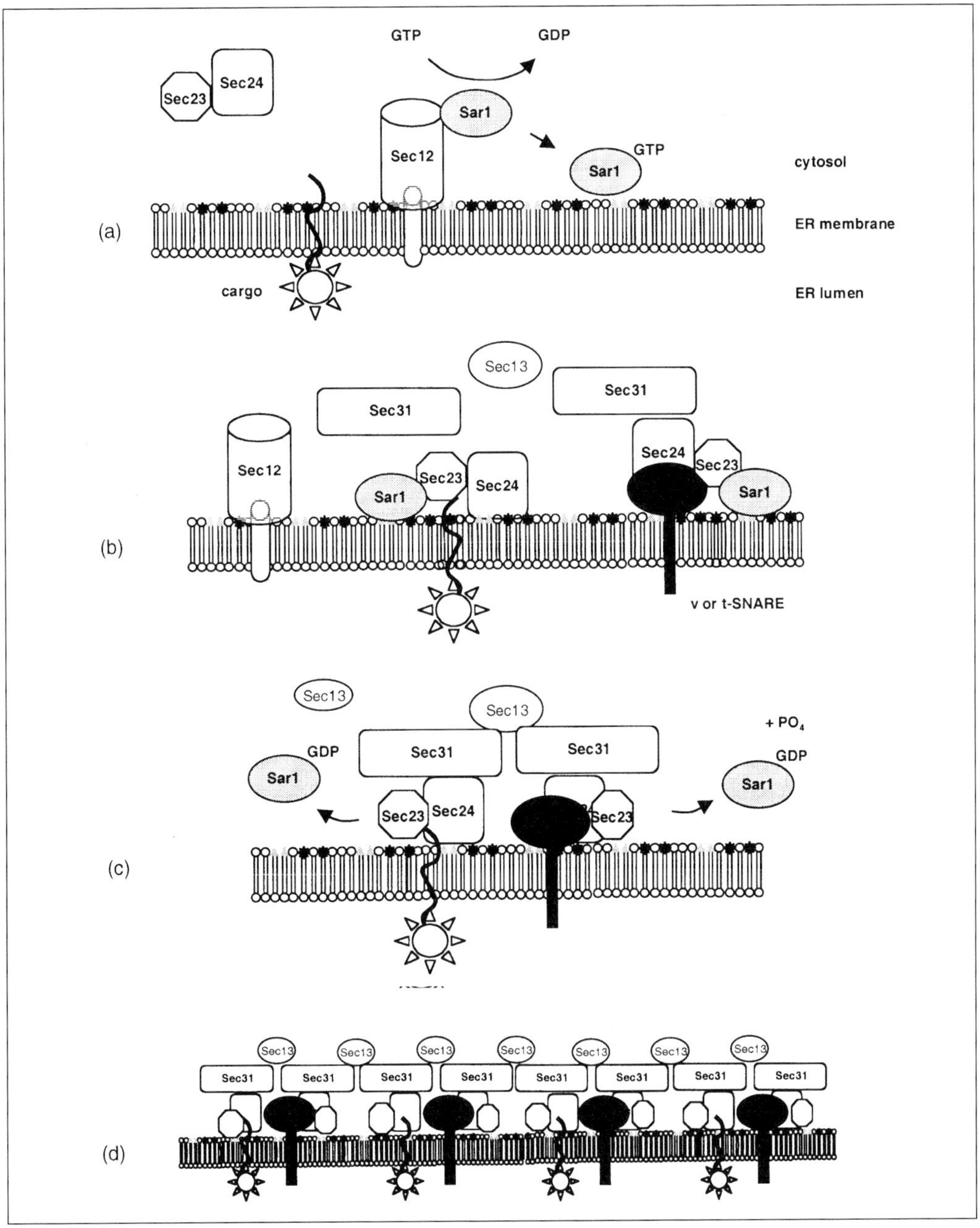

Fig. 6 Regulation of the formation of COPII coats on ER membranes. (a) Sec12p, an integral membrane protein of the ER binds to soluble Sar1p and stimulates the exchange of GDP for GTP on Sar1p. When bound to GTP, Sar1p binds to the ER membrane. (b) Sar1p-GTP recruits a complex of two of the COPII proteins, Sec23p and Sec24p, to the ER membrane. Sec23p and Sec24p bind to cargo and v-SNARE proteins. (c) Sec23p stimulates hydrolysis of GTP on Sar1p, allowing Sar1p to be released from membranes. Two other COPII proteins, Sec31p and Sec13p bind to Sec23p-Sec24p. (d) Repeated rounds of steps (a)-(c) allow COPII subunits to multimerize into a cytosolic protein coat that captures vesicle components and cargo.

for the formation of COPII vesicles *in vitro*, but which is required for packaging certain cargo into the vesicle (113–115). Obviously formation of the coated vesicle must be linked to mechanisms ensuring that the correct cargo is incorporated at the correct location.

10. Conclusion

In the 14 years since the discovery of the first Arf, much has been learned about the structure of the protein, the cellular location of class I and class III Arfs, and the biological pathways in which they engage. The recent discovery of the multidomain structure of Arf GEFs and GAPs indicates that both the mechanisms controlling the GTPase cycle on Arf, and the functions of that cycle, are likely to be more complicated than was previously suspected. Major questions about the functions of Arfs remain. What is the biological role of the class II Arfs? Are the Arf GAP proteins effectors of Arf function and what really controls the cellular locations where Arf is active? What is the relationship between changes in cell cytoskeleton and changes in endosomal membranes controlled by Arf6? What is the role of PLD in membrane traffic? These and other fascinating questions will stimulate future research on these proteins.

References

1. Lee, F. J. *et al.* (1994) Characterization of class II and class III ADP-ribosylation factor genes and proteins in *Drosophila melanogaster*. *J. Biol. Chem.*, **269**, 21555.
2. Stearns, T., Kahn, R. A., Botstein, D., and Hoyt, M. A. (1990) ADP ribosylation factor is an essential protein in *Saccharomyces cerevisiae* and is encoded by two genes. *Mol. Cell Biol.*, **10**, 6690.
3. Brown, H. A. *et al.* (1993) ADP-ribosylation factor, a small GTP-dependent regulatory protein, stimulates phospholipase D activity [see comments]. *Cell*, **75**, 1137.
4. Cockcroft, S. *et al.* (1994) Phospholipase D: a downstream effector of Arf in granulocytes. *Science*, **263**, 523.
5. Radhakrishna, H., Klausner, R. D., and Donaldson, J. G. (1996) Aluminum fluoride stimulates surface protrusions in cells overexpressing the Arf6 GTPase. *J. Cell. Biol.*, **134**, 935.
6. Nakano, A. and Muramatsu, M. (1989) A novel GTP-binding protein, Sar1p, is involved in transport from the endoplasmic reticulum to the Golgi apparatus. *J. Cell. Biol.*, **109**, 2677.
7. Schekman, R. and Orci, L. (1996) Coat proteins and vesicle budding. *Science*, **271**, 1526.
8. Kahn, R. A. and Gilman, A. G. (1984) Purification of a protein cofactor required for ADP-ribosylation of the stimulatory regulatory component of adenylate cyclase by cholera toxin. *J. Biol. Chem.*, **259**, 6228.
9. Kahn, R. A. and Gilman, A. G. (1986) The protein cofactor necessary for ADP-ribosylation of Gs by cholera toxin is itself a GTP binding protein. *J. Biol. Chem.*, **261**, 7906.
10. Moss, J. and Vaughan, M. (1998) Molecules in the Arf orbit. *J. Biol. Chem.*, **273**, 21431.
11. Clark, J. *et al.* (1993) Selective amplification of additional members of the ADP-ribosylation factor (Arf) family: cloning of additional human and *Drosophila* Arf-like genes. *Proc. Natl Acad. Sci., USA*, **90**, 8952.

12. Tsuchiya, M. *et al.* (1991) Molecular identification of ADP-ribosylation factor mRNAs and their expression in mammalian cells. *J. Biol. Chem.*, **266**, 2772.

13. Hosaka, M. *et al.* (1996) Structure and intracellular localization of mouse ADP-ribosylation factors type 1 to type 6 (Arf1-Arf6). *J. Biochem.*, **120**, 813.

14. Radhakrishna, H. and Donaldson, J. G. (1997) ADP-ribosylation factor 6 regulates a novel plasma membrane recycling pathway. *J. Cel. Biol.*, **139**, 49.

15. D'Souza-Schorey, C., Li, G., Colombo, M. I., and Stahl, P. D. (1995) A regulatory role for Arf6 in receptor-mediated endocytosis. *Science*, **267**, 1175.

16. Beraud-Dufour, S. *et al.* (1998) A glutamic finger in the guanine nucleotide exchange factor ARNO displaces Mg^{2+} and the beta-phosphate to destabilize GDP on Arf1. *EMBO J.*, **17**, 3651.

17. Franco, M., Paris, S., and Chabre, M. (1995) The small G-protein ARF1GDP binds to the Gt beta gamma subunit of transducin, but not to Gt alpha GDP–Gt beta gamma. *FEBS Lett.*, **362**, 286.

18. Colombo, M. I. *et al.* (1995) Heterotrimeric G proteins interact with the small Gtpase Arf—possibilities for the regulation of vesicular traffic. *J. Biol. Chem.*, **270**, 24564.

19. Roth, M. G. and Sternweis, P. C. (1997) The role of lipid signaling in constitutive membrane traffic. *Curr. Opinion Cell Biol.*, **9**, 519.

20. Roth, M. G. (1999) Lipid regulators of constitutive membrane traffic. *Trends Cell Biol.*, **9**, 174.

21. Kanoh, H., Williger, B. T., and Exton, J. H. (1997) Arfaptin 1, a putative cytosolic target protein of ADP-ribosylation factor, is recruited to Golgi membranes. *J. Biol. Chem.*, **272**, 5421.

22. Tsai, S. C. *et al.* (1998) Effects of Arfaptin 1 on guanine nucleotide-dependent activation of phospholipase D and cholera toxin by ADP-ribosylation factor. *J. Biol. Chem.*, **273**, 20697.

23. Stearns, T., Willingham, M. C., Botstein, D., and Kahn, R. A. (1990) ADP-ribosylation factor is functionally and physically associated with the Golgi complex. *Proc. Natl Acad. Sci., USA*, **87**, 1238.

24. Gaynor, E. C., Graham, T. R., and Emr, S. D. (1998) COPI in ER/Golgi and intra-Golgi transport—do yeast COPI mutants point the way? *Biochim. Biophys. Acta.*, **1404**, 33.

25. Zhang, C. J. *et al.* (1994) Expression of a dominant allele of human Arf1 inhibits membrane traffic *in vivo*. *J. Cell Biol.*, **124**, 289.

26. Dascher, C. and Balch, W. E. (1994) Dominant inhibitory mutants of Arf1 block endoplasmic reticulum to Golgi transport and trigger disassembly of the Golgi apparatus. *J. Biol. Chem.*, **269**, 1437.

27. D'Souza-Schorey, C. *et al.* (1998) Arf6 targets recycling vesicles to the plasma membrane: insights from an ultrastructural investigation. *J. Cell Biol.*, **140**, 603.

28. Godi, A. *et al.* (1998) ADP ribosylation factor regulates spectrin binding to the Golgi complex. *Proc. Natl Acad. Sci., USA*, **95**, 8607.

29. Serafini, T. *et al.* (1991) ADP-ribosylation factor is a subunit of the coat of Golgi-derived COP-coated vesicles: a novel role for a GTP-binding protein. *Cell*, **67**, 239.

30. Spang, A. *et al.* (1998) Coatomer, Arf1p, and nucleotide are required to bud COPI-coated vesicles from large synthetic liposomes. *Proc. Natl Acad. Sci., USA*, **95**, 11199.

31. Gaynor, E. C. and Emr, S. D. (1997) COPI-independent anterograde transport: cargo-selective ER to Golgi protein transport in yeast COPI mutants. *J. Cell. Biol.*, **136**, 789.

32. Cosson, P. *et al.* (1996) Delta- and zeta-COP, two coatomer subunits homologous to clathrin-associated proteins, are involved in ER retrieval. *EMBO J.*, **15**, 1792.

33. Letourneur, F. *et al.* (1994) Coatomer is essential for retrieval of dilysine-tagged proteins to the endoplasmic reticulum. *Cell*, **79**, 1199.

34. Traub, L. M., Ostrom, J. A., and Kornfeld, S. (1993) Biochemical dissection of AP-1 recruitment onto Golgi membranes. *J. Cell. Biol.*, **123**, 561.

35. Dittie, A. S., Hajibagheri, N., and Tooze, S. A. (1996) The AP-1 adaptor complex binds to immature secretory granules from PC12 cells, and is regulated by ADP-ribosylation factor. *J. Cell Biol.*, **132**, 523.

36. Faundez, V., Horng, J. T., and Kelly, R. B. (1998) A function for the AP3 coat complex in synaptic vesicle formation from endosomes. *Cell*, **93**, 423.

37. Ooi, C. E., Dell'Angelica, E. C., and Bonifacino, J. S. (1998) ADP-Ribosylation factor 1 (Arf1) regulates recruitment of the AP-3 adaptor complex to membranes. *J. Cell Biol.*, **142**, 391.

38. Lenhard, J. M., Kahn, R. A., and Stahl, P. D. (1992) Evidence for ADP-ribosylation factor (Arf) as a regulator of *in vitro* endosome-endosome fusion. *J. Biol. Chem.*, **267**, 13047.

39. West, M. A., Bright, N. A., and Robinson, M. S. (1997) The role of ADP-ribosylation factor and phospholipase D in adaptor recruitment. *J. Cell. Biol.*, **138**, 1239.

40. Traub, L. M. *et al.* (1996) AP-2-containing clathrin coats assemble on mature lysosomes. *J. Cell. Biol.*, **135**, 1801.

41. Barr, F. A. and Huttner, W. B. (1996) A role for ADP-ribosylation factor 1, but not COP I, in secretory vesicle biogenesis from the trans-Golgi network. *FEBS Lett.*, **384**, 65.

42. Chen, Y. G. *et al.* (1997) Phospholipase D stimulates release of nascent secretory vesicles from the trans-Golgi network. *J. Cell Biol.*, **138**, 495.

43. Novick, P. and Zerial, M. (1997) The diversity of Rab proteins in vesicle transport. *Curr. Opinion Cell Biol.*, **9**, 496.

44. Schimmoller, F., Simon, I., and Pfeffer, S. R. (1998) Rab GTPases, directors of vesicle docking. *J. Biol. Chem.*, **273**, 22161.

45. Shome, K., Nie, Y. M., and Romero, G. (1998) ADP-ribosylation factor proteins mediate agonist-induced activation of phospholipase D. *J. Biol. Chem.*, **273**, 30836.

46. Brown, M. T. *et al.* (1998) ASAP1, a phospholipid-dependent Arf GTPase-activating protein that associates with and is phosphorylated by Src. *Mol. Cell. Biol.*, **18**, 7038.

47. Premont, R. T. *et al.* (1998) beta2-adrenergic receptor regulation by GIT1, a G protein-coupled receptor kinase-associated ADP ribosylation factor GTPase-activating protein [In Process Citation]. *Proc. Natl Acad. Sci., USA*, **95**, 14082.

48. Kahn, R. A. *et al.* (1995) Mutational analysis of *Saccharomyces cerevisiae* Arf1. *J. Biol. Chem.*, **270**, 143.

49. Gaynor, E. C., Chen, C. Y., Emr, S. D., and Graham, T. R. (1998) Arf is required for maintenance of yeast Golgi and endosome structure and function. *Mol. Biol. Cell.*, **9**, 653.

50. Chen, C. Y. and Graham, T. R. (1998) An Arf1-delta synthetic lethal screen identifies a new clathrin heavy chain conditional allele that perturbs vacuolar protein transport in *Saccharomyces cerevisiae*. *Genetics*, **150**, 577.

51. Payne, G. S. (1990) Genetic analysis of clathrin function in yeast. *J. Membr. Biol.*, **116**, 93.

52. Graham, T. R. and Krasnov, V. A. (1995) Sorting of yeast alpha-1,3 mannosyltransferase is mediated by a lumenal domain interaction, and a transmembrane domain signal that can confer clathrin-dependent Golgi localization to a secreted protein. *Mol. Biol. Cell.*, **6**, 809.

53. Zhang, C. J., Cavenagh, M. M., and Kahn, R. A. (1998) A family of Arf effectors defined as suppressors of the loss of Arf function in the yeast *Saccharomyces cerevisiae*. *J. Biol. Chem.*, **273**, 19792.

54. Hart, M. J. *et al.* (1998) Direct stimulation of the guanine nucleotide exchange activity of p115 RhoGEF by Galpha13. *Science*, **280**, 2112.

55. Kozasa, T. *et al.* (1998) p115 RhoGEF, a GTPase activating protein for Galpha12 and Galpha13. *Science*, **280**, 2109.

56. Rudge, S. A. *et al.* (1998) Adp-Ribosylation factors do not activate yeast phospholipase Ds but are required for sporulation. *Mol. Biol. Cell*, **9**, 2025.

57. Teal, S. B. *et al.* (1994) An activating mutation in Arf1 stabilizes coatomer binding to Golgi membranes. *J. Biol. Chem.*, **269**, 3135.

58. Aridor, M., Bannykh, S. I., Rowe, T., and Balch, W. E. (1995) Sequential coupling between COPII and COPI vesicle coats in endoplasmic reticulum to Golgi transport. *J. Cell. Biol.*, **131**, 875.

59. Rothman, J. E. and Wieland, F. T. (1996) Protein sorting by transport vesicles. *Science*, **272**, 227.

60. Rothman, J. (1994) Mechanisms of intracellular protein transport. *Nature*, **372**, 55.

61. Peters, P. J. *et al.* (1995) Overexpression of wild-type and mutant Arf1 and Arf6: distinct perturbations of nonoverlapping membrane compartments. *J. Cell. Biol.*, **128**, 1003.

62. Donaldson, J. G., Finazzi, D., and Klausner, R. D. (1992) Brefeldin A inhibits Golgi membrane-catalysed exchange of guanine nucleotide onto Arf protein. *Nature*, **360**, 350.

63. Helms, J. B. and Rothman, J. E. (1992) Inhibition by brefeldin A of a Golgi membrane enzyme that catalyses exchange of guanine nucleotide bound to Arf. *Nature*, **360**, 352.

64. Randazzo, P. A., Yang, Y. C., Rulka, C., and Kahn, R. A. (1993) Activation of ADP-ribosylation factor by Golgi membranes. Evidence for a brefeldin A- and protease-sensitive activating factor on Golgi membranes. *J. Biol. Chem.*, **268**, 9555.

65. Tsai, S. C. *et al.* (1993) Effects of brefeldin A and accessory proteins on association of ADP-ribosylation factors 1, 3, and 5 with Golgi. *J. Biol. Chem.*, **268**, 10820.

66. Lippincott-Schwartz, J. *et al.* (1991) Brefeldin A's effects on endosomes, lysosomes, and the TGN suggest a general mechanism for regulating organelle structure and membrane traffic. *Cell*, **67**, 601.

67. Lippincott-Schwartz, J., Yuan, L. C., Bonifacino, J. S., and Klausner, R. D. (1989) Rapid redistribution of Golgi proteins into the ER in cells treated with brefeldin A: evidence for membrane cycling from Golgi to ER. *Cell*, **56**, 801.

68. Ahmadian, M. R., Mittal, R., Hall, A., and Wittinghofer, A. (1997) Aluminum fluoride associates with the small guanine nucleotide binding proteins. *FEBS Lett.*, **408**, 315.

69. Finazzi, D., Cassel, D., Donaldson, J. G., and Klausner, R. D. (1994) Aluminum fluoride acts on the reversibility of Arf1-dependent coat protein binding to Golgi membranes. *J. Biol. Chem.*, **269**, 13325.

70. Balch, W. E., Kahn, R. A., and Schwaninger, R. (1992) ADP-ribosylation factor is required for vesicular trafficking between the endoplasmic reticulum and the cis-Golgi compartment. *J. Biol. Chem.*, **267**, 13053.

71. Peter, F. *et al.* (1993) Beta-COP is essential for transport of protein from the endoplasmic reticulum to the Golgi *in vitro*. *J. Cell Biol.*, **122**, 1155.

72. Pepperkok, R. *et al.* (1993) Beta-COP is essential for biosynthetic membrane transport from the endoplasmic reticulum to the Golgi complex *in vivo*. *Cell*, **74**, 71.

73. Palmer, D. J. *et al.* (1993) Binding of coatomer to Golgi membranes requires ADP-ribosylation factor. *J. Biol. Chem.*, **268**, 12083.

74. Seaman, M. N. J., Sowerby, P. J., and Robinson, M. S. (1996) Cytosolic and membrane-associated proteins involved in the recruitment of AP-1 adaptors onto the trans-Golgi network. *J. Biol. Chem.*, **271**, 25446.

75. Ostermann, J. *et al.* (1993) Stepwise assembly of functionally active transport vesicles. *Cell*, **75**, 1015.

76. Ktistakis, N. T. *et al.* (1996) Evidence that phospholipase D mediates ADP ribosylation factor-dependent formation of Golgi coated vesicles. *J. Cell. Biol.*, **134**, 295.

77. Chen, Y. G. and Shields, D. (1996) ADP-ribosylation factor-1 stimulates formation of nascent secretory vesicles from the trans-Golgi network of endocrine cells. *J. Biol. Chem.*, **271**, 5297.

78. Zhu, Y., Traub, L. M., and Kornfeld, S. (1998) ADP-ribosylation factor 1 transiently activates high-affinity adaptor protein complex AP-1 binding sites on Golgi membranes. *Mol. Biol. Cell*, **9**, 1323.

79. Chardin, P. *et al.* (1996) A human exchange factor for Arf contains Sec7- and pleckstrin-homology domains. *Nature*, **384**, 481.

80. Peyroche, A., Paris, S., and Jackson, C. L. (1996) Nucleotide exchange on Arf mediated by yeast Gea1 protein. *Nature*, **384**, 479.

81. Morinaga, N., Moss, J., and Vaughan, M. (1997) Cloning and expression of a cDNA encoding a bovine brain brefeldin A-sensitive guanine nucleotide-exchange protein for ADP-ribosylation factor. *Proc. Natl Acad. Sci., USA*, **94**, 12926.

82. Sata, M., Donaldson, J. G., Moss, J., and Vaughan, M. (1998) Brefeldin A-inhibited guanine nucleotide-exchange activity of Sec7 domain from yeast Sec7 with yeast and mammalian ADP ribosylation factors. *Proc. Natl Acad. Sci., USA*, **95**, 4204.

83. Klarlund, J. K. *et al.* (1997) Signaling by phosphoinositide-3,4,5-trisphosphate through proteins containing pleckstrin and Sec7 homology domains. *Science*, **275**, 1927.

84. Pacheco-Rodriguez, G. *et al.* (1998) Guanine nucleotide exchange on ADP-ribosylation factors catalyzed by cytohesin-1 and its Sec7 domain. *J. Biol. Chem.*, **273**, 26543.

85. Klarlund, J. K. *et al.* (1998) Regulation of GRP1-catalyzed ADP ribosylation factor guanine nucleotide exchange by phosphatidylinositol 3,4,5-trisphosphate. *J. Biol. Chem.*, **273**, 1859.

86. Frank, S., Upender, S., Hansen, S. H., and Casanova, J. E. (1998) ARNO is a guanine nucleotide exchange factor for ADP-ribosylation factor 6. *J. Biol. Chem.*, **273**, 23.

87. Frank, S. R., Hatfield, J. C., and Casanova, J. E. (1998) Remodeling of the actin cytoskeleton is coordinately regulated by protein kinase C and the ADP-ribosylation factor nucleotide exchange factor ARNO [In Process Citation]. *Mol. Biol. Cell*, **9**, 3133.

88. Franco, M. *et al.* (1998) ARNO3, a Sec7-domain guanine nucleotide exchange factor for ADP ribosylation factor 1, is involved in the control of Golgi structure and function. *Proc. Natl Acad. Sci., USA*, **95**, 9926.

89. Betz, S. F. *et al.* (1998) Solution structure of the cytohesin-1 (B2–1) Sec7 domain and its interaction with the GTPase ADP ribosylation factor 1. *Proc. Natl Acad. Sci., USA*, **95**, 7909.

90. Mossessova, E., Gulbis, J. M., and Goldberg, J. (1998) Structure of the guanine nucleotide exchange factor Sec7 domain of human ARNO and analysis of the interaction with Arf GTPase. *Cell*, **92**, 415.

91. Cherfils, J. *et al.* (1998) Structure of the Sec7 domain of the Arf exchange factor ARNO. *Nature*, **392**, 101.

92. Goldberg, J. (1998) Structural basis for activation of Arf GTPase: mechanisms of guanine nucleotide exchange and GTP-myristoyl switching. *Cell*, **95**, 237.

93. Paris, S. *et al.* (1997) Role of protein–phospholipid interactions in the activation of Arf1 by the guanine nucleotide exchange factor ARNO. *J. Biol. Chem.*, **272**, 22221.

94. Franco, M., Chardin, P., Chabre, M., and Paris, S. (1996) Myristoylation-facilitated binding of the G protein ARF1GDP to membrane phospholipids is required for its activation by a soluble nucleotide exchange factor. *J. Biol. Chem.*, **271**, 1573.

95. Franco, M., Chardin, P., Chabre, M., and Paris, S. (1995) Myristoylation of ADP-ribosylation factor 1 facilitates nucleotide exchange at physiological Mg^{2+} levels. *J. Biol. Chem.*, **270**, 1337.

96. Cukierman, E., Huber, I., Rotman, M., and Cassel, D. (1995) The Arf1 GTPase-activating protein: zinc finger motif and Golgi complex localization. *Science*, **270**, 1999.

97. Poon, P. P. *et al.* (1996) *Saccharomyces cerevisiae* Gcs1 is an ADP-ribosylation factor GTPase-activating protein. *Proc. Natl Acad. Sci., USA*, **93**, 10074.

98. Huber, I. *et al.* (1998) Requirement for both the amino-terminal catalytic domain and a noncatalytic domain for *in vivo* activity of ADP-ribosylation factor GTPase-activating protein. *J. Biol. Chem.*, **273**, 24786.

99. Aoe, T. *et al.* (1997) The KDEL receptor, ERD2, regulates intracellular traffic by recruiting a GTPase-activating protein for Arf1. *EMBO J.*, **16**, 7305.

100. Aoe, T. *et al.* (1998) Modulation of intracellular transport by transported proteins: insight from regulation of COPI-mediated transport. *Proc. Natl Acad. Sci., USA*, **95**, 1624.

101. Antonny, B. *et al.* (1997) Activation of ADP-ribosylation factor 1 GTPase-activating protein by phosphatidylcholine-derived diacylglycerols. *J. Biol. Chem.*, **272**, 30848.

102. Randazzo, P. A. and Kahn, R. A. (1994) GTP hydrolysis by ADP-ribosylation factor is dependent on both an ADP-ribosylation factor GTPase-activating protein and acid phospholipids. *J. Biol. Chem.*, **269**, 10758 [published erratum appears in *J. Biol. Chem.* (1994) **269** (23), 16519].

103. Greasley, S. E. *et al.* (1995) The structure of rat ADP-ribosylation factor-1 (Arf-1) complexed to GDP determined from two different crystal forms. *Nature, Struct. Biol.*, **2**, 797.

104. Amor, J. C., Harrison, D. H., Kahn, R. A., and Ringe, D. (1994) Structure of the human ADP-ribosylation factor 1 complexed with GDP. *Nature*, **372**, 704.

105. Randazzo, P. A. (1997) Functional interaction of ADP-ribosylation factor 1 with phosphatidylinositol 4,5-bisphosphate. *J. Biol. Chem.*, **272**, 7688.

106. Fensome, A. *et al.* (1996) Arf and PITP restore GTP-gamma-S-stimulated protein secretion from cytosol-depleted Hl60 cells by promoting Pip2 synthesis. *Curr. Biology*, **6**, 730.

107. Liscovitch, M. and Cantley, L. C. (1995) Signal transduction and membrane traffic: the PITP/phosphoinositide connection. *Cell*, **81**, 659.

108. Tuscher, O. *et al.* (1997) Cooperativity of phosphatidylinositol transfer protein and phospholipase D in secretory vesicle formation from the TGN-phosphoinositides as a common denominator. *FEBS Lett.*, **419**, 271.

109. Suchy, S. F., Olivos-Glander, I. M., and Nussabaum, R. I. (1995) Lowe syndrome, a deficiency of phosphatidylinositol 4,5-bisphosphate 5-phosphatase in the Golgi apparatus. *Hum. Molec. Genet.*, **4**, 2245.

110. Bi, K. *et al.* (1998) Positive and negative regulation of COPI coat formation by phospholipase D. *Mol. Biol. Cell*, **9**, 467a.

111. Matsuoka, K. *et al.* (1998) COPII-coated vesicle formation reconstituted with purified coat proteins and chemically defined liposomes. *Cell*, **93**, 263.

112. Matsuoka, K., Morimitsu, Y., Uchida, K., and Schekman, R. (1998) Coat assembly directs v-SNARE concentration into synthetic COPII vesicles. *Mol. Cell*, **2**, 703.

113. Shaywitz, D. A., Espenshade, P. J., Gimeno, R. E., and Kaiser, C. A. (1997) COPII subunit interactions in the assembly of the vesicle coat. *J. Biol. Chem.*, **272**, 25413.

114. Salama, N. R., Chuang, J. S., and Schekman, R. W. (1997) Sec31 encodes an essential component of the COPII coat required for transport vesicle budding from the endoplasmic reticulum. *Mol. Biol. Cell*, **8**, 205.

115. Campbell, J. L. and Schekman, R. (1997) Selective packaging of cargo molecules into endoplasmic reticulum-derived COPII vesicles. *Proc. Natl Acad. Sci., USA*, **94**, 837.

7 | Ran

IAN G. MACARA, AMY BROWNAWELL, AND KATIE WELCH

1. Introduction

Ran is a very unusual GTPase. Unlike all other known members of the Ras super-family of small GTP-binding proteins, Ran is not post-translationally modified. Moreover, its function depends critically on a spatial gradient of the GTP-bound form of Ran. This requirement sets it apart from other small GTPases that act in signal transduction pathways or in vesicle fusion and budding.

Ran is concentrated in the nucleoplasm and plays a central role in active nucleo-cytoplasmic transport through the nuclear pore complex (NPC). It may have additional functions, but these remain to be identified. The vectoriality of transport and the accumulation of cargo in the nuclear or cytoplasmic compartments against a concentration gradient appear to be driven by the Ran–GTP gradient across the NPC. In this limited sense, therefore, nuclear transport behaves like a coupled trans-membrane transport system, similar, for example, to the coupling of glucose accumulation to a sodium ion gradient. At a molecular level, however, Ran-mediated transport is unique, and is based on a complex sequence of protein–protein inter-actions that are controlled by the guanine-nucleotide-bound state of the GTPase. Despite the remarkably rapid progress being made in identifying the players in the nuclear transport machinery, we remain ignorant of many of the fundamental aspects of the transport mechanism.

This chapter will discuss the Ran GTPase and its regulatory factors, the binding proteins that link the GTPase to transport of cargo, and a model for the transport mechanism across the NPC.

2. The Ran GTPase

The Ran open reading frame was first identified in a screen for Ras-like genes expressed in a human teratocarcinoma cell line called TC4 (1). It was independently found as a 25 kDa protein associated with the nuclear protein, RCC1 (regulator of chromosome condensation), and renamed Ran (Ras-like nuclear protein) to reflect its predominantly nuclear localization (Fig. 1) (2, 3). The Ran protein is the most abundant small GTPase in the cell, constituting about 0.4% of the total cell protein

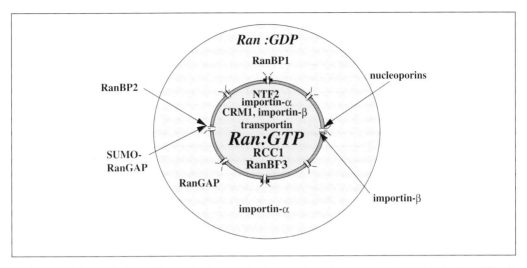

Fig. 1 Subcellular distribution of Ran and other proteins involved in nuclear transport. A unique feature of Ran is the spatial separation of its regulatory factors (RanGAP, RCC1, RanBP1) and of the soluble transport factors. The nucleus and cytosol are separated by a double membrane through which is plugged the nuclear pore complex.

(about 10^7 copies/cell), and is about 80% nuclear. For a mammalian cell, these numbers translate into an intranuclear Ran concentration of 10–20 μM.

Ran genes have been identified in all eukaryotes examined, including fungi, protozoa, higher plants, and metazoa (1, 4–6). The amino-acid sequence conservation is very high (e.g. 88% between human and the budding yeast, *S. cerevisiae*), suggesting that the protein is required for an ancient and essential process. Budding yeast contains two Ran proteins, GSP1 and GSP2, but only GSP1—which is expressed at a level tenfold higher than that of GSP2—is essential for viability. Multiple bands show up on Southern blots of mammalian genomic DNA probed at high stringency with Ran cDNA, but their significance is unclear. There is one report of a testis-specific isoform of Ran in mouse, although it is not known whether the isoform is exclusive to meiotic cells (7).

The crystal structure of Ran in the GDP-bound state is generally similar to that of other small GTPases (8), but the coordination to the Mg^{2+} ion in the guanine nucleotide pocket is different, and the switch 1 region (effector region) of Ran is in a completely different orientation to that found in Ras (see Chapter 9). Overall, the structure suggests that the transition to the GTP-bound state may induce major conformational changes in the protein (9). Uniquely amongst the known small GTPases, the C-terminal region of Ran is believed to undergo such a change. An antipeptide antibody directed against the C-terminal region preferentially recognizes GTP-bound Ran, supporting the idea that exposure of the C terminus is altered in this activated state (10).

The underlying similarities in the structure of the small GTPase family is emphasized by the ability to create constitutively activated or dominant interfering proteins through the substitution of analogous residues in the different family members. Thus dominant interfering mutants of Ran have been created that are resistant to GTP hydrolysis (G19V and Q69L) and which are equivalent to oncogenic Ras mutants (G12V and Q61L). Another dominant interfering Ran mutant (T24N) is equivalent to a Ras mutant, T17N, that sequesters the Ras exchange factor Sos, and blocks activation of endogenous Ras (11–14).

3. Regulating the Ran GTPase cycle

The nucleotide-bound state of Ran is controlled by a guanine nucleotide exchange factor (GEF) and a GTPase activating protein (GAP). Unlike Ras- and Rho-type GTPases, however, for which there are multiple GEFs and GAPs, only a single GEF and a single GAP have been defined unambiguously for Ran.

3.1 RCC1

The mammalian Ran GEF, called RCC1, was initially identified as a protein involved in the regulation of the cell cycle (15). From a set of cell lines with temperature-sensitive cell cycle defects, Nishimoto and colleagues isolated a gene, *RCC1* (regulator of chromosome condensation), that could complement the defect in one of these, tsBN2 (16, 17). *RCC1* isolated from tsBN2 cells contains a point mutation that destabilizes the protein at the non-permissive temperature (18, 19). When tsBN2 cells are switched to the non-permissive temperature during S phase, they undergo premature chromosome condensation, hence the name RCC1 (18, 20). In *Xenopus* oocyte extracts, RCC1 was shown to be required for DNA synthesis (21), but its role in catalysing guanine nucleotide release on Ran remained unsuspected until Ran was isolated by Bischoff and colleagues as an RCC1-associated protein (2, 3, 22).

RCC1 is a nuclear protein of about 45 kDa, that can associate with chromatin and homologues have been found in all eukaryotes examined. In budding yeast the gene is called *PRP20* (or *SRM1* or *MTR1*) (23), in fission yeast, *pim1* (24), and in *Drosophila*, *BJ1* (25). The crystal structure of RCC1 reveals a seven-bladed propellor, each blade consisting of a four-stranded antiparallel β-sheet, reminiscent of the structure of the β subunit of G proteins (26). The RCC1 repeat motif has also been found in several other proteins, including the protein encoded by a gene responsible for retinitis pigmentosa (Rp3) and in p532, a GEF for Arf and Rab GTP-binding proteins, but the significance of this is still unclear.

Mutational analysis suggests that one side of the RCC1 propellor binds Ran and the other binds DNA. Why does a Ran exchange factor need to bind DNA? One interesting possibility is that it permits the re-establishment of a nucleocytoplasmic Ran–GTP gradient after mitosis. The nuclear envelope reforms after M phase by the fusion of vesicles attached to the surface of chromatin. RCC1 attached to the DNA will therefore be retained within the new nucleus and influxing Ran–GDP can be converted to Ran–GTP to power the transport machinery necessary for the establishment of a functional nucleus. Buding yeast, which possesses a closed mitotic cycle, may not require this trick, but the permeability of the nuclear envelope may increase during mitosis and temporarily collapse the Ran gradient. The sequestration of RCC1 during mitosis raises the question as to how RCC1 synthesized during interphase is imported into the nucleus. Although in principle RCC1 may be just small enough to diffuse through the nuclear pores, it seems unlikely that the cell relies on diffusion and a specialized pathway for import may exist.

RCC1 is a potent exchange factor. The half-life for the spontaneous dissociation of GDP from Ran is about 2 hours at 25°C, and this can be accelerated 100 000-fold by RCC1. If the concentration of RCC1 within the nucleus is ~0.5 μM, it can catalyse the exchange of about 1.5×10^6 guanine nucleotides on Ran per second. Assuming one Ran–GTP exchange and hydrolysis is required per transport cycle, and that there are approximately 3000 nuclear pores per cell, then RCC1 is able to cope with ~500 transport events/pore/second. This value provides an upper limit on the rate for all forms of Ran-mediated nuclear import plus export.

3.2 RanGAP

The other half of the regulatory cycle of Ran involves GTP hydrolysis, catalysed by RanGAP, a 65 kDa protein that forms a homodimer (27, 28). The RanGAP gene is also called *RNA1* (*S. cerevisiae*) (29) and *Fug1* (mouse) (30). RanGAP can catalyse the GTP hydrolysis rate on Ran by ~100 000-fold, giving a maximal rate of about 5 per second (31). Since the number of copies of RanGAP per cell (about 3×10^5) is roughly similar to that of RCC1, the GTP hydrolysis capacity appears to nicely balance the exchange capacity.

Although the RanGAP of budding yeast is predominantly cytosolic (32), immuno-localization showed mammalian RanGAP to be concentrated at the nuclear envelope In fact, an anomolously large form of RanGAP (90 kDa) was found to associate with RanBP2, a giant nucleoporin present on the cytoplasmic face of the nuclear pores (Fig. 1). This puzzle was solved by peptide sequencing, which revealed a ubiquitin-like modification of the pore-bound RanGAP (33, 34). The post-translational modification by SUMO-1 (small ubiquitin-related modifier)—also called GMP1 (GAP modifier protein 1)—requires ATP and is mediated by Ubc9, which can also associate with RanBP2 (35–37). SUMO-1 is coupled through an isopeptide link to Lys526 of RanGAP (38).

What is the purpose of this modification? Antibody binding to the pore-associated RanGAP can partially block nuclear import, and since this inhibition cannot be overcome by addition of cytosolic RanGAP, this implies an important function for the modified GAP. However, an alternative explanation for this result is that the antibodies may simply occlude the pore. One likely function of the modification is to ensure rapid hydrolysis of Ran–GTP exiting the pore, so as to increase the efficiency of the transport process. Whatever the case, the SUMO-ation cannot be an absolute requirement for transport, because yeast nuclear pores contain no homologue of RanBP2 and yeast RanGAP is not modified.

What maintains the cytosolic location of RanGAP? In interphase cells, the size of the homodimer is sufficient to prevent diffusion through the pores, but after mitosis, segregation of the RanGAP from the newly created nuclear compartment must occur. Receptor-mediated nuclear export may accomplish this requirement. The yeast RanGAP, RNA1p, has recently been demonstrated to possess several functional nuclear localization signals (NLSs) and nuclear export signals (NESs), and to undergo CRM1-dependent export (39). Similar sequences are present in mammalian RanGAP, but have not yet been shown to be functional (35).

No direct regulation of RCC1 or RanGAP activities has been observed, but the accessibility of Ran to these factors is stringently controlled by other binding proteins, particularly the importin-β family and RanBP1/2. Importin-β and its relatives bind with high affinity to Ran–GTP, blocking RanGAP activity. RanBP1 is a small cytosolic protein, which is described in more detail below, and possesses the unusual ability to function as a co-activator of RanGAP. Moreover, it can reverse the effects of the importin-βs, allowing hydrolysis of the Ran–GTP to proceed. This effect is essential to prevent the cytosolic accumulation of importin-β/Ran–GTP complexes, which would inhibit nuclear transport since they cannot bind cargo.

4. The importin-β family of Ran binding proteins

Importin-β (also called karyopherin-β, p97, and PTAC) was initially isolated as a soluble factor required for NLS-mediated nuclear protein import (40–43). More recently an extended family of proteins related to importin-β has been identified (44, 45). All of these proteins are believed to function as receptors for cargo that transits the NPC. Importin-β binds NLS-cargo via an adaptor protein called importin-α. There are five known isoforms of importin-α in mammalian cells; they have been given a bewildering variety of names, including karyopherin-α1–4, RCH1 (or hSRP1), SRP1–4 (or NPI-1), and Qip1. A more rational nomenclature has been suggested based on phylogeny (46).

Importin-β is a complex, multidomain protein (Fig. 2) with a rich biochemistry. Recent work has shown that in addition to NLS-cargo it mediates the nuclear import of several ribosomal proteins that interact directly with importin-β rather than through an adaptor (47). Thus importin-β is a multifunctional transport receptor, with separate domains for each type of cargo. This added level of complexity may be common to the entire importin-β family.

Importin-α binding requires an intact C terminus on importin-β, while Ran–GTP binds to the N-terminal region of importin-β (48, 49). The interaction with Ran–GTP triggers the dissociation of importin-α (along with any associated cargo) (50–52). Ran–GTP also dissociates ribosomal proteins from importin-β, which bind to a region from about residue 286 to 462 (47) (Fig. 2).

The epitopes on Ran through which it associates with importin-β are separate and distinct from those on Ran that interact with RanBP1 and the three proteins can together form a ternary complex (53). Remarkably, although Ran in the GDP-bound state has negligible affinity for either RanBP1 or importin-β, it is none the less capable of forming a ternary complex with these proteins. The site on importin-β for Ran–GDP/RanBP1 extends further towards the C terminus than that for Ran–GTP

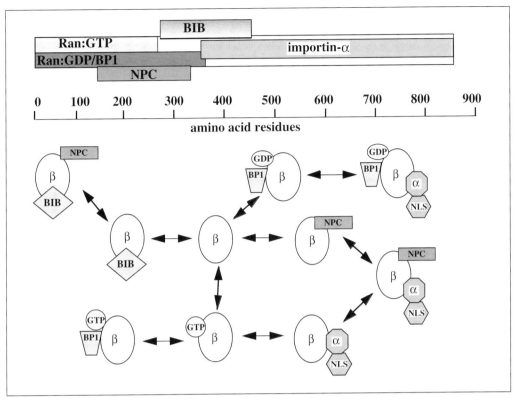

Fig. 2 Domain structure and protein-protein interactions of importin-β. BIB is the basic importin-β binding domain of ribosomal proteins; NPC is the nuclear pore complex; BP1 is RanBP1. Complexes that can be formed *in vitro* using recombinant proteins are also shown. GTP is Ran-GTP; GDP is Ran-GDP; NLS is nuclear cargo. Ran-GTP, but not Ran-GDP/RanBP1, can displace importin-α, BIB-containing proteins, and NPC from importin-β.

and does overlap to some extent with the ribosomal protein and importin-α binding domains, but the RanBP1/Ran–GDP interaction does not disrupt the association with importin-α (48).

Additionally, importin-β possesses a domain that interacts with the nuclear pore complex. This site lies between residues 152 and 352, and the interaction can be disrupted by Ran–GTP binding (42, 49, 54). What components of the NPC comprise this docking site? In nitrocellulose overlay assays, importin-β can bind multiple nucleoporins, particularly those that contain multiple FXFG sequence motifs. Using recombinant yeast proteins, this association was shown to be specific, in that importin-β did not bind nucleoporins containing a different type of repeat (GLFG) (50). But others, using proteins carefully tested for functional competence, were unable to demonstrate *in vitro* binding to the FXFG repeat domain of the p62 nucleoporin (55). A separate study on *S. cerevisiae* provided evidence that yeast importin-β (Kap95) could bind to a GLFG-type nucleoporin (56). Finally, immunoprecipitations from *Xenopus* nuclear envelopes showed that importin-β can associate with

Nup153—a nucleoporin on the internal 'basket' structure of the pore complex—and with Tpr—a non-FXFG protein that is a component of nuclear filaments extending away from the pore (57). While important, these proteins are unlikely to constitute the major docking site for importin-β in intact cells and cannot constitute the sites with which importin-β interacts during transit of the pore. Clearly, much more work will be required to make sense of the interaction of importin-β with the NPC.

The characterization of importin-β was followed by the discovery of a related Ran-binding protein, called transportin (or karyopherin-β2) which is responsible for the import and export of hnRNP1A (58). Transportin interacts directly with hnRNP1A, through a signal sequence called M9 that bears no resemblance to the classical NLS (59). Interestingly, transportin also functions as an import receptor for hnRNP F and perhaps for hRNP D and E, proteins that contain no obvious sequences related to the M9 domain (60).

A measure of the complexity of nuclear transport has been provided by the *S. cerevisiae* genome project, which has revealed a total of 14 open reading frames related to importin-β. Karyopherin β2 appears to be the yeast homologue of transportin (61), while Yrb4p and Pse1p/Kap121 (karyopherins β3 and β4) are likely involved in the import of ribosomal proteins (62). Mtr10p is an importin-β family member that is involved in mRNA export (63) and Kap104p returns mRNA-binding proteins to the nucleus (64). Searches of human EST databases have likewise led to the discovery of a large importin-like family of mammalian proteins. Several of these family members, called RanBP5–RanBP8, have been shown to bind Ran–GTP (44). All are presumed to function as receptors for nuclear cargo. Transportin, importin-β, RanBP5, and RanBP7 can all bind a basic domain (the BIB domain) present in several ribosomal proteins and can function to transport these proteins into the nucleus (47). However, it is quite likely that RanBP5–8, like transportin and importin-β, recognize other, unrelated nuclear targetting signals. Perhaps an archetypal importin that could bind to and transport ribosomal proteins diverged and acquired novel cargo binding cassettes through evolution, as the complexity of the eukaryotic cell increased.

Another group of importin-β family members is responsible for various types of nuclear export. Crm1 (also called exportin1) is required in both yeast and mammalian cells for export of cargo tagged with a leucine-rich nuclear export signal (NES) (65–67). Los1p (in yeast) and exportin-t are responsible for tRNA export (68, 69). Cse1p and CAS are involved in the re-export of importin-α (70). All of these proteins can bind Ran–GTP and all likely possess a similar overall domain organization, with the Ran-binding region being close to the N terminus and the cargo-binding site being C terminal. It remains possible, however, that there are other receptors, structurally unrelated to the importin-β family.

5. The RanBD family of Ran binding proteins

A small, but interesting, family of proteins shares a conserved motif of about 135 amino-acid residues, termed the Ran binding domain (RanBD) (71).

5.1 RanBP1

The best characterized member of this family is RanBP1 (72, 73). Other names include HTF9A in mouse (74), sbp1 in fission yeast (75), and Yrb1p (also CST20) in baker's yeast (76). RanBP1 is a predominantly cytosolic protein containing a single RanBD and a nuclear export signal (NES) and a cytoplasmic retention signal at the carboxy terminus (Fig. 3) (77). RanBP1 binds tightly to Ran–GTP but only weakly to Ran–GDP (K_d ~10 μM) (78). Interestingly, the high-affinity interaction requires the acidic, C-terminal tail of Ran (DEDDDL) (10, 78, 79).

RanBP1 clearly plays an important function in nuclear transport. The depletion of RanBP1 from cytosolic extracts partially inhibits protein nuclear import *in vitro* (80). Temperature-sensitive mutants of Yrb1p are defective in nuclear protein import and mRNA export (81); and deletion of *YRB1* is lethal (81, 82). The injection of RanBP1 into *Xenopus* oocyte nuclei leads to defects in mRNA, tRNA, and snRNA export, while not inhibiting nuclear import (83). However, RanBP1 may possess additional functions. In yeast, the overexpression of Yrb1p leads to chromosome instability and sensitivity to the microtubule-depolymerizing drug, benomyl, implying that it

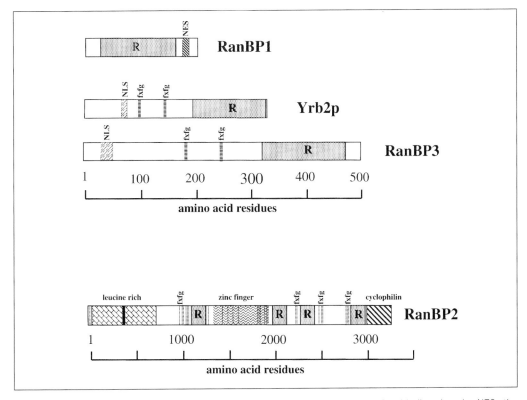

Fig. 3 Doman structures of the RanBD class of Ran-binding proteins. R is the Ran-binding domain; NES, the nuclear export signal; NLS, the nuclear import signal; fxfg, the nucleoporin signature FXFG sequence motif. Note the difference in scale between RanBP2 and the other proteins.

causes a defect in mitotic spindle formation (82). In mouse cells, RanBP1 mRNA expression is repressed in quiescent cells, transcriptionally activated at G_1/S, and peaks in S phase (84). Furthermore the deregulation of RanBP1 expression can cause cell cycle defects (85). None the less, it remains conceivable that these effects are secondary to perturbations in nuclear transport.

5.2 RanBP2

A second member of this family is called RanBP2 or Nup358 (86, 87). No similar gene is present within *S. cerevisiae*, and though of undoubted importance, it cannot be essential to the core functions of Ran in nuclear transport. RanBP2 is a giant protein associated with the cytoplasmic fibrils of the nuclear pores (86–88). It has a complex domain structure (Fig. 3). Spaced along its length are four RanBDs, each very similar to that of RanBP1. Additionally, RanBP2 contains 26 FXFG motifs plus an amino-terminal leucine zipper, zinc-finger repeats, a region homologous to Nup153 (down-stream of the zinc fingers), and a carboxy terminus that is similar to cyclophilin. The function of these domains is unclear, although it is tempting to speculate that the cyclophilin domain might refold large proteins that have traversed the nuclear pores in a partially unfolded state. In support of this hypothesis, the fourth RanBD and the cyclophilin domain are involved in the assembly of functional opsin molecules of the red/green sensitive visual pigment (89). The RanBD binds to opsin and the cyclophilin domain enhances stability of the complex, most likely through a *cis–trans* isomerization. Thus the carboxy terminus of RanBP2 can act as a chaperone and foldase for red/green opsin, and probably for many other poteins. The technical difficulties involved in analysing the function of nuclear pore proteins, and the inability to use the power of yeast genetics, will make functional studies particularly hard. However, a simplified version of RanBP2, called Ranup96, exists in the nematode *C. elegans*, with two RanBDs and several FXFG repeats, which would be accessible to genetics (71). It is likely that a similar gene is also present in *Drosophila*, in which eye-specific promoters might be used to analyse function within the retinal cells without the complications that would arise if mutations in the DRanBP2 prove to be embryonic lethals.

The human *RanBP2* gene is on chromosome 2q11–q13 (90). A related gene, *RanBP2L1*, has been identified within the same chromosomal region, suggesting that a gene cluster of RanBP2s may exist (91). Although a full-length clone encoding *RanBP2L1* has not been isolated, the protein appears to differ from RanBP2 primarily at the carboxy terminus, where it lacks the cyclophilin homology domain.

RanBP1 and RanBP2 can participate in a rich biochemistry with Ran. Perhaps most importantly the RanBD, though inactive alone, can increase by several-fold the ability of RanGAP to catalyse GTP hydrolysis on Ran (28, 73). The mechanism of this co-activation is not yet understood, but likely works through a change in affinity for the RanGAP, rather than through changes in the catalytic mechanism. Additionally, as described above, mammalian RanGAP is modified by conjugation of SUMO-1. The SUMO-ated RanGAP binds to RanBP2, which retains RanGAP on the cytoplasmic side of the nuclear pore. This localization may act to increase nuclear transport efficiency by ensuring a steep Ran–GTP gradient across the pore. However, it could also ensure that a pool of RanGAP is present on the cytoplasmic face of the pores after mitosis, when the gradient must be quickly re-established to permit sorting of all of the nuclear and cytosolic proteins that were mixed after disassembly of the nuclear envelope.

A second property of the RanBDs is to inhibit the ability of RCC1 to dissociate GTP from Ran (71, 92), though during interphase this property would seem to be irrelevant since the proteins are on opposite sides of the nuclear envelope. However, it may be of value during mitosis when the contents of the two compartments become mixed.

RanBDs can also stabilize the Ran–importin-β interaction. Ran–GTP, importin-β. and RanBP1 can bind separately or together in a heterotrimeric complex (93). The sites of interaction on Ran are distinct—and separate from the RanGAP-binding site—and importin-β associates more avidly to a C-terminal deletion mutant of Ran–GTP than with the wild-type protein. The inhibition of RanGAP activity by importin-β can be partially rescued by RanBDs (53) and this effect is enhanced by importin-α, which sequesters the importin-β as it is released from Ran–GTP (94, 95). Similarly, RanBP1 binding to other importin-β family members is required to trigger their dissociation from Ran (44, 69, 70, 96). This permissive role for RanGAP may indeed be the primary function for RanBP1, accounting for its essentiality in yeast: in the absence of RanBP1, import and export complexes would accumulate in dead-end complexes with Ran–GTP, bringing nuclear transport to a grinding halt.

As described above, RanBP1, importin-β, and Ran–GDP can together form a stable heterotrimeric complex (Fig. 2) (48). It will be instructive to study the three-dimensional structure of this complex, as compared to that with Ran–GTP, because an important conformational change in the the heterotrimer undoubtedly accompanies the change in nucleotide state of the Ran. An indication of this change is that whereas Ran–GTP binding to importin-β triggers the dissociation of importin-α (with or without an NLS cargo), Ran–GDP/RanBP1 does not, even though the interaction sites on importin-β for Ran–GDP/RanBP1 and for importin-α overlap.

5.3 RanBP3

All eukaryotic organisms examined contain a third member of the RanBD family, called RanBP3 (in mammals) or Yrb2p (budding yeast). The RanBD of these proteins lacks a substantial fraction of residues that are conserved in RanBP1 and RanBP2, particularly at the carboxy terminus, so that the inter-species similarity in RanBP3s is greater than that between RanBP1 and RanBP3 within any single species (81, 97). This pattern suggests that RanBP3 possesses an evolutionarily conserved function distinct from that of the other family members. Another yeast protein, a nucleoporin called Nup2, also contains a partial RanBD similar to that of Yrb2p and RanBP3, but lacks even more of the consensus domain. It is not clear whether Nup2 can bind Ran directly. Although the proteins give a positive yeast two-hybrid response (98), the interaction might well be indirect.

Yrb2p is a 36 kDa nuclear protein with two putative NLSs and two FXFG motifs at its amino terminus (Fig. 3) (81, 97). Deletion of the *YRB2* gene results in a cold-sensitive phenotype (97, 99). Yrb2p binds only weakly to RanGTP and can co-activate RanGAP, but to a much smaller degree than RanBP1. Δ*yrb2* is synthetically lethal with temperature-sensitive Prp20p or with Δ*rna1* (the yeast RanGAP homologue) (97, 99) and is complemented by overexpresion of Ran (100). These data show that an interaction between Yrb2p and Ran and its regulatory proteins is important to cell survival.

The human protein, RanBP3 has been studied less extensively (101). It exists in several splice variants, although the predominant form, RanBP3b, is 499 residues long. RanBP3c lacks 30 amino acids within the RanBD and RanBP3a possesses a 68-residue insert near the N terminus. RanBP3b possesses a domain structure identical to that of Yrb2p, suggesting that it may be a functional analogue, but overexpression of RanBP3b does not complement the *YRB2* deletion (100). Biochemically, RanBP3b is also similar to Yrb2p (101). Remarkably, however, the mammalian RanBP3 binds very efficiently to yeast Gsp1p–GTP, despite its low affinity for mammalian Ran. Thus, even though the variant RanBD lacks several of the highly conserved motifs found in other RanBDs, it must still possess the epitopes essential for high-affinity interaction with the GTPase. The functional requirement for a low-affinity inter-action may be a consequence of the high (10–20 μM) concentration of Ran within the nucleosol. The nuclear targetting signal of RanBP3 is highly unusual, but possesses some resemblance to the import signal of c-myc, which contains the motif PXXKKXKLD (102). RanBP3 import is mediated by soluble factors in a Ran-dependent pathway, and may bind both importin-α and importin-β (100).

The role of Yrb2p in nuclear transport remains to be established. Deletion of Yrb2p does not affect protein import or mRNA export (97). However, Yrb2p can interact with Xpo1p, the yeast homologue of Crm1 (also called exportin) (103). Nuclear export of proteins containing Leu-rich NESs is reduced in Δyrb2 cells, implicating Yrb2p in this nuclear protein export pathway (103). Whether RanBP3 has a similar function is currently unknown.

6. NTF2/p10

All of the previously described factors recognize the GTP-bound state of Ran. Only one protein is known that prefers RanGDP, and is variously named NTF2 (nuclear transport factor 2), p10, and pp15 (104, 105). NTF2 exists in solution as a dimer and its crystal structure has been solved, both alone and as a complex with Ran–GDP (106, 107). The NTF2 polypeptide chain has an α + β-barrel fold based on a curved β-sheet with a distinctive hydrophobic cavity. A phenylalanine residue in Ran (F72) inserts into this cavity. Additionally, salt bridges are formed between K71 and R76 of Ran with D92/D94 and E42 of NTF2, respectively. NTF2 mutants such as E42 K and

D92/94N, in which the negatively charged residues surrounding the cavity were altered, are defective in Ran–GDP binding (108). Because the interaction interface maintains the positions of key Ran residues involved in binding MgGDP, NTF2 binding may help stabilize the switch state of Ran. The interaction interface with Ran–GDP involves the switch 2 loop of Ran (residues 65–78) with a lesser contribution from the switch 1 loop (residues 39–43). The recent solution to the structure of the Ran–GTP/RanBP1 complex shows that both switch regions undergo extensive conformational changes between the two nucleotide states, which accounts for the binding selectivity of NTF2. NTF2 can also associate with nucleoporin FXFG repeats (104, 109, 110). The site of interaction on NTF2 has not yet been defined, but does not overlap with that for Ran, as the three proteins can form a stable heterotrimer.

The role of NTF2 in nuclear protein import was initially unclear. In one study, NTF2 was purified from *Xenopus laevis* ovarian tissue and shown to be required for the energy-dependent translocation of docked import cargo (105). Independently, HeLa cell cytosol was depleted of import activity by incubation with the nucleoporin p62, and NTF2 was identified as the factor required to complement p62-preabsorbed cytosol. (104). The yeast homologue, Ntf2p, is essential, and genetic evidence also supports its role in nuclear protein import (111). Importantly, the lethality caused by deletion of the *NTF2* gene can be suppressed by overexpression of the yeast Ran, Gsp1p (112), which suggests that the NTF2 protein acts to increase the efficiency of Ran function. In addition, temperature-sensitive mutants of Ntf2p containing single point mutations in highly conserved amino-acid residues show defects in the localization of nuclear proteins, though not in the export of poly(A)$^+$ RNA (111). An E42K mutant of *NTF2* that is defective in Ran binding, but still binds to nucleoporins, is inactive in nuclear transport (108). Finally, overexpression of Ntf2p can rescue temperature-sensitive mutants of Gsp1p that have a decreased affinity for Ntf2p and cause import defects (113). Together, these data suggest that the interaction between Gsp1p and Ntf2p is important for nuclear transport.

But what does NTF2 do? Nuclear transport requires the maintenance of a steep trans-pore Ran–GTP gradient, and microinjection studies showed that in intact cells Ran is rapidly imported into the nucleus. Permeabilized cell assays indicated that the import was not simple diffusion and that NTF2 can function as a Ran import receptor (114, 115). Similar conclusions were reached by Gorlich and colleagues, who showed that cytosol depleted by use of a Ran–GDP affinity matrix was unable to support Ran import into the nucleus. Finally, accumulation of Ran within the nucleus depends on the exchange activity of RCC1 and can be enhanced by the addition of a nuclear Ran–GTP binding protein such as RanBP7 or transportin. Thus a satisfying picture has emerged, in which NTF2 binds Ran–GDP in the cytosol, then by virtue of its ability to interact with nucleoporins permits rapid facilitated diffusion across the nuclear pores. Within the nucleus, RCC1 catalyses GDP/GTP exchange, which releases Ran from the NTF2. Ran–GTP-binding proteins within the nucleus, such as importin-β family proteins and RanBP3, can then sequester the Ran and drive further accumulation by mass action (Fig. 4). This hypothesis explains the requirement for NTF2 in nuclear import and, because Ran can slowly permeate the pores by simple

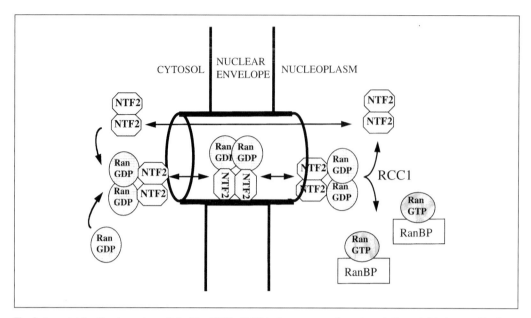

Fig. 4 A model for Ran import mediated by NTF2. NTF2 behaves as a dimer in solution. It binds specifically to Ran-GDP which is present largely in the cytosol. NTF2 also binds nucleoporins, which allows it to shuttle by facilitated diffusion along the inner surface of the nuclear pore complex. Within the nucleosol, RCC1 catalyses nucleotide exchange on Ran, forming Ran-GTP which is released from NTF2. The Ran-GTP is sequestered by nuclear Ran-binding proteins such as importin-β and transportin. NTF2 can return to the cytosol by facilitated diffusion through the pore.

diffusion, may account for the ability of Ran overexpression to suppress the requirement for NTF2 in yeast.

7. Dis3 and other Ran binding proteins

A two-hybrid screen, using human Ran as the bait, isolated a *S. cerevisiae* cDNA encoding the protein Dis3p (116). *DIS3* is essential for viability and its gene product can bind directly to Ran, which enhances nucleotide release by RCC1. The fission yeast *Schizosaccharomyces pombe* homologue of Dis3p is present as a 200 kDa oligomer that includes Spi1p and Pim1p, the *S. pombe* homologues of Ran and RCC1, respectively.

Recently, the human homologue was cloned and shown to rescue partially the temperature sensitive Dis3p mutant in *S. cerevisiae* (117), thus demonstrating a structural and functional conservation from yeast to mammals. Like the *S. cerevisiae* homologue, human Dis3 enhances RCC1-stimulated nucleotide release from Ran in a dose-dependent manner and binds Ran in both the GTP- and GDP-bound states. Dis3 turns out to be identical to Rrp44 (118), which is a component of the exosome, a nuclear RNA processing complex. The exosome consists of at least five proteins, four of which are homologous to previously characterized bacterial $3' \rightarrow 5'$ exoribonucleases. Rrp44 is homologous to RNase II, and possesses a processive $3' \rightarrow 5'$ exoribonuclease activity *in vitro* (118). The function of Ran, if any, in regulating exosome function remains to be determined. Dis3/Rrp44 may, however, provide a link between RNA processing and Ran-dependent RNA export.

Two other proteins reported to interact with Ran have been identified, Gtr1p (119, 120) and RanBPM (which apparently associates with the centrosome), although it is not clear whether these interactions are direct. None the less, this association with RanBPM is of considerable interest, as it may provide a long-sought link between Ran and control of the cell cycle. It has been difficult to provide unambiguous evidence of a role for Ran distinct from that in nucleocytoplasmic transport, but the Ran(T24N) mutant can block activation of p34cdc2 kinase in *Xenopus* oocyte mitotic extracts that lack a nuclear envelope (121, 122), suggesting a role in regulating entry into mitosis.

Additional evidence for alternate functions of Ran and RCC1 has emerged from genetic studies in *S. pombe.* Disruption of Ran GTP/GDP cycling using either *pim1* (RCC1) or *rna1* (RanGAP) temperature-sensitive mutants results in cell cycle arrest at the end of mitosis, but has no effect on cells in G_2 or on the kinetics of entry into mitosis, and the *pim1* mutation does not interfere with NLS-mediated nuclear protein import (75, 123). Despite these provocative hints, definitive and unambiguous evidence for functions distinct from nuclear transport remains to be established. There is no question, however, that any such evidence would provide exciting new insights into the control of the cell cycle.

8. Ran and nucleocytoplasmic transport

How is the Ran–GTP gradient across the nuclear pore maintained and how is this gradient coupled to nuclear cargo import and export? We have answered the first question in a previous section, where evidence was presented that NTF2 is a nuclear import factor for Ran–GDP. To answer the second question we propose a general model for nuclear import (Fig. 5a). In the first step, a transport complex is assembled in the cytoplasm. This complex consists, at a minimum, of importin-β, importin-α (for NLS-mediated import), and cargo. However, we argue that importin-β is associated in the cytosol with RanBP1 and Ran–GDP and that the import complex will therefore also contain these components. Neither RanBP1 nor Ran–GDP are thought to accompany the complex through the pore, but are displaced by the interaction of importin-β with nucleoporin components. The importin-β/importin-α/cargo complex can then flicker from side to side of the pore by facilitated diffusion along the inner pore surface.

In the third step, we suggest that the import complex can freely dissociate from the pore, but will quickly rebind unless it interacts with Ran–GTP. This interaction triggers the release of importin-α and the NLS cargo from importin-β, and terminates the import reaction. This mechanism does not require a specific internal docking site from which Ran–GTP releases the import complex. Rather, the Ran-dependent dissociation can take place within the nucleoplasm. However, given the high concentration of Ran within the nucleus (about 10 μM) the complex is highly likely to encounter Ran–GTP before it can diffuse far from the exit point of the pore.

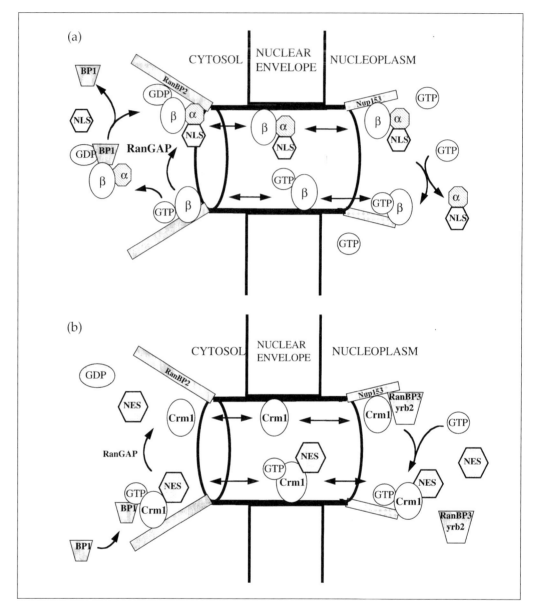

Fig. 5 (a) A model for Ran-mediated import of NLS cargo. Ran-GDP can form a complex with RanBP1, importin-β plus importin-α, and NLS cargo. On docking at the pore, the RanBP1 and Ran-GDP are released. The importin-β shuttles through the pore by facilitated diffusion. If it encounters Ran-GTP in the nucleus, the importin-α and cargo are released. Importin-β then returns to the cytosol as a complex with Ran-GTP, where RanGAP in the presence of either RanBP1 or RanBP2 triggers GTP hydrolysis leading to dissociation of this complex. (b) A model for Ran-mediated export of NES cargo. Ran-GTP can form a complex with the export receptor, Crm1, plus cargo. It binds nucleoporins and shuttles through the pore by facilitated diffusion. In the cytosol, RanBP1 or RanBP2 release the complex from the nucleoporins, and with RanGAP hydrolyse the GTP, triggering dissociation of the complex. The released Crm1 shuttles back through the pore where a RanBP3/Ran-GTP complex may catalyse release into the nucleosol.

Finally, the Ran–GTP/importin-β complex can also interact with nucleoporins on the internal surface of the pore, and slide back by facilitated diffusion towards the cytoplasmic compartment. If it dissociates on the cytoplasmic side of the pore, it will likely bind to RanBP2, where the associated RanGAP can hydrolyse the GTP. The Ran–GDP/importin-β complex could then dissociate from the RanBP2 or reform an import complex directly, with importin-α and NLS cargo.

This model predicts that importin-β will be associated predominantly with the nuclear pores rather than free in the cytosol or nucleoplasm—a distribution that has been confirmed in many laboratories. Secondly, it suggests that importin-β when not associated with cargo can move by facilitated diffusion through the pore in the absence of energy. Again, this type of transport has been observed for both importin-β and transportin (42, 124). Thirdly, it suggests that cytoplasmic Ran–GDP may not be essential for nuclear import, because the import complex may be able to associate directly with the nucleoporins within the pores. It also argues that the energy requirement for accumulation of cargo against a concentration gradient is a reflection of the necessity for recycling the importin-β, rather than for the release of the cargo into the nucleoplasm. Finally, it suggests that RanBP1 may have two important functions: to trigger release of the importin-β from the pore into the cytoplasm and to permit establishment of a new import complex.

It seems reasonable to suppose a fundamental unity in nuclear transport pathways mediated by different members of the importin-β family and this unity is supported by the experimental evidence. For example, collapse of the trans-pore Ran–GTP gradient inhibits both nuclear import and the export of proteins and RNA (83). Moreover, an importin-β fragment (45–462) that cannot be dissociated from the pores by Ran–GTP also inhibits multiple nuclear import and export pathways (49).

In the light of this information, how is NES-mediated nuclear export accomplished? A model based on that of Mattaj and colleagues is shown in Fig. 5b (66, 125). The NES receptor, Crm1 or exportin, associates with NES cargo in the nucleus. This association requires Ran–GTP. As for import, we propose that the Crm1/Ran–GTP/cargo complex can flicker by facilitated diffusion across the pore, through association of the Crm1 with nucleoporins on the internal wall of the pore. If the complex emerges into the cytoplasic compartment, dissociation is triggered by the combined action of RanGAP and RanBP2 or RanBP1, thereby releasing the NES cargo. Free Crm1 will re-equilibrate back across the pore by facilitated diffusion, and we suggest that RanBP3/Yrb2p in complex with Ran–GTP may trigger release of Crm1 into the nucleus, where it can bind cargo.

The vectoriality of import and export is therefore determined by the inverse effects of Ran–GTP on the association of the cargo with its receptor, but the fundamental mechanism of each type of transport is similar.

9. Conclusions

Five years ago, virtually nothing was known about the molecular mechanisms of nucleocytoplasmic transport. Today, a satisfying picture is emerging into which each of the many factors involved in transport can be placed. Ran has emerged as the kingpin of an ancient and efficient machine for communication between the two major compartments of the eukaryotic cell. Early in evolution, the nuclear pores may have functioned simply to permit facilitated diffusion of nucleic acids and proteins between these compartments. Indeed, such systems may still exist within the cell, such as that which permits transport of the hnRNP K protein, and which is independent of soluble factors. The Ran–GTP gradient solved the problem of permitting active transport between the nucleus and cytosol, without the necessity of redesigning the pores. Layers of complexity could then be added through divergence of the receptors, the creation of adaptor proteins, and the addition of multiple types of binding site on each receptor.

Within the near future cargoes for all of the known importin receptors are likely to be identified. Several new members of the importin family probably await discovery. Crystal structures for all of the major components in various complexes are likely to be determined within the next 2–3 years, although it will be technically more difficult to answer questions concerning the exact nature of the import complex that traverses the pore. For example, does it contain Ran–GDP or is Ran–GDP released prior to entry? The regulation of the various nuclear transport pathways remains largely unexplored. For instance, are specific pathways shut down during apoptosis, differentiation, or during the cell cycle? Are certain pathways tissue specific? It will also be important to settle the difficult question of whether Ran has functions other than those directly involved in transport. However, the major challenge for the future is to determine the structure and design of the nuclear pore complex, to elucidate the specifics of its interaction with transport factors and to understand its dynamics. Given the size of the pores, this task is likely to occupy us for several decades.

References

1. Drivas, G. T., Shih, A., Coutavas, E., Rush, M. G., and D'Eustachio, P. (1990) Characterization of four novel ras-like genes expressed in a human teratocarcinoma cell line. *Mol. Cell. Biol.*, **10**, 1793.
2. Bischoff, F. R. and Ponstingl, H. (1991) Mitotic regulator protein RCC1 is complexed with a nuclear ras-related polypeptide. *Proc. Natl Acad. Sci., USA*, **88**, 10830.
3. Bischoff, F. R., Maier, G., Tilz, G., and Ponstingl, H. (1990) A 47-kDa human nuclear protein recognized by antikinetochore autoimmune sera is homologous with the protein encoded by RCC1, a gene implicated in onset of chromosome condensation. *Proc. Natl Acad. Sci., USA*, **87**, 8617.
4. Belhumeur, P., Lee, A., Tam, R., DiPaolo, T., Fortin, N., and Clark, M. W. (1993) GSP1 and GSP2, genetic suppressors of the prp20-1 mutant in *Saccharomyces cerevisiae*: GTP-binding proteins involved in the maintenance of nuclear organization. *Mol. Cell. Biol.*, **13**, 2152.
5. Dontfraid, F. F. and Chakrabarti, D. (1994) Cloning and expression of a cDNA encoding the homologue of Ran/TC4 GTP-binding protein from *Plasmodium falciparum. Biochem. Biophys. Res. Commun.*, **201**, 423.
6. Bush, J. and Cardelli, J. (1993) Molecular cloning and DNA sequence of a *Dictyostelium*

cDNA encoding a Ran/TC4 related GTP binding protein belonging to the ras super-family. *Nucl. Acids Res.*, **21**, 1675.

7. Coutavas, E. E., Hsieh, C. M., Ren, M., Drivas, G. T., Rush, M. G., and D'Eustachio, P. D. (1994) Tissue-specific expression of Ran isoforms in the mouse. *Mamm. Genome*, **5**, 623.

8. Wittinghofer, A. and Pai, E. F. (1991) The structure of Ras protein: a model for a universal molecular switch. *Trends Biochem. Sci.*, **16**, 382.

9. Scheffzek, K., Klebe, C., Fritz-Wolf, K., Kabsch, W., and Wittinghofer, A. (1995) Crystal structure of the nuclear Ras-related protein Ran in its GDP- bound form. *Nature*, **374**, 378.

10. Richards, S. A., Lounsbury, K. M., and Macara, I. G. (1995) The C terminus of the nuclear RAN/TC4 GTPase stabilizes the GDP-bound state and mediates interactions with RCC1, RAN-GAP, and HTF9A/RANBP1. *J Biol. Chem.*, **270**, 14405.

11. Palacios, I., Weis, K., Klebe, C., Mattaj, I. W., and Dingwall, C. (1996) RAN/TC4 mutants identify a common requirement for snRNP and protein import into the nucleus. *J. Cell Biol.*, **133**, 485.

12. Lounsbury, K. M., Richards, S. A., Carey, K. L., and Macara, I. G. (1996) Mutations within the Ran/Tc4 GTPase—effects on regulatory factor interactions and subcellular local-ization. *J. Biol. Chem.*, **271**, 32834.

13. Ren, M., Coutavas, E., D'Eustachio, P., and Rush, M. G. (1994) Effects of mutant Ran/TC4 proteins on cell cycle progression. *Mol. Cell. Biol.*, **14**, 4216.

14. Murphy, G. A., Moore, M. S., Drivas, G., Delaossa, P. P., Villamarin, A., Deustachio, P., and Rush, M. G. (1997) A T42a Ran mutation – differential interactions with effectors and regulators, and defect in nuclear protein import. *Mol. Biol. Cell*, **8**, 2591.

15. Seki, T., Hayashi, N., and Nishimoto, T. (1996) RCC1 in the Ran pathway. *J. Biochem.*, **120**, 207.

16. Nishimoto, T., Eilen, E., and Basilico, C. (1978) Premature of chromosome condensation in a ts DNA – mutant of BHK cells. *Cell*, **15**, 475.

17. Ohtsubo, M., Okazaki, H., and Nishimoto, T. (1989) The RCC1 protein, a regulator for the onset of chromosome condensation locates in the nucleus and binds to DNA. *J. Cell. Biol.*, **109**, 1389.

18. Uchida, S., Sekiguchi, T., Nishitani, H., Miyauchi, K., Ohtsubo, M., and Nishimoto, T. (1990) Premature chromosome condensation is induced by a point mutation in the hamster RCC1 gene. *Mol. Cell. Biol.*, **10**, 577.

19. Ohtsubo, M., Yoshida, T., Seino, H., Nishitani, H., Clark, K. L., Sprague, G. F. Jr, Frasch, M., and Nishimoto, T. (1991) Mutation of the hamster cell cycle gene RCC1 is comple-mented by the homologous genes of *Drosophila* and *S. cerevisiae*. *EMBO J.*, **10**, 1265.

20. Nishitani, H., Ohtsubo, M., Yamashita, K., Iida, H., Pines, J., Yasudo, H., Shibata, Y., Hunter, T., and Nishimoto, T. (1991) Loss of RCC1, a nuclear DNA-binding protein, un-couples the completion of DNA replication from the activation of cdc2 protein kinase and mitosis. *EMBO J.*, **10**, 1555.

21. Dasso, M., Nishitani, H., Kornbluth, S., Nishimoto, T., and Newport, J. W. (1992) RCC1, a regulator of mitosis, is essential for DNA replication. *Mol. Cell. Biol.*, **12**, 3337.

22. Bischoff, F. R. and Ponstingl, H. (1991) Catalysis of guanine nucleotide exchange on Ran by the mitotic regulator RCC1. *Nature*, **354**, 80.

23. Kadowaki, T., Goldfarb, D., Spitz, L. M., Tartakoff, A. M., and Ohno, M. (1993) Regulation of RNA processing and transport by a nuclear guanine nucleotide release protein and members of the Ras superfamily. *EMBO J.*, **12**, 2929.

24. Matsumoto, T. and Beach, D. (1991) Premature initiation of mitosis in yeast lacking RCC1 or an interacting GTPase. *Cell*, **66**, 347.

25. Frasch, M. (1991) The maternally expressed *Drosophila* gene encoding the chromatin-binding protein BJ1 is a homolog of the vertebrate gene Regulator of Chromatin Condensation, RCC1. *EMBO J.*, **10**, 1225.

26. Sondek, J., Bohm, A., Lambright, D. G., Hamm, H. E., and Sigler, P. B. (1996) Crystal structure of a G-protein beta gamma dimer at 2.1A resolution (see comments). *Nature*, **379**, 369 (published erratum appears in *Nature* (1996) **379**, (6568), 847).

27. Bischoff, F. R., Krebber, H., Kempf, T., Hermes, I., and Ponstingl, H. (1995) Human RanGTPase-activating protein RanGAP1 is a homologue of yeast Rna1p involved in mRNA processing and transport. *Proc. Natl Acad. Sci., USA*, **92**, 1749.

28. Bischoff, F. R., Klebe, C., Kretschmer, J., Wittinghofer, A., and Ponstingl, H. (1994) RanGAP1 induces GTPase activity of nuclear Ras-related Ran. *Proc. Natl Acad. Sci., USA*, **91**, 2587.

29. Traglia, H. M., Atkinson, N. S., and Hopper, A. K. (1989) Structural and functional analyses of *Saccharomyces cerevisiae* wild-type and mutant RNA1 genes. *Mol. Cell. Biol.*, **9**, 2989.

30. Becker, J., Melchior, F., Gerke, V., Bischoff, F. R., Ponstingl, H., and Wittinghofer, A. (1995) RNA1 encodes a GTPase-activating protein specific for Gsp1p, the Ran/TC4 homologue of *Saccharomyces cerevisiae*. *J. Biol. Chem.*, **270**, 11860.

31. Klebe, C., Bischoff, F. R., Ponstingl, H., and Wittinghofer, A. (1995) Interaction of the nuclear GTP-binding protein Ran with its regulatory proteins RCC1 and RanGAP1. *Biochemistry*, **34**, 639.

32. Hopper, A. K., Traglia, H. M., and Dunst, R. W. (1990) The yeast RNA1 gene product necessary for RNA processing is located in the cytosol and apparently excluded from the nucleus. *J. Cell Biol.*, **111**, 309.

33. Matunis, M. J., Coutavas, E., and Blobel, G. (1996) A novel ubiquitin-like modification modulates the partitioning of the Ran-GTPase-activating protein RanGAP1 between the cytosol and the nuclear pore complex. *J. Cell Biol.*, **135**, 1457.

34. Mahajan, R., Delphin, C., Guan, T. L., Gerace, L., and Melchior, F. (1997) A small ubiquitin-related polypeptide involved in targeting RanGAP1 to nuclear pore complex protein RanBP2. *Cell*, **88**, 97.

35. Matunis, M. J., Wu, J. A., and Blobel, G. (1998) Sumo-1 modification and its role in targeting the Ran GTPase-activating protein, RanGAP1, to the nuclear pore complex. *J. Cell Biol.*, **140**, 499.

36. Saitoh, H., Sparrow, D. B., Shiomi, T., Pu, R. T., Nishimoto, T., Mohun, T. J., and Dasso, M. (1997) Ubc9p and the conjugation of Sumo-1 to RanGAP1 and RanBP2. *Curr. Biol.*, **8**, 121.

37. Lee, G. W., Melchior, F., Matunis, M. J., Mahajan, R., Tian, Q. S., and Anderson, P. (1998) Modification of Ran GTPase-activating protein by the small ubiquitin-related modifier SUMO-1 requires Ubc9, an E2-type ubiquitin-conjugating enzyme homologue. *J. Biol. Chem.*, **273**, 6503.

38. Mahajan, R., Gerace, L., and Melchior, F. (1998) Molecular characterization of the SUMO-1 modification of RanGAP1 and its role in nuclear envelope association. *J. Cell Biol.*, **140**, 259.

39. Feng, W., Benko, A. L., Lee, J.-H., Stanford, D. R., and Hopper, A. K. (1998) Antagonistic effects of NES and NLS motifs determine *S. cerevisiae* Rna1p subcellular distribution. *J. Cell. Sci.*, **112**, 339.

40. Chi, N. C., Adam, E. J., and Adam, S. A. (1995) Sequence and characterization of cytoplasmic nuclear protein import factor p97. *J. Cell Biol.*, **130**, 265.

41. Gorlich, D., Kostka, S., Kraft, R., Dingwall, C., Laskey, R. A., Hartmann, E., and Prehn, S. (1995) Two different subunits of importin cooperate to recognize nuclear localization signals and bind them to the nuclear envelope. *Curr. Biol.*, **5**, 383.

42. Kose, S., Imamoto, N., Tachibana, T., Shimamoto, T., and Yoneda, Y. (1997) Ran-unassisted nuclear migration of a 97-Kd component of nuclear pore-targeting complex. *J. Cell Biol.*, **139**, 841.

43. Radu, A., Blobel, G., and Moore, M. S. (1995) Identification of a protein complex that is required for nuclear protein import and mediates docking of import substrate to distinct nucleoporins. *Proc. Natl Acad. Sci., USA*, **92**, 1769.

44. Gorlich, D., Dabrowski, M., Bischoff, F. R., Kutay, U., Bork, P., Hartmann, E., Prehn, S., and Izaurralde, E. (1997) A novel class of RanGTP binding proteins. *J. Cell Biol.*, **138**, 65.

45. Wozniak, R. W., Rout, M. P., and Aitchison, J. D. (1998) Karyopherins and Kissing Cousins. *Trends Cell Biol.*, **8**, 184.

46. Malik, H. S., Eickbush, T. H., and Goldfarb, D. S. (1997) Evolutionary specialization of the nuclear targeting apparatus. *Proc. Natl Acad. Sci., USA*, **94**, 13738.

47. Jakel, S. and Gorlich, D. (1998) Importin b, transportin, RanBP5 and RanBP7 mediate nuclear import of ribosomal proteins in mammalian cells. *EMBO J.*, **17**, 4491.

48. Chi, N. C., Adam, E. J. H., and Adam, S. A. (1997) Different binding domains for Ran-GTP and Ran-GDP/RanBP1 on nuclear import factor P97. *J. Biol. Chem.*, **272**, 6818.

49. Kutay, U., Izaurralde, E., Bischoff, F. R., Mattaj, I. W., and Gorlich, D. (1997) Dominant-negative mutants of importin-beta block multiple pathways of import and export through the nuclear pore complex. *EMBO J.*, **16**, 1153.

50. Rexach, M. and Blobel, G. (1995) Protein import into nuclei: association and dissociation reactions involving transport substrate, transport factors, and nucleoporins. *Cell*, **83**, 683.

51. Moroianu, J., Blobel, G., and Radu, A. (1996) Nuclear protein import: Ran-GTP dissociates the karyopherin alphabeta heterodimer by displacing alpha from an overlapping binding site on beta. *Proc. Natl Acad. Sci., USA*, **93**, 7059.

52. Gorlich, D., Pante, N., Kutay, U., Aebi, U., and Bischoff, F. R. (1996) Identification of different roles for RanGDP and RanGTP in nuclear protein import. *EMBO J.*, **15**, 5584.

53. Lounsbury, K. M. and Macara, I. G. (1997) Ran-binding protein 1 (RanBP1) forms a ternary complex with Ran and karyopherin beta and reduces Ran GTPase-activating protein (RanGAP) inhibition by karyopherin beta. *J. Biol. Chem.*, **272**, 551.

54. Chi, N. C. and Adam, S. A. (1997) Functional domains in nuclear import factor P97 for binding the nuclear localization sequence receptor and the nuclear pore. *Mol. Biol. Cell*, **8**, 945.

55. Percipalle, P., Clarkson, W. D., Kent, H. M., Rhodes, D., and Stewart, M. (1997) Molecular interactions between the importin alpha/beta heterodimer and proteins involved in vertebrate nuclear protein import. *J. Mol. Biol.*, **266**, 722.

56. Iovine, M. K., Watkins, J. L., and Wente, S. R. (1995) The GLFG repetitive region of the nucleoporin Nup116p interacts with Kap95p, an essential yeast nuclear import factor. *J. Cell Biol.*, **131**, 1699.

57. Shah, S., Tugendreich, S., and Forbes, D. (1998) Major binding sites for the nuclear import receptor are the internal nucleoporin Nup153 and the adjacent nuclear filament protein Tpr. *J. Cell Biol.*, **141**, 31.

58. Pollard, V. W., Michael, W. M., Nakielny, S., Siomi, M. C., Wang, F., and Dreyfuss, G. (1996) A novel receptor-mediated nuclear protein import pathway. *Cell*, **86**, 985.

59. Michael, W. M., Choi, M. Y., and Dreyfuss, G. (1995) A nuclear export signal in Hnrnp A1—a signal-mediated, temperature-dependent nuclear protein export pathway. *Cell*, **83**, 415.

60. Siomi, M. C., Eder, P. S., Kataoka, N., Wan, L. L., Liu, Q., and Dreyfuss, G. (1997)

Transportin-mediated nuclear import of heterogeneous nuclear Rnp proteins. *J. Cell Biol.*, **138**, 1181.

61. Bonifaci, N., Moroianu, J., Radu, A., and Blobel, G. (1997) Karyopherin beta-2 mediates nuclear import of a mRNA binding protein. *Proc. Natl Acad. Sci., USA*, **94**, 5055.

62. Rout, M. P., Blobel, G., and Aitchison, J. D. (1997) A distinct nuclear import pathway used by ribosomal proteins. *Cell*, **89**, 715.

63. Pemberton, L. F., Rosenblum, J. S., and Blobel, G. (1997) A distinct and parallel pathway for the nuclear import of an mRNA-binding protein. *J. Cell Biol.*, **139**, 1645.

64. Aitchison, J. D., Blobel, G., and Rout, M. P. (1996) Kap104p: a karyopherin involved in the nuclear transport of messenger RNA binding proteins. *Science*, **274**, 624.

65. Stade, K., Ford, C. S., Guthrie, C., and Weis, K. (1997) Exportin 1 (Crm1p) is an essential nuclear export factor. *Cell*, **90**, 1041.

66. Fornerod, M., Ohno, M., Yoshida, M., and Mattaj, I. W. (1997) Crm1 is an export receptor for leucine-rich nuclear export signals. *Cell*, **90**, 1051.

67. Ossarehnazari, B., Bachelerie, F., and Dargemont, C. (1997) Evidence for a role of Crm1 in signal-mediated nuclear protein export. *Science*, **278**, 141.

68. Arts, G. J., Fornerod, M., and Mattaj, I. W. (1998) Identification of a nuclear export receptor for Trna. *Curr. Biol.*, **8**, 305.

69. Kutay, U., Lipowsky, G., Izaurralde, E., Bischoff, F. R., Schwarzmaier, P., Hartmann, E., and Gorlich, D. (1998) Identification of a Trna-specific nuclear export receptor. *Mol. Cell*, **1**, 359.

70. Kutay, U., Bischoff, F. R., Kostka, S., Kraft, R., and Gorlich, D. (1997) Export of importin alpha from the nucleus is mediated by a specific nuclear transport factor. *Cell*, **90**, 1061.

71. Beddow, A. L., Richards, S. A., Orem, N. R., and Macara, I. G. (1995) The Ran/TC4 GTPase-binding domain: identification by expression cloning and characterization of a conserved sequence motif. *Proc. Natl Acad. Sci., USA*, **92**, 3328.

72. Coutavas, E., Ren, M., Oppenheim, J. D., D''Eustachio, P., and Rush, M. G. (1993) Characterization of proteins that interact with the cell-cycle regulatory protein Ran/TC4. *Nature*, **366**, 585.

73. Lounsbury, K. M., Beddow, A. L., and Macara, I. G. (1994) A family of proteins that stabilize the Ran/TC4 GTPase in its GTP-bound conformation. *J. Biol. Chem.*, **269**, 11285.

74. Bressan, A., Somma, M. P., Lewis, J., Santolamazza, C., Copeland, N. G., Gilbert, D. J., Jenkins, N. A., and Lavia, P. (1991) Characterization of the opposite-strand genes from the mouse bidirectionally transcribed HTF9 locus. *Gene*, **103**, 201.

75. He, X., Hayashi, N., Walcott, N. G., Azuma, Y., Patterson, T. E., Bischoff, F. R., Nishimoto, T., and Sazer, S. (1998) The identification of cDNAs that affect the mitosis-to-interphase transition in *Schizosaccharomyces pombe*, including sbp1, which encodes a spi1p-GTP-binding protein. *Genetics*, **148**, 645.

76. Butler, G. and Wolfe, K. H. (1994) Yeast homolog of mammalian ran binding-protein-1. *Biochim. Biophys. Acta-Gene Structure and Expression*, **1219**, 711.

77. Richards, S. A., Lounsbury, K. M., Carey, K. L., and Macara, I. G. (1996) A nuclear export signal is essential for the cytosolic localization of the Ran binding protein, RanBP1. *J. Cell Biol.*, **134**, 1157.

78. Kuhlmann, J., Macara, I., and Wittinghofer, A. (1997) Dynamic and equilibrium studies on the interaction of Ran with its effector, RanBP1. *Biochemistry*, **36**, 12027.

79. Ren, M., Villamarin, A., Shih, A., Coutavas, E., Moore, M. S., LoCurcio, M., Clarke, V., Oppenheim, J. D., D'Eustachio, P., and Rush, M. G. (1995) Separate domains of the Ran GTPase interact with different factors to regulate nuclear protein import and RNA processing. *Mol. Cell. Biol.*, **15**, 2117.

80. Chi, N. C., Adam, E. J. H., Visser, G. D., and Adam, S. A. (1996) Ranbp1 stabilizes the interaction of Ran with P97 in nuclear protein import. *J. Cell Biol.*, **135**, 559.

81. Schlenstedt, G., Wong, D. H., Koepp, D. M., and Silver, P. A. (1995) Mutants in a yeast Ran binding protein are defective in nuclear transport. *EMBO J.*, **14**, 5367.

82. Ouspenski, II, Mueller, U. W., Matynia, A., Sazer, S., Elledge, S. J., and Brinkley, B. R. (1995) Ran-binding protein-1 is an essential component of the ran rcc1 molecular switch system in budding yeast. *J. Biol. Chem.*, **270**, 1975.

83. Izaurralde, E., Kutay, U., Vonkobbe, C., Mattaj, I. W., and Gorlich, D. (1997) The asymmetric distribution of the constituents of the Ran system is essential for transport into and out of the nucleus. *EMBO J.*, **16**, 6535.

84. Guarguaglini, G., Battistoni, A., Pittoggi, C., Dimatteo, G., Difiore, B., and Lavia, P. (1997) Expression of the murine RanBP1 and Htf9-C genes is regulated from a shared bidirectional promoter during cell cycle progression. *Biochem. J.*, **325**, 277.

85. Battistoni, A., Guarguaglini, G., Degrassi, F., Pittoggi, C., Palena, A., Dimatteo, G., Pisano, C., Cundari, E., and Lavia, P. (1997) Deregulated expression of the RanBP1 gene alters cell cycle progression in murine fibroblasts. *J. Cell Sci.*, **110**, 2345.

86. Yokoyama, N., Hayashi, N., Seki, T., Pante, N., Ohba, T., Nishii, K., Kuma, K., Hayashida, T., Miyata, T., Aebi, U., *et al.* (1995) A giant nucleopore protein that binds Ran/TC4. *Nature*, **376**, 184.

87. Wu, J., Matunis, M. J., Kraemer, D., Blobel, G., and Coutavas, E. (1995) Nup358, a cytoplasmically exposed nucleoporin with peptide repeats, Ran-GTP binding sites, zinc fingers, a cyclophilin A homologous domain, and a leucine-rich region. *J. Biol. Chem.*, **270**, 14209.

88. Wilken, N., Senecal, J. L., Scheer, U., and Dabauvalle, M. C. (1995) Localization of the Ran-GTP binding protein RanBP2 at the cytoplasmic side of the nuclear pore complex. *Eur. J. Cell Biol.*, **68**, 211.

89. Ferreira, P. A., Nakayama, T. A., Pak, W. L., and Travis, G. H. (1996) Cyclophilin-related protein RanBP2 acts as chaperone for red/green opsin. *Nature*, **383**, 637.

90. Krebber, H., Bastians, H., Hoheisel, J., Lichter, P., Ponstingl, H., and Joos, S. (1997) Localization of the gene encoding the Ran-binding protein RanBP2 to human chromosome 2q11–q13 by fluorescence *in situ* hybridization. *Genomics*, **43**, 247.

91. Nothwang, H. G., Rensing, C., Kubler, M., Denich, D., Brandl, B., Stubanus, M., Haaf, T., Kurnit, D., and Hildebrandt, F. (1998) Identification of a novel Ran binding protein 2 related gene (RANBP2L1) and detection of a gene cluster on human chromosome 2q11–q12. *Genomics*, **47**, 383.

92. Bischoff, F. R., Krebber, H., Smirnova, E., Dong, W., and Ponstingl, H. (1995) Co-activation of RanGTPase and inhibition of GTP dissociation by Ran-GTP binding protein RanBP1. *EMBO J.*, **14**, 705.

93. Lounsbury, K. M., Richards, S. A., Perlungher, R. R., and Macara, I. G. (1996) Ran binding domains promote the interaction of Ran with p97/beta- karyopherin, linking the docking and translocation steps of nuclear import. *J. Biol. Chem.*, **271**, 2357.

94. Floer, M., Blobel, G., and Rexach, M. (1997) disassembly of RanGTP-karyopherin beta complex, an intermediate in nuclear protein import. *J. Biol. Chem.*, **272**, 19538.

95. Bischoff, F. R. and Gorlich, D. (1997) RanBP1 is crucial for the release of RanGTP from importin beta-related nuclear transport factors. *FEBS Lett.*, **419**, 249.

96. Schlenstedt, G., Smirnova, E., Deane, R., Solsbacher, J., Kutay, U., Gorlich, D., Ponstingl, H., and Bischoff, F. R. (1997) Yrb4p, a yeast Ran-GTP-binding protein involved in import of ribosomal protein l25 into the nucleus. *EMBO J.*, **16**, 6237.

97. Noguchi, E., Hayashi, N., Nakashima, N., and Nishimoto, T. (1997) Yrb2p, a Nup2p-related yeast protein, has a functional overlap with Rna1p, a yeast Ran-GTPase-activating protein. Mol. Cell. Biol., **17**, 2235.

98. Dingwall, C., Kandels-Lewis, S., and Seraphin, B. (1995) A family of Ran binding proteins that includes nucleoporins. *Proc. Natl Acad. Sci., USA*, **92**, 7525.

99. Taura, T., Schlenstedt, G., and Silver, P. A. (1997) Yrb2p is a nuclear protein that interacts with Prp20p, a yeast Rcc1 homologue. *J. Biol. Chem.*, **272**, 31877.

100. Welch, K. A. and Macara, I. G., unpublished.

101. Mueller, L., Cordes, V. C., Bischoff, F. R., and Ponstingl, H. (1998) Human RanBP3, a group of nuclear RanGTP binding proteins. *FEBS Lett.*, **427**, 330.

102. Makkerh, J. P. S., Dingwall, C., and Laskey, R. A. (1996) Comparative mutagenesis of nuclear localization signals reveals the importance of neutral and acidic amino acids. Curr Biol, 6, 1025.

103. Taura, T., Krebber, H., and Silver, P. A. (1998) A member of the Ran-binding protein family, Yrb2p, is involved in nuclear protein export. *Proc. Natl Acad. Sci., USA*, **95**, 7427.

104. Paschal, B. M. and Gerace, L. (1995) Identification of NTF2, a cytosolic factor for nuclear import that interacts with nuclear pore complex protein p62. *J. Cell Biol.*, **129**, 925.

105. Moore, M. S. and Blobel, G. (1994) Purification of a Ran-interacting protein that is required for protein import into the nucleus. *Proc. Natl Acad. Sci., USA*, **91**, 10212.

106. Bullock, T. L., Clarkson, W. D., Kent, H. M., and Stewart, M. (1996) The 1.6 angstroms resolution crystal structure of nuclear transport factor 2 (NTF2). *J. Mol. Biol.*, **260**, 422.

107. Stewart, M., Kent, H. M., and McCoy, A. J. (1998) Structural basis for molecular recognition between nuclear transport factor 2 (Ntf2) and the GDP-bound form of the Ras-family GTPase Ran. *J. Mol. Biol.*, **277**, 635.

108. Clarkson, W. D., Corbett, A. H., Paschal, B. M., Kent, H. M., McCoy, A. J., Gerace, L., Silver, P. A., and Stewart, M. (1997) Nuclear protein import is decreased by engineered mutants of nuclear transport factor 2 (Ntf2) that do not bind GDP-Ran. *J. Mol. Biol.*, **272**, 716.

109. Clarkson, W. D., Kent, H. M., and Stewart, M. (1996) Separate binding sites on nuclear transport factor 2 (Ntf2) for GDP-Ran and the phenylalanine-rich repeat regions of nucleoporins p62 and Nsp1p. *J. Mol. Biol.*, **263**, 517.

110. Nehrbass, H. and Blobel, G. (1996) Role of the nuclear transport factor p10 in nuclear transport. *Science*, **272**, 120.

111. Corbett, A. H. and Silver, P. A. (1996) The Ntf2 gene encodes an essential, highly conserved protein that functions in nuclear transport *in vivo*. *J. Biol. Chem.*, **271**, 18477.

112. Paschal, B. M., Fritze, C., Guan, T., and Gerace, L. (1997) High levels of the GTPase Ran/Tc4 relieve the requirement for nuclear protein transport factor 2. *J. Biol. Chem.*, **272**, 21534.

113. Wong, D. H., Corbett, A. H., Kent, H. M., Stewart, M., and Silver, P. A. (1997) interaction between the small GTPase Ran/Gsp1p and Ntf2p is required for nuclear transport. *Mol. Cell. Biol.*, **17**, 3755.

114. Smith, A., Brownawell, A., and Macara, I. G. (1998) Nuclear import of Ran:GDP is mediated by NTF2. *Curr. Biol.*, **8**, 1403.

115. Ribbeck, K., Lipowsky, G., M., K. H., Stewart, M., and Gorlich, D. (1998) NTF2 mediates nuclear import of Ran. *EMBO J.*, **17**, 6587.

116. Noguchi, E., Hayashi, N., Azuma, Y., Seki, T., Nakamura, M., Nakashima, N., Yanagida, M., He, X., Mueller, U., Sazer, S., and Nishimoto, T. (1996) Dis3, implicated in mitotic control, binds directly to Ran and enhances the Gef activity of Rcc1. *EMBO J.*, **15**, 5595.

117. Shiomi, T., Fukushima, K., Suzuki, N., Nakashima, N., Noguchi, E., and Nishimoto, T. (1998) Human dis3p, which binds to either GTP- or GDP-Ran, complements *Saccharomyces cerevisiae* dis3. *J. Biochem.*, **123**, 883.

118. Mitchell, P., Petfalski, E., Shevchenko, A., Mann, M., and Tollervey, D. (1997) The exosome: a conserved eukaryotic RNA processing complex containing multiple 3′→5′ exoribonucleases. *Cell*, **91**, 457.

119. Nakashima, N., Hayashi, N., Noguchi, E., and Nishimoto, T. (1996) Putative GTPase Gtr1p genetically interacts with the RanGTPase cycle in *Saccharomyces cerevisiae*. *J. Cell Sci.*, **109**, 2311.

120. Hirose, E., Nakashima, N., Sekiguchi, T., and Nishimoto, T. (1998) Raga is a functional homologue of *S. cerevisiae* Gtr1p involved in the Ran/Gsp1-GTPase pathway. *J. Cell Sci.*, **111**, 11.

121. Kornbluth, S., Dasso, M., and Newport, J. (1994) Evidence for a dual role for TC4 protein in regulating nuclear structure and cell cycle progression. *J. Cell Biol.*, **125**, 705.

122. Dasso, M., Seki, T., Azuma, Y., Ohba, T., and Nishimoto, T. (1994) A mutant form of the Ran/TC4 protein disrupts nuclear function in *Xenopus laevis* egg extracts by inhibiting the RCC1 protein, a regulator of chromosome condensation. *EMBO J.*, **13**, 5732.

123. Sazer, S. (1996) The search for the primary function of the Ran GTPase continues. *Trends Cell Biol.*, **6**, 81.

124. Nakielny, S. and Dreyfuss, G. (1997) Import and export of the nuclear protein import receptor transportin by a mechanism independent of GTP hydrolysis. *Curr. Biol.*, **8**, 89.

125. Mattaj, I. and Englmeier, L. (1998) Nucleocytoplasmic transport: the soluble phase. *Ann. Rev. Biochem.*, **67**, 265.

8 | GTPases in protein translocation and protein elongation

DOUGLAS M. FREYMANN AND PETER WALTER

1. Introduction

Each family of GTPases builds upon a structural core in distinct ways to effect the GTPase switch (1), by which the protein undergoes a transition between 'active' and 'inactive' states (2–4). In this chapter we describe the structure and function of two GTPase families not covered in depth elsewhere in this volume: the GTPase subunits of the signal recognition particle (SRP) and SRP receptor (SR), which mediate co-translational protein targeting (5), and the elongation factors in protein synthesis, taking the GTPases EF-Tu and EF-G as representatives of the family (6–11). We cannot hope to do justice to the breadth of each field, but focus on orienting the reader to the role of the GTPase in the biochemistry of each system.

2. GTPases in SRP-mediated protein targetting

The signal recognition particle (SRP) is a ribonucleoprotein that catalyses the co-translational targeting of secreted and membrane proteins to the endoplasmic reticulum in mammalian cells and to the plasma membrane in bacteria (5, 12). SRP binds to signal sequences as they emerge as part of nascent polypeptide chains from the ribosome. During this interaction, SRP communicates with the ribosome (Fig. 1). Upon binding of a signal sequence, translation elongation stalls or pauses transiently, presumably to ascertain that the length of the nascent protein is kept as short as possible prior to its targeting to the membrane. As a next step, SRP interacts with the SRP receptor (SR) in the ER membrane. This interaction releases SRP from the ribosome/nascent chain complex. Translation resumes as the ribosome becomes bound to a protein translocation channel through which the protein chain is fed across the lipid bilayer. Thus, SRP and SRP receptor function as 'initiation factors' for protein translocation. The key features of this process are thought to be evolutionarily conserved: homologues of SRP and SR are found in all living cells analysed to date and are thought to function similarly. During cotranslational protein translocation, an

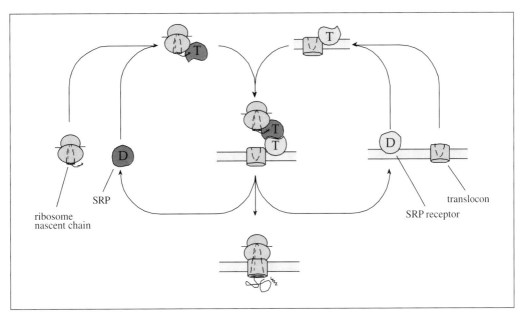

Fig. 1 SRP-dependent protein targetting. The SRP recognizes the hydrophobic signal sequence as it emerges from the ribosome. Interaction between the SRP and ribosome arrests translation. The SRP recognizes a membrane-associated receptor (SR) in a process requiring GTP binding. GTP hydrolysis accompanies release of the polypeptide to the translocation apparatus of the membrane, and SRP dissociates from the translating ribosome complex. In prokaryotes the cycle apparently does not involve translation arrest. T, GTP-bound state; D, GDP-bound state. (Figure adapted from (5) with permission.)

unfolded polypeptide chain is channelled across the membrane, through a narrow, water-filled pore, as it exits the ribosome. The strict coupling between protein synthesis and translocation ensures that the protein cannot fold or misfold in the cytosol. Cotranslational translocation therefore has conceptual advantages over post-translational translocation modes, which are likely to impose many more constraints on the passenger proteins to retain them in a translocation-competent state.

Mammalian SRP is a ribonucleoprotein comprising six polypeptides and one RNA (13, 14). Signal-sequence recognition is mediated by the SRP54 subunit (15, 16), which forms the 'business end' of the particle and also binds directly to SRP RNA. In addition, SRP54 is a GTPase, as was first deduced from its sequence after it was determined by cDNA cloning (17, 18). Intriguingly, the GTPase domain of SRP54 was found to be closely related to a GTPase domain in the SRP receptor α subunit (SRα). Sequence homology of the SRP and SR GTPase domains to other GTPases is restricted to the discrete boxes that comprise the GTP-binding site, and distinct conserved features of the SRP54 and SRα sequences suggest that they constitute a new family in the GTPase superfamily (Fig. 2). This poses the exciting questions of why, mechanistically, an otherwise unique GTPase module is repeated twice in the same cellular pathway and how such a system might have evolved. There is currently no satisfying answer to either of these questions.

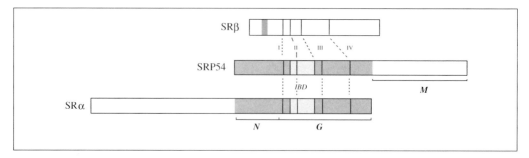

Fig. 2 Conserved sequence motifs of the SRP family of GTPases. The positions of the conserved sequence boxes in the SRP family of GTPases are indicated. SRP54 and SRα (and the prokaryotic Ffh and FtsY) include a unique ~50 amino-acid insertion, the I-box domain (IBD). The IBD is positioned between the motifs I and III, and contains the sequence motif II implicated in interaction between SRP and SR. SRβ does not have the IBD, and is more structurally related to the small GTPases such as Ras. The GTPase domain (G) contains the highly conserved motifs (I, II, III, and IV) found in all GTPases (see Chapter 9). (Figure adapted from (5) with permission.)

GTP binding and hydrolysis accompany interaction of SRP54 and SR receptor (19). Binding of SRP to SR causes SRP to dissociate from the signal sequence and the ribosome (20). Indeed, SR was first purified by following this activity in a functional assay (21, 22). Biochemical studies have suggested, however, that only GTP binding, but not GTP hydrolysis, is required for the initial steps in the targeting reaction. Thus GTP binding is thought to accompany targeting and GTP hydrolysis to cause the subsequent release of the SRP from the membrane (23) (Fig. 1).

Both SRP54 and SRα are evolutionarily conserved. The prokaryotic homologues, called Ffh and FtsY, show very similar properties to their mammalian counterparts, and have been used extensively to characterize their interactions biochemically. Like mammalian SRP, Ffh is also an RNA-binding protein that is complexed with *E. coli* 4.5S RNA (24). However, 4.5S RNA lacks a domain present in mammalian SRP RNA, the Alu domain, that is required for SRP's elongation arrest activity (25). Ffh and 4.5S RNA are the only subunits of a stable ribonucleoprotein in the bacterial cytosol that functions as a streamlined, 'minimal' SRP. The bacterial SRP receptor is a single polypeptide chain that is either soluble or loosely associated with the bacterial inner membrane (26–28).

Besides having a far more complex SRP, the story in eukaryotic cells is further complicated as the SRP receptor comprises two distinct subunits (29). The SRα subunit is anchored to the membrane via the β subunit (SRβ), which contains a bona fide transmembrane domain. Intriguingly, the β subunit is also a GTPase, although it is not a member of the SRP GTPase family and shows more sequence similarity to other small GTPases, such as Ras and Arf (30). Mutations in the GTPase domain of SRβ disrupt SR function *in vivo* (31). However, the function of the SRβ GTPase domain in the targeting pathway, is still unknown.

Originally, the proposal that Ffh and FtsY function in prokaryotic secretion met with considerable controversy (32, 33). Consequently, a substantial effort has been directed towards demonstrating a role in protein translocation for these components.

Disruption of *E. coli* SRP by expression of a dominant lethal 4.5S RNA mutant (24) or depletion of the Ffh protein, weakly impairs translocation of periplasmic proteins but has profound effects on the insertion of integral membrane proteins (34–36). Similarly, FtsY has been shown to be essential for biogenesis of membrane proteins (37). As expected, the GTPase activity of Ffh is essential (38), as is binding and hydrolysis of GTP by FtsY (39). The prokaryotic SRP has been reviewed recently (40–42).

The extreme functional conservation between the prokaryotic and eukaryotic systems is best demonstrated by the fact that Ffh can substitute for some functions of SRP54 *in vitro* (43). Recently, Ffh and FtsY together have been shown to substitute efficiently for their eukaryotic counterparts in *in vitro* translocation assays (44). Thus prokaryotic Ffh can recognize efficiently signal sequences, eukaryotic ribosomes, and ER translocation sites, and induces the formation of a fully functional ribosome/membrane junction. Targetting in this assay was shown to be strictly cotranslational. Consistent with its streamlined structure, prokaryotic SRP lacks translation arrest activity, which on mammalian SRP maps to the Alu-domain of SRP RNA that is absent in *E. coli* 4.5S RNA. An important feature in this experiment was to replace both components of the targetting machinery, since prokaryotic SRP does not work with mammalian SR and vice versa. Consistent with this notion, the converse attempt to rescue translocation in a bacterial Ffh conditional mutant using mammalian SRP54 failed (45).

Ffh proteins from a variety of bacteria, including *E. coli*, *Bacillus subtilus*, and *Mycoplasma mycoides*, have been shown to posses intrinsic GTPase activities (46, 47). GTP hydrolysis by Ffh, however, is significantly stimulated by interaction with its receptor, FtsY (48). Indeed, it was shown that the two GTPases, Ffh and FtsY, interact directly and act as regulatory proteins for each other. *In vitro* experiments, which took advantage of a mutation that changed the specificity of FtsY from GTP to xanthosine triphosphate (XTP), demonstrated that FtsY stimulated hydrolysis of GTP by Ffh/4.5S RNA in a reaction that required XTP, and that, reciprocally, Ffh/4.5S stimulated the hydrolysis of XTP by FtsY in a reaction that required GTP (49). Thus, just as the structures of the two GTPases are similar, so are their enzymatic properties, both acting as GTP-dependent GTPase activating proteins (GAPs) for each other. To our knowledge, this reciprocally symmetrical interaction is unique among known GTPases. In evolutionary terms, this reaction of the two closely related proteins may have evolved after gene duplication from an originally homotypic interaction, although it is difficult to envision a role for such an ancestral homotypic interaction in protein targetting.

In addition to the SR, other ligands of SRP also affect its nucleotide binding and hydrolysis properties. A role of the ribosome to stimulate GTP binding, for example, was suggested in experiments investigating GTP binding and hydrolysis by SRP54 (50), indicating that ribosome-bound SRP may be set to its activated, GTP-bound state. However, these data are currently at odds with results that suggest that signal-sequence binding prevents SRP from assuming the GTP-bound conformation (51, 52).

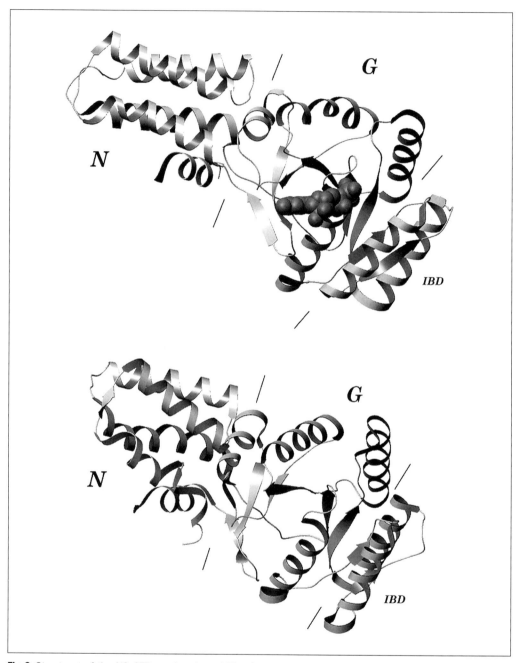

Fig. 3 Structures of the NG GTPase domains of Ffh and FtsY. The GTPase of the SRP comprises ~290 amino acids in two structural domains. The N domain is an α-helical bundle that rests end-on against one side of the GTPase domain opposite the phosphate-binding loops of the active site and the IBD. The NG domains from the two proteins, *T. aquaticus* Ffh (top) and *E. coli* FtsY (bottom), have similar structure, although the N domain of the FtsY is shorter and apparently more disordered. The position of bound GDP in the Ffh NG domain is indicated. Protein databank IDs 1ng1 and 1ftsY (53, 70).

A significant advance in our understanding of the SRP GTPases has come from recent crystallographic work. In particular, the structures of *Thermus aquaticus* Ffh and the GTPase of *E. coli* FtsY have been determined (53–55) (Fig. 3). In full agreement with the domain structure proposed from sequence alignments, Ffh has three structural domains, termed 'N', 'G', and 'M'; similarly, FtsY contains an N (amino-terminal) and a G domain, both structurally very similar to those of Ffh. The G domain is a classical GTPase-fold, with a unique insertion of approximately 50 amino acids between the motifs I and III which extends the core β-sheet by two strands. This domain, which is characteristic for the SRP GTPase subfamily, is termed the 'I-box domain' (IBD) (Fig. 2).

The structure of the intact Ffh shows that the M domain includes a large hydrophobic pocket that is proposed to be the signal-sequence binding site (55). The M domain was previously shown to be responsible for signal-sequence binding to SRP (56). It is characterized by its remarkable abundance in methionine residues (hence its name (17)), and mutational studies of the M domain have helped to map residues involved in signal-sequence binding (57). From the structure it is now clear that the methionine side-chains line the putative signal-sequence binding pocket. In this way, a fluid and hydrophobic environment is formed that has sufficient plasticity to bind structurally divergent signal sequences. Such a model of signal-sequence recognition was originally suggested in the methionine bristle hypothesis (17). Interestingly the M domain is only loosely associated with the N and G domains of the (apo) protein in the crystal, suggesting that the domains might undergo substantial rearrangements during recognition and binding of signal sequences to allow for interdomain communication. Indeed, the N and G domains appear to contribute to signal-sequence binding, as shown by mutational studies with SRP54 which showed that disruption of residues at the interface between the N and G domains produced defects in signal-sequence recognition (58). Chemical modification of cysteine residues in the NG domains of SRP54 blocks signal-sequence binding and, conversely, the nucleotide-bound state of the NG domains affects signal-sequence binding (59).

The M domain also binds to SRP RNA. A helix-turn-helix motif located opposite to the signal-sequence-binding pocket positions positively charged amino-acid side-chains that are thought to mediate RNA binding. Studies in *B. subtilus* have helped to map the residues (60). Binding of Ffh to 4.5S RNA induces a conformational change in the RNA, as assayed by fluorescence measurements (61). The structures of the SRP RNA and 4.5S RNA are the subject of current studies (61–64), but the mechanistic role for SRP RNA is still unknown. SRP RNA is not required for interaction with signal sequences (65), and hence does not simply serve as a structural scaffold for the M domain. It has been shown, however, that binding 4.5S RNA stabilizes the M domain (66). An attractive hypothesis is that SRP RNA might communicate with the ribosome through base-pairing interactions with ribosomal RNA. Conformational changes in SRP induced by signal-sequence binding could then be passed on to the RNA, which in turn could regulate SRP's interaction with the ribosome.

The biochemical and structural studies of Ffh and FtsY suggest that significant conformational changes must occur as the proteins cycle through the steps of occupancy with different guanine nucleotides. From the biochemistry it is clear that a major function of the conformational changes is to modulate the interaction between the two components. SRP and SRP receptor interact in their GTP-bound states, but have only a low affinity when empty or GDP-bound. The notion that SRP and SR primarily interact through their respective NG domains (54) is supported by the observation that SRP RNA and M domain are not required for stimulation of GTPase activity *in vitro* (67). Furthermore, the NG domains of FtsY can function to direct *E. coli* SRP to the membrane even when fused to an unrelated membrane protein (68).

An additional feature distinguishing SRP GTPases from other GTPases is their comparatively low affinity for GDP. The existence of a stable 'empty state' in the structure of the apo Ffh NG domain from *T. aquaticus* is consistent with biochemical evidence in the eukaryotic system (51, 52), which also suggested that the apo-state is a functionally significant intermediate. After completion of a targetting cycle, SRP and SR presumably release GDP spontaneously, i.e. without the need for an external exchange factor. This led to the notion that SRP GTPases have an 'intrinsic exchange factor', and it has been suggested that the unique IBD might confer this property upon SRP GTPases (69). However, this view is not supported by the implications of the recent structures of the GDP- and Mg^{2+}GDP-bound forms of NG from *T. aquaticus* (70). Rather, these structures suggest that the low affinity for nucleotide results from a number of different elements in the structure. These include:

- the stabilization of the apo-state by a hydrogen-bonding network of active-site side-chains;
- a flexible closing loop;
- flexibility in the NG domain interface which is coupled to a rearrangement in the box IV guanine-binding site; and
- hydrogen bonding through a conserved glutamine side-chain to the β-phosphate of GDP in the absence of Mg^{2+}, which potentially represents the stabilization of an intermediate in product release (70).

Several proteins with similar GTPase structures, although seemingly unrelated to the SRP, have been identified by sequence comparisons. These include PilA, a transcriptional regulator that controls pilin gene expression and comprises two domains. Its C-terminal domain is a GTPase (71) that shows substantial sequence conservation with the G domain, but not the N domain, of Ffh. PilA appears to be phosphorylated (72). The sequence of FlhF, a protein associated with flagellar cellular motility identified in *B. subtilus* and *Treponema pallidium*, includes a domain with putative structural homology to Ffh and FtsY (73, 74). The functions of these proteins, and the basis for their apparent structural relationship to the SRP GTPases, are not understood.

The GTPases of SRP/SR-mediated protein targeting present a fascinating puzzle, as both the biochemical logic and the structural nature of the interaction between the two proteins has yet to be understood. That GTP hydrolysis can function to give unidirectionality to the interactions in the protein targeting reaction provides a basis for rationalizing the mechanism. But it remains unclear why a mechanism utilizing hydrolysis of *two* or, in the eukaryotic case, even three GTP molecules has evolved. This question is particularly fascinating in light of that fact that the two GTPases are structurally related and act as mutual GTPase activators, suggesting that there is an intrinsic symmetry in their interaction.

3. GTPases in protein elongation

The ribosome translates the genetic information encoded in the messenger RNA into the sequence of the polypeptide by polymerizing amino acids carried to the ribosome by tRNAs. During the complex multistep process of translation, GTPases play central roles at several stages: during initiation, elongation, and release of the nascent protein. The most well studied of these translation GTPases are the prokaryotic elongation factors, EF-Tu and EF-G (6–11, 75, 76). The elongation factors have been studied for many years (77), and their biochemical and functional properties are the subject of many excellent reviews. We will briefly review the structural biology of the two proteins, and refer the reader to more comprehensive recent reviews of EF-Tu (78) and the ribosome in translation (79).

There are currently several related models for ribosome function (79, 80). Simply put, it is thought that tRNAs interact with three sites on the ribosome during the elongation cycle, binding first to the A site, translocating to the P site following polymerization, and being released from the E site. Two sites on the ribosome are always occupied, and the two elongation-factor GTPases are essential in driving and regulating this process. An overview of the roles of EF-Tu and EF-G in the ribosomal elongation cycle is given in Fig. 4. The two GTPases interact alternately at a common site on the ribosome to catalyse the structural interconversions that carry the ribosome through the elongation cycle.

The role of EF-Tu in the GTP-bound conformation is to carry the aminoacylated tRNA to the A site of the translating ribosome. Following recognition of the proper codon/anticodon interaction, the ribosome stimulates the GTPase activity of the EF-Tu, generating the GDP complex which has low affinity for the tRNA and which then can release it and dissociate from the ribosome. It is thought that in addition to mediating the initial interaction between the tRNA and the ribosome, the hydrolysis of GTP by EF-Tu serves as a timer to enhance the fidelity of translation (81). The aminoacylated tRNA, now in the A site, accepts the growing polypeptide from the P site. EF-Tu thus participates in the first step taken by the tRNA during protein synthesis. Once free of the ribosome, tightly bound GDP is released from EF-Tu by interaction with the exchange factor, EF-Ts, which then allows re-binding of GTP and continuation of the catalytic cycle.

Subsequent to the transpeptidation of the aminoacylated tRNA, the translocation of the tRNA-polypeptide to the P site is promoted by the GTP-bound EF-G. This step requires the coordinate movement of two tRNAs from pretranslocational to post-translocational positions in the P and E sites of the ribosome. It is thought that on binding of the GTP EF-G complex, the GTP is hydrolysed (82), generating an activated conformation of the EF-G, which promotes the transition between the two

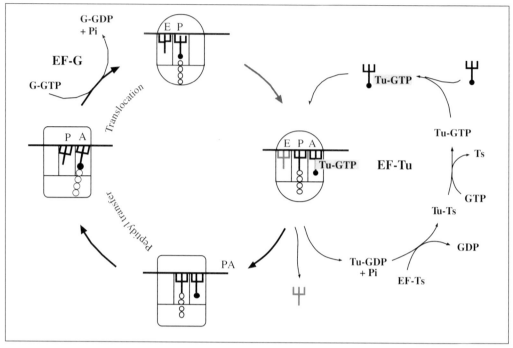

Fig. 4 Elongation factors EF-Tu and EF-G in translation elongation cycle. The ribosome is shown alternating between pre- and post-translocational states. EF-Tu in the GTP conformation carries the aminoacylated tRNA to the A site of the translating ribosome. Following proper recognition of the mRNA codon by the anticodon, the ribosome stimulates the GTPase of EF-Tu, yielding the GDP complex with low affinity for the tRNA. EF-Tu dissociates, and interaction with the exchange factor, EF-Ts, is necessary to release the bound GDP from EF-Tu so that GTP can re-bind. The aminoacylated tRNA in the A site accepts peptidyl transfer from the growing polypeptide in the P site. EF-G then promotes translocation of the tRNA polypeptide to the P site, and a shift of the ribosome along the mRNA. Prior to or during the transition to the post-translocational state the EF-G-bound GTP is hydrolysed, and subsequently EF-G can dissociate. The empty tRNA in the E site is released as translation of the mRNA continues. (Figure adapted from (6) with permission.)

ribosomal states. In the post-translocational state the GDP-bound EF-G can dissociate from the ribosome. The empty tRNA in the E site is released, either prior to or coordinately with the EF-Tu-catalysed binding of another aminoacylated tRNA at the A site, and translation of the mRNA continues. Thus EF-G catalyses the second step in the cycle, and the two GTPases together coordinate cycles of tRNA binding and translocation by catalysing complementary transitions in the state of the ribosome (post- to pretranslocational, and pre- to post-translocational) (6).

It appears clear that the basic features of this translational cycle are highly conserved through evolution. The eukaryotic proteins that correspond to the bacterial proteins EF-Tu and EF-G are termed EF-1α and EF-2 (83), and elongation factor homologues have been identified from species in all kingdoms. The structural and functional features of the eukaryotic translation factor EF-1α have been reviewed (83). Differences in the structure and biochemistry of the eukaryotic proteins presumably reflect the more complex mechanisms of eukaryotic translation (83).

Crystallographic studies of the EF-Tu have been in progress for a long time (84, 85) and have recently yielded a series of spectacular structures (7). The structures of EF-Tu with bound GDP and GMP-PNP (a GTP analogue) are shown in Fig. 5. EF-Tu comprises three domains, of which domain 1 is structurally homologous to the GTPase domain of Ras (85). Domain 2 has a β-barrel structure and is connected to domain 1 by a flexible linker peptide. In the GDP-bound form of EF-Tu the three domains are loosely associated, in a sort of triangle, with domains 2 and 3 interacting across a tight interface, and domains 1 and 3 across a relatively weak interface (86) (Fig. 5). These interactions produce, in the centre of the protein, a relatively large hole. The interaction between domains 1 and 3 is centred near the α2-helix, the switch 2 helix of the GTPase fold.

The structures of *T. aquaticus* and *T. thermophilus* EF-Tu with the bound GTP analogue GMP-PNP reveal a much more compact structure (87, 88) (Fig. 5). Domains 2 and 3 are reoriented, apparently in response to the creation of a new interaction surface on domain 1 which accompanies a shift in the orientation and position of the domain 1 switch 2 helix. The shift in the helix itself is due to small but well-defined structural shifts that occur locally in response to binding of the nucleoside triphosphate analogue (87). The change in the orientations of domains 2 and 3 has the effect of eliminating the hole in the centre of the structure (87, 88) and creating what is the binding cleft for the aminoacyl-tRNA. In the *E. coli* EF-Tu, an α- to β-conformational change in the effector loop, or switch 1 region, has been observed between the GDP- and GTP-bound forms (89, 90). The structural changes that accompany nucleotide binding to domain 1 are therefore coupled to large domain shifts in the protein that create (or destroy) the binding site for the aminoacyl-tRNA; they provide a beautiful example of the amplification of small differences in the protein interactions with GDP and GTP to effect a GTPase switch.

The structure of the second elongation factor, EF-G, has been determined both in complex with GDP (91, 92) and without bound nucleotide (93). The protein is highly elongated, and comprises five domains (Fig. 6). Domain 1 is structurally homologous to the GTPases of Ras and EF-Tu. Domains 2, 3, and 5 are clustered around domain 1, while domain 4 extends away from the other domains so that the protein extends over ~120 Å (93). Although domain 3 is poorly defined in the crystal structure, its position between the tightly associated domain pairs 1/2 and 4/5 suggests a role as a hinge between the two (92). The structure of the GTP-bound conformation of EF-G is unknown. It has been proposed that the GTP-bound state of EF-G might mimic the GDP-bound state of EF-Tu, and vice versa (11, 94); however, solution studies suggest that a large structural change corresponding to that of the GTP- and GDP-bound states of EF-Tu does not occur in EF-G (95).

These structures reveal the relationship of the EF-Tu and EF-G to the GTPase superfamily, identify a common structural core, and suggest how binding and hydrolysis of GTP are coupled to structural change in the elongation-factor GTPases. Interestingly, the two proteins EF-Tu and EF-G share structural homology beyond the GTPase domain, as structure-based sequence alignment of the translation GTPases identifies an extensive consensus structural unit comprising domains 1 and

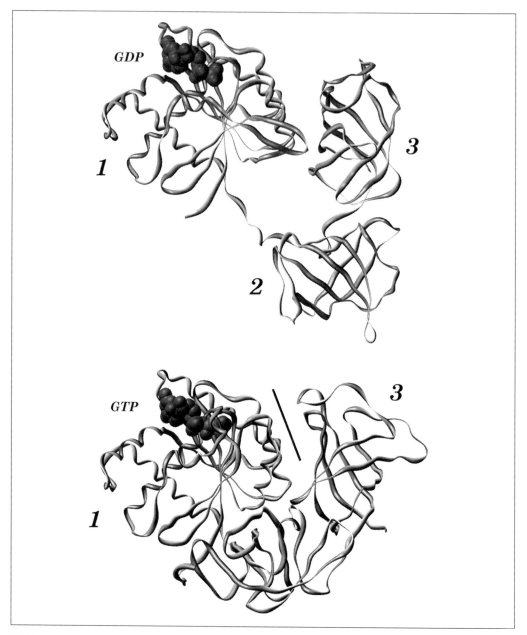

Fig. 5 Structure of EF-Tu. The protein comprises ~400 amino acids organized into three domains. The structures of EF-Tu with bound Mg^{2+}GDP (top) and Mg^{2+}GTP (bottom) are shown. On binding GTP (GTP analogue), a shift in the position and orientation of the α2-helix of domain 1 creates a new interaction surface (indicated by a line) which allows the structural rearrangement of domains 2 and 3. The binding site for aminoacylated tRNA in the GTP complex is at the interface of the three domains; it is destroyed by the remarkably large rearrangement of domains 2 and 3 following hydrolysis of the bound GTP. Interaction with the ribosome stimulates the GTPase activity; however, the mechanism by which the interaction is communicated to loops I and II ('GAP' region) of the GTPase domain 1 is unknown. Protein databank IDs 1eft and 1tui (88, 90). (Figure adapted from (7) with permission.)

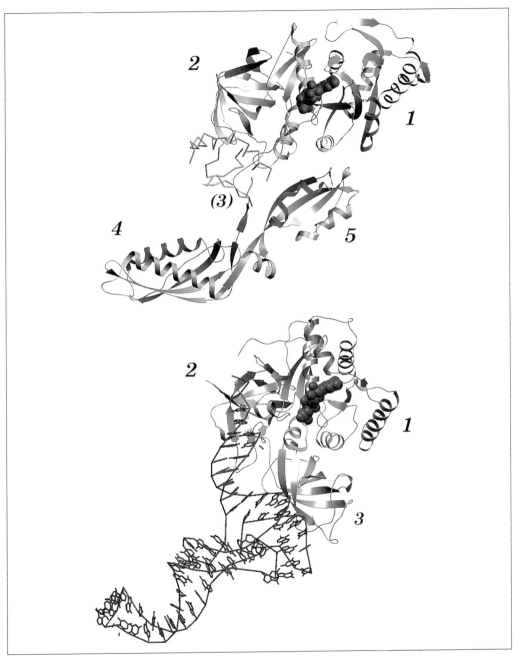

Fig. 6 Structural homology between EF-Tu/tRNA and EF-G. A ribbon drawing of the structure of the GDP complex of EF-G is shown (top). It is similar to that of the apo protein; the structure of the GTP complex is unknown. EF-G (top) is a large protein, ~700 amino acids, organized into five domains. Domain 3 is poorly defined in the crystal structure, and only fragments are shown in the figure. The structure of the EF-Tu ternary complex with GTP analogue and tRNA (bottom) is similar in dimension and shape to the structure of EF-G. The anticodon of the tRNA is at the bottom of the figure. Protein databank IDs 2efg and 1ttt (92, 96). (Figure adapted from (96).)

2 (75). It was proposed that the common ancestor of the translation GTPases contained two domains, and that its structural persistence reflects interaction with the common GTPase effector site of the ribosome (75).

Two additional pieces of the puzzle are supplied by the structure of the ternary complex of *E. coli* EF-Tu, yeast Phe-tRNA, and the non-hydrolysable GTP analogue GMP-PNP (9, 96) (Fig. 6). The structure of the complex shows that the tRNA binds in a cleft between the three domains of the GTP form of EF-Tu, with the anticodon end of the tRNA extending away from the protein (97). The binding site is in precisely the region of the protein that undergoes structural change in the GDP form, thus explaining how tRNA is released following GTP hydrolysis at the ribosome. Consistent with its function as a transporter of disparate aminoacyl-tRNAs, the protein recognizes common features of the RNA; its contacts with the tRNA are located at the aminoacylated 3′ CCA end, the 5′ phosphate, and one side of the tRNA T helix (9). The switch 1 and 2 regions of the EF-Tu GTPase contribute to the interaction with tRNA. In other GTPases these regions are involved in interaction with effectors; however, their conformations in the ternary complex do not change appreciably from the RNA-free structure.

Remarkably, the shape of the ternary protein/tRNA complex is similar to that of the five-domain EF-G (96, 98) (Fig. 6). Thus, in addition to the common domains 1 and 2 noted above, which are in similar conformation, the additional domains 3–5 of EF-G appear to mimic the positions and shape of the acceptor and anticodon helices of the tRNA (96, 99). The overall shape of the two protein complexes suggests that they recognize similar structures on the ribosome. Indeed this makes sense, as the proteins bind at overlapping sites on the ribosome and both the GDP complex of EF-G and the ternary GMP-PNP/EF-Tu/aminoacylated tRNA complex would be expected to interact specifically with the post-translocational state of the ribosome (96). Furthermore, it is consistent with the notion that there is one ribosomal centre, which activates the GTPase activities of both proteins by a common mechanism (96), and that the structural events catalysed by EF-G and EF-Tu at the ribosome have a similar mechanism (100). Interestingly, this thinking can be extended to the structure and function of the translation release factor RF-3, which is also a GTPase (96), as sequence conservation between the release factors and the C terminus of EF-G are consistent with a role for tRNA mimicry in release factor function (101).

The elongation factors are brought to the ribosome in an 'activated' GTP-bound form; following their release as the GDP complexes, the two elongation factors have different strategies for nucleotide exchange. EF-G rapidly releases bound GDP, so it does not require an external exchange factor and can readily rebind GTP from the cellular pool (2). The structure of EF-G reveals an insertion in the GTPase domain 1, termed the G′ subdomain, which interacts with the nucleotide-binding site and may serve as an internal exchange factor (93). In contrast, EF-Tu requires a third elongation factor, EF-Ts, to promote nucleotide exchange (Fig. 4). The structures of complexes of EF-Tu and its exchange factor, from both *T. thermophilus* (102), and *E. coli* (103), have been determined and reveal that EF-Ts interacts over two extensive surfaces of EF-Tu. Interestingly, in the *T. thermophilus* structure the corresponding EF-Ts surfaces are supplied by a symmetrical dimer, while in *E. coli* they are supplied by the monomeric, but structurally homologous, protein (102). The EF-Tu/EF-Ts interaction disrupts residues that contribute to the magnesium ion coordination site, and also more directly distorts the backbone of the phosphate-binding loop, so facilitating release of bound $Mg^{2+}GDP$ (102, 103).

A key question unanswered by the structures is, of course, how interaction with the ribosome stimulates the GTPase activity of the EF-Tu and EF-G, and thus turns 'off' the 'GTPase switch' in each case. The ribosome can be thought to oscillate between two states, a post-translocational state, which serves as the effector for EF-Tu, and a pretranslocational state, which serves as the effector for EF-G (104) (Fig. 4). The ribosomal-binding sites for EF-Tu and EF-G (and for IF-2 and release factors RF-1 and -2) overlap, and the interactions have been mapped to a region of the 50S subunit defined by the L7/L12 stalk, proteins L10 and L11, and the α-sarcin loop (105). Both EF-Tu and EF-G interact with the α-sarcin loop of the 23S rRNA (106, 107), and EF-G binding to the sarcin/ricin domain of *E. coli* 23S rRNA can be mimicked using a 27-mer oligoribonucleotide derived from 23S rRNA (108). There is also evidence for functional interaction between EF-Tu and the 16S rRNA (109). It appears that only one GTP is hydrolysed by EF-Tu per aminoacyl-tRNA bound and peptide bond formed (110); however, this has been the subject of some discussion as hydrolysis of two GTPases per cycle has been measured in other laboratories (111, 112).

A single GTPase centre on the ribosome may well stimulate the GTPase activities of both EF-Tu and EF-G by a common, or structurally homologous mechanism. Biochemical studies suggest that codon–anticodon recognition provides the specific signal for GTP hydrolysis by EF-Tu on the ribosome (113), and this is consistent with a proof-reading mechanism that utilizes control of GTPase activation (114). However, the nature of the activating interaction remains unknown. Kinetic studies of EF-Tu have identified a conformational change following codon recognition and preceding GTP hydrolysis on the ribosome (115–117). The close structural relationship between the tRNA and the GTPase domain 1 suggests that ribosome recognition of the anticodon could readily be communicated with the GTPase domain of EF-Tu by some conformational mechanism. However, it remains to be seen whether the ribosome could instead form direct interactions with the GTPase domain analogous to those observed in other GTPase/GAP complexes (118). Mutagenic and proteolytic studies have shown that disrupting specific residues of the EF-Tu domain 1 effector, the switch 1 and 2 loops that contribute to both the GTPase active site and to tRNA binding (96), can uncouple the ribosome-stimulated GTPase activation from RNA recognition and nucleotide binding (119, 120). Mutations of domain 2 residues G222 and G280 have long been known (121), and the mutation G222D has been shown to inhibit the codon-induced structural transitions of tRNA and EF-Tu on the ribosome (122). Thus both domains 1 and 2 of EF-Tu have been implicated in the mechanism of GTPase activation, but the specific details of the ribosome interaction with each protein remain unknown.

GTPases structurally related to the elongation factors function in translation initiation and termination (123, 124). The *E. coli* translation initiation factor IF-2 has been characterized biochemically; its purified G domain displays GTP binding (125) and GTP hydrolysis activities (126). Other proteins are known which have similar or specialized functions (124). Thus, for example, in *E. coli* SelB, an elongation factor, replaces EF-Tu to incorporate a rare amino acid, selenocysteine, into a small, specific subset of proteins. The GTP activity of SelB is modulated by both the ribosome and by selenocysteine-inserting RNA elements (127) and the protein has structural homology with EF-Tu over three of its four domains (128–130).

The crystal structures of EF-Tu and EF-G provide a framework for understanding the mechanism by which each of the proteins utilizes binding and hydrolysis of GTP to interact with tRNA and the ribosome to promote translation. However, an understanding of their direct interactions with the ribosome is missing. Biochemical studies have begun to map these interactions but recent results raise the exciting possibility that they will soon be visualized directly. The position of the EF-Tu/tRNA complex on the ribosome A site has been already been visualized at 18 Å resolution (131), and recently the position of elongation factor G on the ribosome has been visualized using three-dimensional cryoelectron microscopy (132).

4. Conclusions

Interestingly, both the SRP/SR and EF-Tu/EF-G systems comprise GTPase pairs that have structural homology beyond the core GTPase domain. That structurally homologous N and G domains are shared by the SRP and SR GTPases, and structurally homologous domains 1 and 2 are shared by the elongation-factor GTPases presumably reflects the requirement of each pair to interact with a common macromolecular partner. In the case of the elongation factors, this is almost certainly the ribosome. In the case of the SRP GTPases the putative common partner is unknown, but could comprise the SRP itself (or its RNA), the signal peptide, the translocation apparatus, or, intriguingly, the ribosome as well. Could there be additional connections between the two GTPase systems? The NG domains of FtsY are preceded by a highly acidic domain of unknown function. Might this domain structurally mimic SRP RNA bound to Ffh, just as domains 3–5 of EF-G mimic aminoacyl-tRNA bound to EF-Tu? The prokaryotic 4.5S RNA, the homologue of eukaryotic SRP RNA, binds to Ffh to form the prokaryotic SRP. However, 4.5S RNA has also been proposed to have a role in translation (133), and in cell extracts, EF-G was shown to bind to 4.5S RNA with apparent specificity (134). Moreover, SRP can functionally engage with ribosomes only at a specific step in the elongation cycle (135), indicating that SRP, like elongation factors, discriminates the conformational states of the ribosome. These intriguing parallels, together with the extreme evolutionary conservation of either system, suggest that there may be significant connections that remain to be appreciated and incorporated into our current models.

References

1. Bourne, H. R., Sanders, D. A., and McCormick, F. (1990) The GTPase superfamily: a conserved switch for diverse cell functions. *Nature*, **348**, 125.

2. Bourne, H. R., Sanders, D. A., and McCormick, F. (1991) The GTPase superfamily: conserved structure and molecular mechanism. *Nature*, **349**, 117.

3. Kjeldgaard, M., Nyborg, J., and Clark, B. (1996) The GTP binding motif: variations on a theme. *FASEB J.*, **10**, 1347.

4. Schweins, T. and Wittinghofer, A. (1994) Structures, interactions and relationships. *Curr. Biol.*, 4, 547.

5. Walter, P. and Johnson, A. (1994) Signal sequence recognition and protein targeting to the endoplasmic reticulum membrane. *Annu. Rev. Cell Biol.*, **10**, 87.

6. Nierhaus, K. H. (1996) An elongation factor turn-on. *Nature*, **379**, 491.

7. Sprinzl, M. (1994) Elongation factor Tu: a regulatory GTPase with an integrated effector. *Trends Biochem. Sci.*, **19**, 245.

8. Weijland, A., Harmark, K., Cool, R. H., Anborgh, P. H., and Parmeggiani, A. (1992) Elongation factor Tu: a molecular switch in protein biosynthesis. *Mol. Microbiol.*, **6**, 683.

9. Clark, B. and Nyborg, J. (1997) The ternary complex of EF-Tu and its role in protein biosynthesis. *Curr. Opinion Struct. Biol.*, **7**, 110.

10. Abel, K. and Jurnak, F. (1996) A complex profile of protein elongation: translating chemical energy into molecular movement. *Structure*, **4**, 229.

11. Liljas, A. and Garber, M. (1995) Ribosomal proteins and elongation factors. *Curr. Opinion Struct. Biol.*, **5**, 721.

12. Lutcke, H. (1995) Signal recognition particle (SRP), a ubiquitous initiator of protein translocation. *Eur. J. Biochem.*, **228**, 531.

13. Walter, P. and Blobel, G. (1980) Purification of membrane-associated protein complex required for protein translocation across the endoplasmic reticulum. *Proc. Natl Acad. Sci. USA*, **77**, 7112.

14. Walter, P. and Blobel., G. (1982) Signal recognition particle contains a 7S RNA essential for protein translocation across the endoplasmic reticulum. *Nature*, **299**, 691.

15. Kurzchalia, T. V., Wiedmann, M., Girshovich, A. S., Bochkareva, E. S., Bielka, H., and Rapoport, T. A. (1986) The signal sequence of nascent preprolactin interacts with the 54K polypeptide of the signal recognition particle. *Nature*, **320**, 634.

16. Krieg, U. C., Walter, P., and Johnson, A. E. (1986) Photocrosslinking of the signal sequence of nascent preprolactin to the 54-kilodalton polypeptide of the signal recognition particle. *Proc. Natl Acad. Sci., USA*, **83**, 8604.

17. Bernstein, H. D., Poritz, M. A., Strub, K., Hoben, P. J., Brenner, S., and Walter, P. (1989) Model for signal sequence recognition from amino-acid sequence of 54K subunit of signal recognition particle. *Nature*, **340**, 482.

18. Römisch, K., Webb, J., Herz, J., Prehn, S., Frank, R., Vingron, M., and Dobberstein, B. (1989) Homology of the 54K protein of signal recognition particle, docking protein, and two E. coli proteins with putative GTP-binding domains. *Nature*, **340**, 478.

19. Connolly, T. and Gilmore, R. (1993) GTP hydrolysis by complexes of the signal recognition particle and the signal recognition particle receptor. *J. Cell Biol.*, **123**, 799.

20. Gilmore, R. and Blobel, G. (1983) Transient involvement of signal recognition particle and its receptor in the microsomal membrane prior to protein translocation. *Cell*, **35**, 677.

21. Gilmore, R., Walter, P., and Blobel, G. (1982) Protein translocation across the endoplasmic reticulum. II. Isolation and characterization of the signal recognition particle receptor. *J. Cell Biol.*, **95**, 470.

22. Gilmore, R., Blobel, G., and Walter, P. (1982) Protein translocation across the endoplasmic reticulum. I. Detection in the microsomal membrane of a receptor for the signal recognition particle. *J. Cell Biol.*, **95**, 463.

23. Rapiejko, P. J. and Gilmore, R. (1994) Signal sequence recognition and targeting of ribosomes to the endoplasmic reticulum by the signal recognition particle do not require GTP. *Mol. Biol. Cell*, **5**, 887.

24. Poritz, M. A., Bernstein, H. D., Strub, K., Zopf, D., Wilhelm, H., and Walter, P. (1990) An *E. coli* ribonucleoprotein containing 4.5S RNA resembles mammalian signal recognition particle. *Science*, **250**, 1111.

25. Poritz, M. A., Strub, K., and Walter, P. (1988) Human SRP RNA and *E. coli* 4.5S RNA contain a highly homologous structural domain. *Cell*, **55**, 4.

26. Luirink, J., ten Hagen-Jongman, C., van der Weijden, C., Oudega, B., High, S., Dobberstein, B., and Kusters, R. (1994) An alternative protein targeting pathway in *Escherichia coli*: studies on the role of FtsY. *EMBO J.*, **13**, 2289.

27. de Leeuw, E., Poland, D., Mol, O., Sinning, I., ten Hagen-Jongman, C., Oudega, B., and Luirink, J. (1997) Membrane association of FtsY, the *E. coli* SRP receptor. *FEBS Lett.*, **416**, 225.

28. Ladefoged, S. and Christiansen, G. (1997) A GTP-binding protein of *Mycoplasma hominis*: a small sized homolog to the signal recognition particle receptor FtsY. *Gene*, **201**, 37.

29. Tajima, S., Lauffer, L., Rath, V., and Walter, P. (1986) The signal recognition particle receptor is a complex that contains two distinct polypeptide chains. *J. Cell Biol.*, **103**, 1167.

30. Miller, J., Tajima, S., Lauffer, L., and Walter, P. (1995) The beta subunit of the signal recognition particle receptor is a transmembrane GTPase that anchors the alpha subunit, a peripheral membrane GTPase, to the endoplasmic reticulum membrane. J. Cell Biol., **128**, 273.

31. Ogg, S. C., Barz, W. P., and Walter, P. (1998) A functional GTPase domain, but not its transmembrane domain, is required for function of the SRP receptor beta-subunit. *J. Cell Biol.*, **142**, 341.

32. Bassford, P., Beckwith, J., Ito, K., Kumamoto, C., Mizushima, S., Oliver, D., Randall, L., Silhavy, T., Tai, P. C. and Wickner, B. (1991) The primary pathway of protein export in *E. coli. Cell*, **65**, 367.

33. Beckwith, J. (1991) 'Sequence-gazing?'. *Science*, **251**, 1161.

34. Ulbrandt, N., Newitt, J., and Bernstein, H. (1997) The *E. coli* signal recognition particle is required for the insertion of a subset of inner membrane proteins. *Cell*, **88**, 187.

35. MacFarlane, J. and Muller, M. (1995) Functional integration of a polytopic membrane protein of *E. coli* requires the bacterial signal recognition particle. *Biochem. Soc. Trans.*, **23**, 560S.

36. de Gier, J., Mansournia, P., Valent, Q., Phillips, G., Luirink, J., and von Heijne, G. (1996) Assembly of a cytoplasmic membrane protein in *Escherichia coli* is dependent on the signal recognition particle. *FEBS Lett.*, **399**, 307.

37. Seluanov, A. and Bibi, E. (1997) FtsY, the prokaryotic signal recognition particle receptor homologue, is essential for biogenesis of membrane proteins. *J. Biol. Chem.*, **272**, 2053.

38. Samuelsson, T., Olsson, M., Wikstrom, P., and Johansson, B. (1995) The GTPase activity of the *Escherichia coli* Ffh protein is important for normal growth. *Biochim. Biophys. Acta*, **1267**, 83.

39. Kusters, R., Lentzen, G., Eppens, E., van Geel, A., van der Weijden, C., Wintermeyer, W., and Luirink, J. (1995) The functioning of the SRP receptor FtsY in protein-targeting in *E. coli* is correlated with its ability to bind and hydrolyse GTP. *FEBS Lett.*, **372**, 253.

40. Luirink, J. and Dobberstein, B. (1994) Mammalian and *Escherichia coli* signal recognition particles. *Mol. Microbiol.*, **11**, 9.

41. de Gier, J., Valent, Q., Von Heijne, G., and Luirink, J. (1997) The *E. coli* SRP: preferences of a targeting factor. *FEBS Lett.*, **408**, 1.

42. Bernstein, H. D. (1998) Protein targeting: Getting into the groove. *Curr. Biol.*, **8**, R715.

43. Bernstein, H. D., Zopf, D., Freymann, D. M., and Walter, P. (1993) Functional substitution of the signal recognition particle 54-kDa subunit by its *Escherichia coli* homolog. *Proc. Natl Acad. Sci., USA*, **90**, 5229.

44. Powers, T. and Walter, P. (1997) Co-translational protein targeting catalyzed by the *Escherichia coli* signal recognition particle and its receptor. *EMBO J.*, **16**, 4880.

45. Patel, S. and Austen, B. (1996) Substitution of fifty four homologue (Ffh) in *Escherichia coli* with the mammalian 54-kDa protein of signal-recognition particle. *Eur. J. Biochem.*, **238**, 760.

46. Nakamura, K., Nishiguchi, M., Honda, K., and Yamane, K. (1994) The *Bacillus subtilis* SRP54 homologue, Ffh, has an intrinsic GTPase activity and forms a ribonucleoprotein complex with small cytoplasmic RNA *in vivo*. *Biochem. Biophys. Res. Commun.*, **199**, 1394.

47. Samuelsson, T. and Olsson, M. (1993) GTPase activity of a bacterial SRP-like complex. *Nucl. Acids Res.*, **21**, 847.

48. Miller, J., Bernstein, H., and Walter, P. (1994) Interaction of *E. coli* Ffh/4.5S ribonucleoprotein and FtsY mimics that of mammalian signal recognition particle and its receptor. *Nature*, **367**, 657.

49. Powers, T. and Walter, P. (1995) Reciprocal stimulation of GTP hydrolysis by two directly interacting GTPases. *Science*, **269**, 1422.

50. Bacher, G., Lutcke, H., Jungnickel, B., Rapoport, T. A., and Dobberstein, B. (1996) Regulation by the ribosome of the GTPase of the signal-recognition particle during protein targeting. *Nature*, **381**, 248.

51. Miller, J., Wilhelm, H., Gierasch, L., Gilmore, R., and Walter, P. (1993) GTP binding and hydrolysis by the signal recognition particle during initiation of protein translocation. *Nature*, **366**, 351.

52. Rapiejko, P. J. and Gilmore, R. (1997) Empty site forms of the SRP54 and SR alpha GTPases mediate targeting of ribosome-nascent chain complexes to the endoplasmic reticulum. *Cell*, **89**, 703.

53. Montoya, G., Svensson, C., Luirink, J., and Sinning, I. (1997) Crystal structure of the NG domain from the signal-recognition particle receptor FtsY. *Nature*, **385**, 365.

54. Freymann, D. M., Keenan, R. J., Stroud, R. M., and Walter, P. (1997) Structure of the conserved GTPase domain of the signal recognition particle. *Nature*, **385**, 361.

55. Keenan, R. J., Freymann, D. M., Walter, P., and Stroud, R. M. (1998) Crystal structure of the signal sequence binding subunit of the signal recognition particle. *Cell*, **94**, 181.

56. Zopf, D., Bernstein, H., Johnson, A., and Walter, P. (1990) The methionine-rich domain of the 54 kd protein subunit of the signal recognition particle contains an RNA binding site and can be crosslinked to a signal sequence. *EMBO J.*, **9**, 4511.

57. Takamatsu, H., Bunai, K., Horinaka, T., Oguro, A., Nakamura, K., Watabe, K., and Yamane, K. (1997) Identification of a region required for binding to presecretory protein in *Bacillus subtilis* Ffh, a homologue of the 54-kDa subunit of mammalian signal recognition particle. *Eur. J. Biochem.*, **248**, 575.

58. Newitt, J. and Bernstein, H. (1997) The N-domain of the signal recognition particle 54-kDa subunit promotes efficient signal sequence binding. *Eur. J. Biochem.*, **245**, 720.

59. Lütcke, H., High, S., Römisch, K., Ashford, A. J., and Dobberstein, B. (1992) The methionine-rich domain of the 54 kDa subunit of signal recognition particle is sufficient for the interaction with signal sequences. *EMBO J.*, **11**, 1543.

60. Kurita, K., Honda, K., Suzuma, S., Takamatsu, H., Nakamura, K., and Yamane, K. (1996) Identification of a region of *Bacillus subtilis* Ffh, a homologue of mammalian SRP54 protein, that is essential for binding to small cytoplasmic RNA. *J. Biol. Chem.*, **271**, 13140.

61. Lentzen, G., Dobberstein, B., and Wintermeyer, W. (1994) Formation of SRP-like particle induces a conformational change in *E. coli* 4.5S RNA. *FEBS Lett.*, **348**, 233.

62. Schmitz, U., Freymann, D., James, T., Keenan, R., Vinayak, R., and Walter, P. (1996) NMR studies of the most conserved RNA domain of the mammalian signal recognition particle (SRP). *RNA*, **2**, 1213.

63. Gowda, K., Chittenden, K., and Zwieb, C. (1997) Binding site of the M-domain of human protein SRP54 determined by systematic site-directed mutagenesis of signal recognition particle RNA. *Nucl. Acids Res.*, **25**, 388.

64. Lentzen, G., Moine, H., Ehresmann, C., Ehresmann, B., and Wintermeyer, W. (1996) Structure of 4.5S RNA in the signal recognition particle of *Escherichia coli* as studied by enzymatic and chemical probing. *RNA*, **2**, 244.

65. Bunai, K., Takamatsu, H., Horinaka, T., Oguro, A., Nakamura, K., and Yamane, K. (1996) *Bacillus subtilis* Ffh, a homologue of mammalian SRP54, can intrinsically bind to the precursors of secretory proteins. *Biochem. Biophys. Res. Commun.*, **227**, 762.

66. Zheng, N. and Gierasch, L. (1997) Domain interactions in *E. coli* SRP: stabilization of M domain by RNA is required for effective signal sequence modulation of NG domain. *Mol. Cell*, **1**, 79.

67. Macao, B., Luirink, J., and Samuelsson, T. (1997) Ffh and FtsY in a *Mycoplasma mycoides* signal-recognition particle pathway: SRP RNA and M domain of Ffh are not required for stimulation of GTPase activity *in vitro*. *Mol. Microbiol.*, **24**, 523.

68. Zelazny, A., Seluanov, A., Cooper, A., and Bibi, E. (1997) The NG domain of the prokaryotic signal recognition particle receptor, FtsY, is fully functional when fused to an unrelated integral membrane polypeptide. *Proc. Natl Acad. Sci., USA*, **94**, 6025.

69. Moser, C., Mol, O., Goody, R., and Sinning, I. (1997) The signal recognition particle receptor of *Escherichia coli* (FtsY) has a nucleotide exchange factor built into the GTPase domain. *Proc. Natl Acad. Sci., USA*, **94**, 11339.

70. Freymann, D. M., Keenan, R. J., Stroud, R. M., and Walter, P. (1999) Functional changes in the structure of the SRP GTPase on binding GDP and Mg^{2+}GDP, *Nature Structural Biology*, **6**, 793.

71. Arvidson, C. and So, M. (1995) The *Neisseria* transcriptional regulator PilA has a GTPase activity. *J. Biol. Chem.*, **270**, 26045.

72. Taha, M. and Giorgini, D. (1995) Phosphorylation and functional analysis of PilA, a protein involved in the transcriptional regulation of the pilin gene in *Neisseria gonorrhoeae*. *Mol. Microbiol.*, **15**, 667.

73. Carpenter, P., Hanlon, D., and Ordal, G. (1992) flhF, a *Bacillus subtilis* flagellar gene that encodes a putative GTP-binding protein. *Mol. Microbiol.*, **6**, 2705.

74. Hardham, J., Frye, J., Young, N., and Stamm, L. (1997) Identification and sequences of the *Treponema pallidum* flhA, flhF, and orf304 genes. *DNA Seq.*, **7**, 107.

75. Ævarsson, A. (1995) Structure-based sequence alignment of elongation factors Tu and G with related GTPases involved in translation. *J. Mol. Evol.*, **41**, 1096.

76. Nyborg, J. and Kjeldgaard, M. (1996) Elongation in bacterial protein biosynthesis. *Curr. Opinion Biotechnol.*, **7**, 369.

77. Lucas-Lenard, J. and Lipmann, F. (1966) Separation of three microbial amino acid polymerization factors. *Proc. Natl Acad. Sci., USA*, **55**, 1562.

78. Krab, I. M. and Parmeggiani, A. (1998) EF-Tu, a GTPase odyssey. *Biochim. Biophys. Acta*, **1443**, 1.

79. Green, R. and Noller, H. F. (1997) Ribosomes and translation. *Annu. Rev. Biochem.*, **66**, 679.

80. Neirhaus, K. H. (1990) The allosteric three-site model for the ribosomal elongation cycle: features and future. *Biochemistry*, **29**, 4997.

81. Thompson, R. C. and Karim, A. M. (1982) The accuracy of protein biosynthesis is limited by its speed: high fidelity selection by ribosomes of aminoacyl-tRNA ternary complexes containing GTP[gamma S]. *Proc. Natl Acad. Sci., USA*, **79**, 4922.

82. Rodnina, M. V., Savelsbergh, A., Katunin, V. I., and Wintermeyer, W. (1997) Hydrolysis of GTP by elongation factor G drives tRNA movement on the ribosome [see comments]. *Nature*, **385**, 37.

83. Negrutskii, B. and Elqskaya, A. (1998) Eukaryotic translation elongation factor 1 alpha: structure, expression, functions, and possible role in aminoacyl-tRNA channeling. *Prog. Nucl. Acid Res. Mol. Biol.*, **60**, 47.

84. la Cour, T. F., Nyborg, J., Thirup, S., and Clark, B. F. (1985) Structural details of the binding of guanosine diphosphate to elongation factor Tu from *E. coli* as studied by X-ray crystallography. *EMBO J.*, **4**, 2385.

85. Jurnak, F. (1985) Structure of the GDP domain of EF-Tu and location of amino acids homologous to ras oncogene proteins. *Science*, **230**, 32.

86. Kjeldgaard, M. and Nyborg, J. (1992) Refined structure of elongation factor Tu from *Escherichia coli*. *J. Mol. Biol.*, **223**, 721.

87. Berchtold, H., Reshetnikova, L., Reiser, C., Schirmer, N., Sprinzl, M., and Hilgenfeld, R. (1993) Crystal structure of active elongation factor Tu reveals major domain rearrangements. *Nature*, **365**, 126.

88. Kjeldgaard, M., Nissen, P., Thirup, S., and Nyborg, J. (1993) The crystal structure of elongation factor EF-Tu from *Thermus aquaticus* in the GTP conformation. *Structure*, **1**, 35.

89. Abel, K., Yoder, M., Hilgenfeld, R., and Jurnak, F. (1996) An alpha to beta conformational switch in EF-Tu. *Structure*, **4**, 1153.

90. Polekhina, G., Thirup, S., Kjeldgaard, M., Nissen, P., Lippmann, C., and Nyborg, J. (1996) Helix unwinding in the effector region of elongation factor EF-Tu-GDP. *Structure*, **4**, 1141.

91. al-Karadaghi, S., Aevarsson, A., Garber, M., Zheltonosova, J., and Liljas, A. (1996) The structure of elongation factor G in complex with GDP: conformational flexibility and nucleotide exchange. *Structure*, **4**, 555.

92. Czworkowski, J., Wang, J., Steitz, T. A., and Moore, P. B. (1994) The crystal structure of elongation factor G complexed with GDP, at 2.7 Å resolution. *EMBO J.*, **13**, 3661.

93. Ævarsson, A., Brazhnikov, E., Garber, M., Zheltonosova, J., Chirgadze, Y., al-Karadaghi, S., Svensson, L., and Liljas, A. (1994) Three-dimensional structure of the ribosomal translocase: elongation factor G from *Thermus thermophilus*. *EMBO J.*, **13**, 3669.

94. Liljas, A., Ævarsson, A., al-Karadaghi, S., Garber, M., Zheltonosova, J., and Brazhnikov, E. (1995) Crystallographic studies of elongation factor G. *Biochem. Cell Biol.* **73**, 1209.

95. Czworkowski, J. and Moore, P. (1997) The conformational properties of elongation factor G and the mechanism of translocation. *Biochemistry*, **36**, 10327.

96. Nissen, P., Kjeldgaard, M., Thirup, S., Polekhina, G., Reshetnikova, L., Clark, B. F. C., and Nyborg, J. (1995) Crystal structure of the ternary complex of Phe-tRNAphe, EF-Tu, and a GTP analog. *Science*, **270**, 1464.

97. Nissen, P., Kjeldgaard, M., Thirup, S., Clark, B., and Nyborg, J. (1996) The ternary complex of aminoacylated tRNA and EF-Tu-GTP. Recognition of a bond and a fold. *Biochimie*, **78**, 921.

98. Nyborg, J., Nissen, P., Kjeldgaard, M., Thirup, S., Polekhina, G., Clark, B., and Reshetnikova, L. (1997) Macromolecular mimicry in protein biosynthesis. *Fold Des.*, **2**, S7.

99. Nyborg, J., Nissen, P., Kjeldgaard, M., Thirup, S., Polekhina, G., and Clark, B. (1996) Structure of the ternary complex of EF-Tu: macromolecular mimicry in translation. *Trends Biochem. Sci.*, **21**, 81.

100. Wilson, K. S. and Noller, H. F. (1998) Molecular movements inside the translational engine. *Cell*, **92**, 337.

101. Ito, K., Ebihara, K., Uno, M., and Nakamura, Y. (1996) Conserved motifs in prokaryotic and eukaryotic polypeptide release factors: tRNA-protein mimicry hypothesis. *Proc. Natl Acad. Sci., USA*, **93**, 5443.

102. Wang, Y., Jiang, Y., Meyering-Voss, M., Sprinzl, M., and Sigler, P. (1997) Crystal structure of the EF-TuEF-Ts complex from *Thermus thermophilus*. *Nat. Struct. Biol.*, **4**, 950.

103. Kawashima, T., Berthet-Colominas, C., Wulff, M., Cusack, S., and Leberman, R. (1996) The structure of the *Escherichia coli* EF-TuEF-Ts complex at 25 A resolution. *Nature*, **379**, 511.

104. Mesters, J. R., Potapov, A. P., de Graaf, J. M., and Kraal, B. (1994) Synergism between the GTPase activities of EF-TuGTP and EF-GGTP on empty ribosomes. Elongation factors as stimulators of the ribosomal oscillation between two conformations. *J. Mol. Biol.*, **242**, 644.

105. Noller, H. F. (1991) Ribosomal RNA and translation. *Annu. Rev. Biochem.*, **60**, 191.

106. Sköld, S. E. (1983) Chemical crosslinking of elongation factor G to the 23S RNA in 70S ribosomes from *Escherichia coli*. *Nucl. Acids Res.*, **11**, 4923.

107. Moazed, D., Robertson, J. M., and Noller, H. F. (1988) Interaction of elongation factors EF-G and EF-Tu with a conserved loop in 23S RNA. *Nature*, **334**, 362.

108. Munishkin, A. and Wool, I. (1997) The ribosome-in-pieces: binding of elongation factor EF-G to oligoribonucleotides that mimic the sarcin/ricin and thiostrepton domains of 23S ribosomal RNA. *Proc. Natl Acad. Sci., USA*, **94**, 12280.

109. Powers, T. and Noller, H. (1993) Evidence for functional interaction between elongation factor Tu and 16S ribosomal RNA. *Proc. Natl Acad. Sci., USA*, **90**, 1364.

110. Rodnina, M. and Wintermeyer, W. (1995) GTP consumption of elongation factor Tu during translation of heteropolymeric mRNAs. *Proc. Natl Acad. Sci., USA*, **92**, 1945.

111. Weijland, A. and Parmeggiani, A. (1994) Why do two EF-Tu molecules act in the elongation cycle of protein biosynthesis? *Trends Biochem. Sci.*, **19**, 188.

112. Scoble, J., Bilgin, N., and Ehrenberg, M. (1994) Two GTPs are hydrolysed on two molecules of EF-Tu for each elongation cycle during code translation. *Biochimie*, **76**, 59.

113. Rodnina, M., Pape, T., Fricke, R., Kuhn, L., and Wintermeyer, W. (1996) Initial binding of the elongation factor TuGTPaminoacyl-tRNA complex preceding codon recognition on the ribosome. *J. Biol. Chem.*, **271**, 646.

114. Powers, T. and Noller, H. (1994) The 530 loop of 16S rRNA: a signal to EF-Tu? *Trends Genet.*, **10**, 27.

115. Rodnina, M., Fricke, R., Kuhn, L., and Wintermeyer, W. (1995) Codon-dependent conformational change of elongation factor Tu preceding GTP hydrolysis on the ribosome. *EMBO J.*, **14**, 2613.

116. Rodnina, M., Pape, T., Fricke, R., and Wintermeyer, W. (1995) Elongation factor Tu, a GTPase triggered by codon recognition on the ribosome: mechanism and GTP consumption. *Biochem. Cell Biol.*, **73**, 1221.

117. Kalbitzer, H., Feuerstein, J., Goody, R., and Wittinghofer, A. (1990) Stereochemistry and lifetime of the GTP hydrolysis intermediate at the active site of elongation factor Tu from *Bacillus stearothermophilus* as inferred from the 17O-55Mn superhyperfine interaction. *Eur. J. Biochem.*, **188**, 355.

118. Scheffzek, K., Ahmadian, M. R., Kabsch, W., Wiesmüller, L., Lautwein, A., Schmitz, F., and Wittinghofer, A. (1997) The Ras–RasGAP complex: structural basis for GTPase activation and its loss in oncogenic ras mutants. *Science*, **277**, 333.

119. Zeidler, W., Schirmer, N. K., Egle, C., Ribeiro, S., Kreutzer, R., and Sprinzl, M. (1996) Limited proteolysis and amino acid replacements in the effector region of *Thermus thermophilus* elongation factor Tu. *Eur. J. Biochem.*, **239**, 265.

120. Jonak, J., Anborgh, P. H., and Parmeggiani, A. (1994) Histidine-118 of elongation factor Tu: its role in aminoacyl-tRNA binding and regulation of the GTPase activity. *FEBS Lett.*, **343**, 94.

121. Swart, G. W. M., Parmeggiani, A., Kraal, B., and Bosch, L. (1987) Effect of the mutation glycine-222→aspartic acid on the functions of elongation factor Tu. *Biochemistry*, **26**, 2047.

122. Vorstenbosch, E., Pape, T., Rodnina, M. V., Kraal, B., and Wintermeyer, W. (1996) The G222D mutation in elongation factor Tu inhibits the codon-induced conformational changes leading to GTPase activation on the ribosome. *EMBO J.*, **15**, 6766.

123. Schmitt, E., Guillon, J., Meinnel, T., Mechulam, Y., Dardel, F., and Blanquet, S. (1996) Molecular recognition governing the initiation of translation in *Escherichia coli*. A review. *Biochimie*, **78**, 543.

124. Laalami, S., Grentzmann, G., Bremaud, L., and Cenatiempo, Y. (1996) Messenger RNA translation in prokaryotes: GTPase centers associated with translational factors. *Biochimie*, **78**, 577.

125. Vachon, G., Laalami, S., Grunberg-Manago, M., Julien, R., and Cenatiempo, Y. (1990) Purified internal G-domain of translational initiation factor IF-2 displays guanine nucleotide binding properties. *Biochemistry*, **29**, 9728.

126. Severini, M., Spurio, R., La Teana, A., Pon, C., and Gualerzi, C. (1991) Ribosome-independent GTPase activity of translation initiation factor IF2 and of its G-domain. *J. Biol. Chem.*, **266**, 22800.

127. Huttenhofer, A. and Bock, A. (1998) Selenocysteine inserting RNA elements modulate GTP hydrolysis of elongation factor SelB. *Biochemistry*, **37**, 885.

128. Bock, A., Hilgenfeld, R., Tormay, P., Wilting, R., and Kromayer, M. (1997) Domain structure of the selenocysteine-specific translation factor SelB in prokaryotes. *Biomed. Environ. Sci.*, **10**, 125.

129. Kromayer, M., Wilting, R., Tormay, P., and Bock, A. (1996) Domain structure of the prokaryotic selenocysteine-specific elongation factor SelB. *J. Mol. Biol.*, **262**, 413.

130. Hilgenfeld, R., Bock, A., and Wilting, R. (1996) Structural model for the selenocysteine-specific elongation factor SelB. *Biochimie*, **78**, 971.

131. Stark, H., Rodnina, M. V., Rinke-Appel, J., Brimacombe, R., Wintermeyer, W., and van Heel, M. (1997) Visualization of elongation factor Tu on the *Escherichia coli* ribosome. *Nature*, **389**, 403.

132. Agrawal, R., Penczek, P., Grassucci, R., and Frank, J. (1998) Visualization of elongation factor G on the *Escherichia coli* 70S ribosome: the mechanism of translocation. *Proc. Natl Acad. Sci., USA*, **95**, 6134.

133. Brown, S. (1989) Time of action of 4.5S RNA in *Escherichia coli* translation. *J. Mol. Biol.*, **209**, 79.

134. Shibata, T., Fujii, Y., Nakamura, Y., Nakamura, K., and Yamane, K. (1996) Identification of protein synthesis elongation factor G as a 4.5 S RNA-binding protein in *Escherichia coli*. *J. Biol. Chem.*, **271**, 13162.

135. Ogg, S. and Walter, P. (1995) SRP samples nascent chains for the presence of signal sequences by interacting with ribosomes at a discrete step during translation elongation. *Cell*, **81**, 1075.

9 | The functioning of molecular switches in three dimensions

ALFRED WITTINGHOFER

1. Introduction

Two major groups of GTP-binding proteins have been identified. The first is rather small and includes FtsZ and tubulin, both of which polymerize to form fibre-like structures. The second is a major class and consists of many families or superfamilies sharing common sequence motifs and a common core structure. These are referred to as GTP-binding proteins or GTPases and they act as regulatory molecules in all eukaryotic cells (1). The common core structure is relatively well conserved within but not necessarily between families. The major families are described in the first eight chapters of this book and include the protein biosynthesis factors such as elongation factors Tu (EF-Tu) and G (EF-G) and their mammalian homologues EF-1α and EF-2, the superfamily of Ras-related proteins, the α subunits of heterotrimeric G proteins, the signal recognition particle and its receptor together with the bacterial counterparts Ffh and FtsY, and the dynamin/Mx type of proteins (2, 3). In addition to these two groups there are a few GTP-specific metabolic enzymes, such as adenylosuccinate synthetase or PEP-carboxykinase, but these have no sequence conservation either between each other or to the proteins mentioned above, and will not be considered here any further. Although the protein biosynthesis factors were the first GTP-binding proteins to be discovered and many ground-breaking biochemical and structural studies have been carried out on these, this chapter will focus primarily on small GTP-binding proteins and G proteins since these are more closely related to each other and their biology in eukaryotic cells has been studied intensively in recent years. Structural aspects of GTP-binding proteins have been dealt with in earlier reviews (4, 5).

2. General properties of GTP-binding proteins

2.1 Biochemistry

Classical GTP-binding proteins contain a set of conserved sequence elements, which are (in order of appearance in the sequence): **GxxxxGKS/T**, F/Y, **T**, **DxxGQ/H/T**,

T/NKxD, C/SAK/L/T (invariant residues in bold). Most of the proteins bind guanine nucleotides with very high affinity, having dissociation constants in the picomolar to nanomolar range. Due to the high affinity, they always contain bound nucleotide in a 1:1 molar ratio and are fairly unstable in the absence of nucleotide. The β-phosphate is believed to be required for tight binding, although this has been shown formally only for Ras, where GMP binds with a drastically reduced affinity (6). All GTP-binding proteins bind guanine nucleotides with high specificity, since slight modifications of the guanine base show drastically reduced affinity and adenine nucleotides are bound very weakly. In the case of Ras, the affinity for ATP is six orders of magnitude less than for GTP, although the reasons for this are not completely understood (6). GTP-binding proteins hydrolyse GTP rather slowly, with rate constants in the order of $0.002–10$ min^{-1}, and they do not show any steady-state turnover of GTP since one product of the reaction, GDP, stays tightly bound. The cycling of GTP-binding proteins is controlled *in vivo* by regulatory factors: GEFs (guanine nucleotide exchange factors) increase the dissociation of tightly bound nucleotides such as GDP (but also of GTP) and GAPs (GTPase-activating proteins) activate the GTPase cleavage reaction such that the half-life of the GTP-bound state is reduced to less than seconds.

2.2 General features of the G domain

The first three-dimensional structure obtained for a GTP-binding protein was the 6Å resolution structure of trypsinized elongation factor Tu from *Escherichia coli* (7). Higher-resolution data for EF-Tu were reported later, but this still did not allow complete chain tracing. Help came from the discovery of Ras, a protein that contained all the hallmarks of a GTP-binding protein and yet was only 21 kDa in size, thereby establishing that the residues responsible for nucleotide binding would probably be contained in a single domain. This spurred further model building on EF-Tu and led to the prediction that EF-Tu and Ras contain a domain with similar structure (8, 9), now called the G domain, and eventually led to a complete chain tracing for the bacterial elongation factor (10).

Although Ras was originally proposed to have the same topology as EF-Tu, the first three-dimensional structure of Ras suggested it had a somewhat different fold. Later, however, it was established that it does, in fact, have the same set of secondary elements in the same topological order (11–15). Since Ras is the smallest known and most well-documented G-domain-containing protein, it will be described here as the minimal scaffold for the GTP-binding proteins, where all other proteins can be regarded as variations and extensions of this canonical G-domain structure.

As is common for most nucleotide binding proteins, the G domain is an α,β protein. It consists of a six-stranded β-sheet, which is mostly parallel except for one strand located at one edge and running antiparallel to the others, and five helices located on both sides of the sheet. The β-sheet has a 231456 topology, as shown diagrammatically in the topological diagram of Fig. 1. The structures of all the regulatory GTP-binding proteins determined so far contain the same G-domain structure, except SRP and SRP receptor, which have a similar but not identical topology.

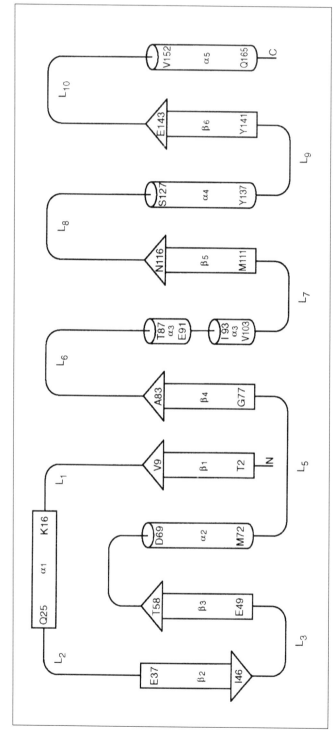

Fig. 1 Topology of the G domain, using the numbering from Ras•GppNHp (Brookhaven database: pdb-file 5p21). β-strands and their direction are indicated by arrows, helices by cylinders, connecting sequences are loops. All these secondary structural elements are numbered according to their appearance in the sequence. According to their appearance in the structure, β-strands are numbered 2,3,1,4,5,6.

2.3 The sequence elements

Although some amino-acid sequence conservation had been noticed before the first three-dimensional structure was available (16), the exact role of these conserved sequence elements could not be fully appreciated. It is now clear that the conserved residues are invariably involved in binding of guanine nucleotides, hydrolysis of GTP or controlling the conformational switch. They are referred to as PM and G motifs, describing their involvement in either binding to phosphate/Mg^{2+} ion or to the guanine base, respectively (17). The network of interactions between nucleotide and the conserved sequence elements is represented in Fig. 2, taking Ras as the model.

2.3.1 The PM1 motif ([10]GxxxxGKS/T)

This forms a loop (L1 in Ras), also called the P-loop, which is involved in binding the charged phosphate groups. Using a totally conserved lysine, it forms a ring-like structure that wraps tightly around the β-phosphate of GDP/GTP. The main-chain nitrogen atoms of residues 13–16 point towards the negatively charged phosphates and, together with the side-chain of Lys16, create a positively polarized environment. The serine/threonine residue following the lysine coordinates to the important cofactor Mg^{2+} in both the GDP- and GTP-bound state. Its substitution by Ala or Asn

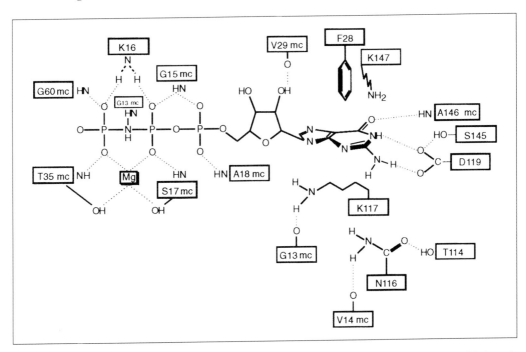

Fig. 2 Role of conserved sequence elements, exemplified on the Ras•GppNHp structure. Phe28 is only conserved as F/Y in Ras-like GTP-binding proteins. Although Ala146 is totally conserved in the G3 (G5, see ref. 4) motif, its strong main chain interaction with the O6 substituent on the guanine base, believed to be responsible for most of the specificity of guanine nucleotide binding, is not universally recognized in all structures.

(which renders the protein a dominant-negative inhibitor of normal Ras) severely modifies the interaction between protein, metal ion, and nucleotide (18, 19). As a result of the weak affinity to guanine nucleotides, this mutant protein has a relatively high affinity (compared to GDP/GTP) for guanine nucleotide exchange factors.

2.3.2 The PM2 motif (^{35}T)

Although various consensus motifs have been proposed around this conserved threonine, none of these residues, except Thr itself, are actually conserved. The Thr is a direct ligand of Mg^{2+} in the GTP-bound state, it binds the γ-phosphate of GTP and is a key residue that triggers the conformational change after GTP hydrolysis (20). Mutation of Thr35 to alanine in Ras reduces the affinity for nucleotide, and the mutant protein can no longer be activated by GAP, probably due to the incorrect coordination of Mg^{2+} (19).

2.3.3 The PM3 motif (^{57}DxxGQ/H/T)

The aspartic acid is involved in binding Mg^{2+} via a water molecule. Glycine is coordinated via a main-chain H-bond to the γ-phosphate, another interaction crucial for the conformational change after GTP hydrolysis. Most GTP-binding proteins described here contain a Gln residue crucial for GTP hydrolysis, although it is replaced by Thr in Rap and by His in elongation factors.

2.3.4 The G1 motif (F/Y)

This residue is only conserved in small, Ras-related GTP-binding proteins. It is situated perpendicular to the guanine base and creates strong hydrophobic interactions between protein and base together with the conserved lysine from the G2 motif (below). Mutating this residue to leucine weakens the nucleotide binding drastically by increasing the dissociation rate.

2.3.5 The G2, G4 motif (^{116}N/TKxD)

The guanine base of the nucleotide is situated in a hydrophobic pocket and specific hydrophobic contacts, ionic interactions, and hydrogen bonds stabilize its binding. Both the Asn and Lys are involved in linking together various subregions of the nucleotide-binding site. In addition, they appear to make weak hydrogen bonds to the O6-oxygen and N7-nitrogen of the guanine base and the O6-oxygen of the ribose. The aspartic acid is involved in a double hydrogen bond to the guanine base and is one of the elements responsible for the high specificity of the protein for guanine nucleotides. Many residues in the G2 motif have been mutated in Ras-related proteins. The effect of the mutations is invariably to increase the dissociation rate of nucleotides and such mutants are permanently activated due to the fast exchange reaction. Mutating the Asp to Asn reduces the affinity for guanine nucleotide several hundredfold, but shows a similarly increased affinity for xanthosine nucleotides (21), since binding of the latter preserves the double hydrogen-bond system. This type of mutants has been useful to study the role of EF-Tu, Ras, Rab5, and FtsY.

2.3.6 The G3 motif ([145]S/CAK/L/T)

Except for the alanine, this motif is not totally conserved except in the Ras super-family. It seems to have a helper function in the binding of the guanine base because its side-chains do not participate in any direct interaction with the base. Instead the Lys/Leu is involved in stabilizing the position of Phe28 (the G1 motif) by a hydro-phobic interaction and Ser145 forms a hydrogen bond with Asp119. The Ala146 main-chain nitrogen makes a strong hydrogen bond with O6 of the base and this interaction is another factor responsible for the specificity of guanine nucleotide binding because adenine would not be accommodated for steric reasons.

3. Structures of small GTP-binding proteins

3.1 Ras

Ras is a central regulator of many cellular processes, such as growth, differentiation, cell cycle, and apoptosis (see Chapter 3). The study of its structure and function was sparked by the discovery that it is an oncogene frequently found in human tumours and that the activation of the normal Ras gene involves a point mutation in codons at either of two positions, Gly12 in the P-loop (PM1 motif) or the conserved Gln61 in the PM3 motif. Ras is 189 residues long and contains a CaaX box motif at the C-terminal end which is post-translationally modified by farnesylation, proteolysis, and carboxymethylation. Full-length Ras has proved difficult to crystallize and structural analysis has been carried out on the actual G domain, residues 1–171 (13, 14) or 1–166 (11, 12). The 1–166 fragment has been shown to retain all the biochemical properties of the intact molecule. However, the C-terminal 23 amino acids and their post-translation modification are necessary to localize the protein to the plasma mem-brane and for full biological activity. In the case where full-length Ras could be crystallized, the C-terminal end was found to be disordered (14).

Ras has been crystallized in many forms, complexed with GDP (13), GppNHp (11, 12), mGppNHp, a fluorescent analogue (22), $GppCH_2p$ (13) and indirectly also with GTP, using a caged-GTP analogue and generating the Michaelis–Menten complex with flash photolysis of the crystals (23). The latter approach also allowed analysis of the GTPase reaction in the crystal. The high-resolution structure of Ras was instru-mental in defining the general rules and requirements for all regulatory GTP-binding proteins (Fig. 3). Most important was the discovery that the structural differences between the GTP-bound and the GDP-bound forms (13, 23) are localized to two regions of the protein, subsequently termed the switch 1 and switch 2 regions (13). It could also be shown that the trigger for the conformational change involved residues Thr35 and Gly60 from switch 1 and 2 respectively, totally invariant in GTP-binding proteins, which form main chain hydrogen bonds to the γ-phosphate (20). After GTP hydrolysis there is a structural rearrangement into a new conformation, mediated by the loss of the interactions of Thr35 and Gly60. The concept has emerged that the binding of the switch regions to the γ-phosphate is like loading a spring and that this is released after GTP hydrolysis (24). Since these two residues are totally invariant in

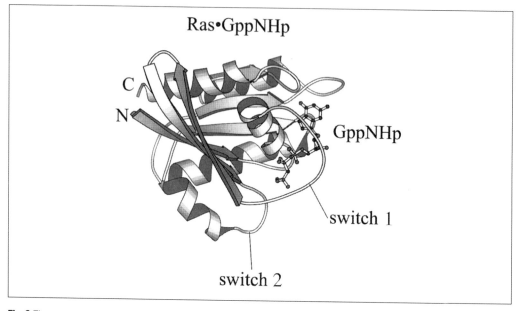

Fig. 3 The structure of the canonical G-domain fold, exemplified with Ras•GppNHp (pdb-file 5p21).

all GTP-binding proteins, it was predicted that the mechanism of the conformational switch would be universal (20) although many modifications of this mechanism have been discovered since.

The three-dimensional structures of the many oncogenic and non-oncogenic Ras mutants, such as G12V, G12R, G12P, G12D, Q61L, Q61H and D38E, complexed to either GTP analogues or GDP have also been determined (14, 25, 26) and they are very similar to that of wild-type Ras. Significant effects of the mutations on the three-dimensional structure are confined to the active site around the γ-phosphate. In the case of the Gln61 mutants, it seems that the absence of the catalytically important glutamine residue *per se* and not any change in structure is responsible for the reduced GTPase. In the G12R mutation, the long side-chain of arginine pushes the nucleophilic water and glutamine 61 out of the way. In G12V the bifurcated aliphatic side-chain of valine disturbs the catalytic configuration of Gly60 and Gln61 (25). In G12D the carboxylate side-chain of Asp12 makes hydrogen bonds with the γ-phosphate, Gln61 and Tyr32 (26). However, from these structural studies and the biochemical analysis of the Ras proteins, a consistent explanation for the loss of the GTPase activity could not be put forward.

Ras has also been studied extensively by magnetic resonance techniques. Even before the complete structure was solved, preliminary assignments of key residues allowed conformational changes in the switch 1 and 2 regions to be monitored and structural assignments to be made (15, 27–29). Eventually the complete structure of Ras in solution in both the GDP-bound (30) and GppNHp-bound forms (31) was solved by nuclear magnetic resonance (NMR) and this agreed with the structure determined by X-ray crystallography. NMR also revealed additional dynamic elements of the structure, which are mobile on the nanosecond time scale. In addition, loops L1, L2, L4 in the GppNHp or GTPγS conformation were found to exist in two or more slowly interconverting conformations (called regional polysterism) (31, 32). The dynamic behaviour of the switch 1 region was also demonstrated by phosphorus-NMR, where the β-phosphate of Ras•GppNHp shows two chemical shifts that interconvert slowly at 4 °C and rapidly at 30 °C. This is due to Tyr32 being located either close to, or away from, the nucleotide and, interestingly, effectors and GAPs interact with only one of these two conformations (33, 34). Electron paramagnetic resonance (EPR) studies using electron spin-echo envelope modulation (ESEEM) suggested that the interaction of Thr35 with Mn^{2+} ion (simulating the Mg^{2+} ion), which is seen in the crystal structure of all GTP-binding proteins, might be very weak in solution (35–37).

3.2 Rap1A and Rap2

The structure of Rap1A, a close relative of Ras, in the GppNHp form has been solved in a complex with the putative Ras-effector c-Raf-1 (38), which will be described later. The structure of Rap•GppNHNp is mostly indistinguishable from that of Ras•GppNHp.

Rap2 has been solved in the GDP, GTPγS, and the GTP-bound form (39). The low GTPase rate of Rap2 has made it possible to crystallize directly the GTP-bound form and solve the structure, not via the caged-GTP approach reported by Schlichting *et al.* (23). Conformational differences between the GDP- and GTP-bound forms of Rap2 were again restricted to the switch 1 and switch 2 regions, but they were less extensive than in Ras or other GTP-binding proteins (see below). When comparing different structures in the switch 2 region, it was noted that there are large differences in the GDP-bound forms, whereas the GTP-bound forms of different GTP-binding proteins (including G_{α} and elongation factor Tu) are much more similar. This is presumably due to the conserved hydrogen bonding network between the γ-phosphate and the invariant Thr and Gly residues. In the Rap2A•GTP structure, Tyr32 was located close to the nucleoside triphosphate making a direct hydrogen bond to the γ-phosphate and to the amide nitrogen of Gly13, whereas in the GTPγS complex, Tyr32 was pointing away from the phosphates. In the Rap1A•GppNHp•RafRBD complex Tyr32 of Rap1A is located close to the γ-phosphate and can make a water-mediated contact (38); it is directly coordinated in the mutant Raps•GppNHp•Raf complex (40), in a similar manner to interactions observed in the P-NMR spectrum of Ras•GppNHp (34). It has been argued that the interaction of the Tyr32-OH group with the γ-phosphate oxygen in the GTP- but not the GTPγS-form points to a role of the phenolic oxygen in the GTPase reaction. However, mutational studies of Ras have shown that Tyr32 can be replaced by Trp or Phe without a significant effect on the intrinsic or the GAP-mediated GTPase rate, and in the Ras–RasGAP structure Tyr32 points away from the reaction centre (41), making a role of Tyr32 in GTP hydrolysis, at least for Ras, unlikely.

3.3 Arf

Arf (for <u>A</u>DP <u>r</u>ibosylation <u>f</u>actor) is thought to recruit coat proteins to the membranes of developing vesicles (see Chapter 6). Arf is unique among Ras-related GTP-binding proteins in that it has an N-terminal extension of 14 residues (compared to Ras), modified on the N-terminal Gly by a myristoyl group. GDP/GTP exchange involves the insertion into and interaction of the myristoylated N-terminus with membranes.

The structure of non-myristoylated mammalian Arf-1 (rat and human) in the GDP-form was solved in two laboratories using three crystal forms (42, 43) and showed numerous unique structural features not found in other Ras-related proteins (Fig. 4). The G-domain fold contained an extra short β-strand in the effector region instead of the effector loop. The N-terminus is a nine-residue amphipathic helix located parallel to helix α5 in a groove where it is stabilized by hydrophobic interactions. Another distinguishing feature is the third β-strand containing the totally invariant Asp residue of the PM2 motif, which in all other GTP-binding proteins is involved in Mg^{2+} binding. In Arf the location of this β-strand is out of register by two residues with respect to Ras, such that Asp67 is coordinated to Mg^{2+} via a water molecule from a different orientation when compared to all other GTP-binding proteins. In the structure of human, but not rat, Arf-1, a glutamic acid residue (Glu54) is directly involved in Mg^{2+} coordination, creating a very unusual sevenfold coordination. This discrepancy between rat Arf-1 and other GTP-binding proteins could be due to the high Ca^{2+} concentration in the crystallization medium, since Ca^{2+} is known to show

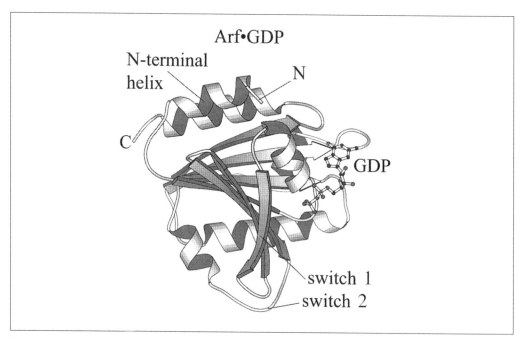

Fig. 4 Arf•GDP structure (coordinates from J. Goldberg) shows the additional N-terminal helix and the extra β-strand (b2E) in the switch 1 region. The N-terminal myristoyl group protein normally present on Arf was absent.

sevenfold or even eightfold coordination geometry and may have taken the position of Mg^{2+}. Another distinguishing feature of the Arf•GDP structure is that, due to the different conformation of the effector region, the invariant Thr residue (PM2), which is thought to be involved in binding γ-phosphate and Mg^{2+} in the GTP-bound structures, is far away from the Mg^{2+} ion and the putative γ-phosphate binding site.

3.3.1 The Arf•GTP structure and the N-myristoyl switch

Since the myristoylated full-length Arf could not be crystallized in the GTP-bound form, the N-terminally truncated form was used to obtain crystals and to solve the structure (Fig. 5) (44). It showed a large structural rearrangement compared to Arf•GDP (Fig. 4) such that there was now a remarkable similarity of Arf•GTP to the canonical triphosphate structure of Ras with a similar location of the switch 1 and 2 regions (11–13). Thus the extra β-strand in switch 1 of Arf•GDP disappears and Thr48, homologous to Thr35 of Ras, now interacts with the γ-phosphate and Mg^{2+}. Furthermore, Gly70 (Gly60 in Ras) is coordinated to the γ-phosphate and Gln71 is close to the nucleophilic water as expected.

The conformational switch of Arf must be more extensive than Ras since it also involves the N-terminal helix and the myristoyl group, which become exposed allowing their interaction with membranes. The structure of Arf•GTP leads to a change in the register between β-strands 2 and 3 by two residues, thus forming a

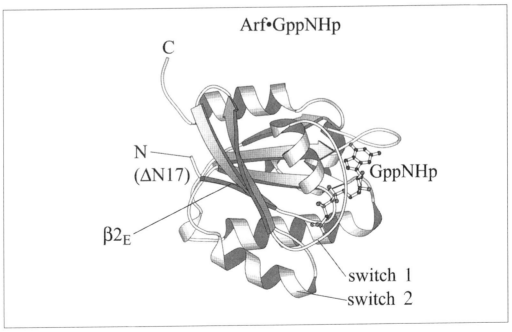

Fig.5 Arf•GppNHp (coordinates from J. Goldberg) structure shows the very dramatic change in the switch 1 region, with the unravelling of the extra β-strand such that it looks like Ras•GppNHp. The protein was crystallized without the N-terminal 17 residues that form the helical extension. β2$_E$ is an extra β-strand in the effector region.

more extensive interface between the two strands. The change in β-sheet register also changes the position of residues in switch 1 and 2, which are displaced towards the active site, and places Asp67 (Asp57 of Ras) so that it coordinates Mg^{2+} from a normal location.

Biochemical experiments had shown that the GDP–GTP exchange reaction on Arf only takes place in the presence of phospholipids and that in the GTP-bound form the hydrophobic portion of the N-terminal helix becomes accessible (45). The structure of Arf•GTP supplies an explanation for the change in the location of the N-terminal helix and the myristoyl group. In Arf•GDP a continuous hydrophobic patch is formed by the extra β-strand and the following β-turn, and residues completely conserved in all Arfs constitute a binding site for the N-terminal α helix and, presumably, the myristoyl group. In Arf•GTP this patch is no longer available, thus exposing the helix and lipid group for interaction with the membrane. This in turn means that the unusual conformation of Arf•GDP with the change in β-strands 2 and 3 is stabilized by the N-terminal extension and is destabilized after the γ-phosphate enters and induces changes in β2 and β3. Thus, Arf constitutes a particular variation of the canonical switch mechanism of GTP-binding proteins called the myristoyl switch (44).

3.4   Ran

Ran (for <u>Ra</u>s-related protein in the <u>n</u>ucleus) is the only nuclear GTP-binding protein described so far, and the only one that is not N- or C-terminally modified by hydrophobic residues (see Chapter 7). It constitutes another variation of the Ras-related protein make-up since it has a long C-terminal extension that ends in a highly conserved strongly acidic DEDDDL motif. The Ran•GDP structure was solved by X-ray crystallography by Scheffzek *et al.* (46). As with the N-terminal extension of Arf, part of the C-terminal extension of Ran is an α-helix, which is located in a groove close to helix α5 of the G domain (Fig. 6). It is connected to the main body of the G-domain via an unstructured linker, and is followed by the invariant DEDDDL motif. Another feature common to Arf and Ran is the presence of an extra β-strand in the effector region (called $β2_E$ to keep the proper numbering of the canonical G-domain fold) when compared to the effector loop seen in Ras. As in Arf, Thr42, corresponding to Thr35 in Ras, is far away from the nucleotide binding site and it was expected that if Ran•GTP were to adopt the canonical Ras•GTP structure, it would need to undergo a large conformational change (46–48). Another observation different from earlier GDP-bound structures was that residue E69 (corresponding to E62 in Ras) formed a water-mediated ligand interaction with Mg^{2+}. Glu62 is totally conserved in all Ras-related proteins, although no function (except the one found in Ran•GDP) has yet been assigned to it.

3.4.1   Ran•GTP and the C-terminal switch

The RanGTP structure has been determined indirectly by solving the complex between a Ran-binding domain of RanBP2 and RanGppNHp (49). There is indeed a

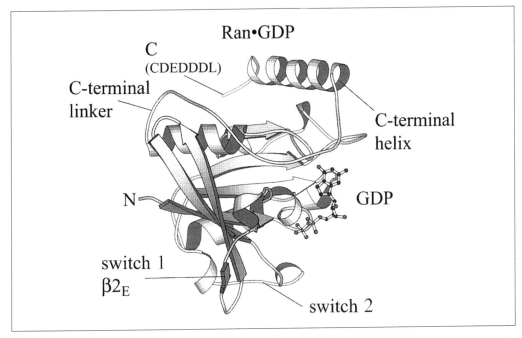

Fig 6 Ran•GDP shows a similar modification of the switch 1 region as Arf•GDP. In addition Ran has a long C-terminal extension which forms a linker, a long α-helix, an invariant DEDDDL motif, which is very important for its function.

large conformational change in the effector region such that the extra β-strand (β2$_E$) disappears and adopts a conformation quite similar to that of Ras. Phe35 is now perpendicular to the guanine base and Thr42 contacts the Mg^{2+} and the γ-phosphate as expected from the other triphosphate structures. Even more strikingly, there is a large conformational change involving the C-terminal extension of Ran, see Fig. 21. The structure suggests that the major driving force of this molecular switch is the steric clash of the C-terminal linker region of Ran with the new conformation of the effector loop when Ran binds GTP, mostly involving residues 32–34 from the effector loop and 182–184 from the linker. That the DEDDDL motif is an important element of the C-terminal switch derives from the observation that these residues stabilize the GDP-bound and destabilize the GTP-bound conformation. It appears that in the GTP-bound form, the acidic tail of Ran makes rather unfavourable interactions (50) and that these are relieved by binding to the RanBP module (see below) which contains a basic patch of residues designed to bind the DEDDDL motif.

3.5 Cdc42/Rho/Rac

The Rho proteins Rho, Rac, and Cdc42 constitute a distinct subfamily of small GTP-binding proteins (see Chapter 4). The major distinguishing feature of Rho proteins is the presence of a 13-amino-acid insertion located between β5 and helix 4 (17). This insertion was shown to be involved in the interaction of Rac/Cdc42 with two downstream targets, the NADPH oxidase complex and IQGAP, whereas other effector interactions are not affected. The special feature of Rac and Cdc42 proteins is the presence of a Thr instead of Asn in the NKxD motif, but since Asn does not directly participate in the interaction with nucleotide this is not expected to severely effect the binding.

The structure of human Rac1•GppNHp (Fig.7) showed that the insertion consists of two helices, a short 3_{10} helix and an α-helix followed by a loop (51). Excluding this insertion, Ras•GppNHp and Rac•GppNHp are very similar with an rms deviation of 1.25 Å. Other possibly important differences from the Ras structure are the flexibility of the switch 1 region and the inability to assign specifically to Thr 35 the role of a γ-phosphate hydrogen bond interaction partner. Gln61 in switch 2 was very well defined, with a hydrogen bond from its side- and main-chain nitrogen to a putative catalytic water molecule. The Mg^{2+} ion coordination was found to be exactly as in Ras, with two phosphates, two Thr side-chains, and two waters as ligands.

The Rho structure was solved in the GDP (52) and GTP analogue (53) forms by X-ray crystallography. Both show the characteristic insertion helix, but are otherwise very similar to the Ras structure, with a few characteristic differences. In the GDP-bound form, the coordination of the Mg^{2+} ion is different from that in Ras in that, in addition to the canonical β-phosphate and Ser/Thr-OH, a main chain carbonyl

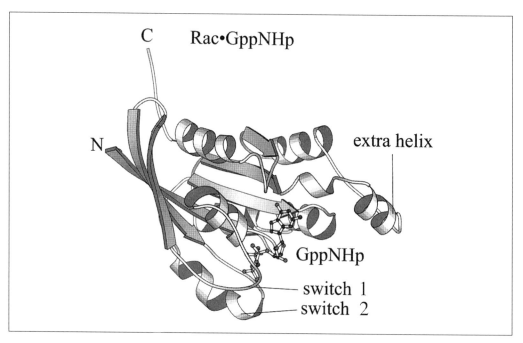

Fig. 7 Rac•GppNHp (pdb-file 1MH1) showing its modification to the canonical G domain, an α-helical insertion between β-strand 5 and helix α4.

from Thr37 (Thr35 in Ras) also interacts with the metal ion, which along with three water molecules completes the first coordination sphere. As in Ras, but unlike Rac•GppNHp, the effector region of Rho–GDP is well defined. In the triphosphate structure, the rms deviation to Ras is 1.68 Å for 163 common atoms. Differences in the structure of the effector region between Ras and Rho (and others) may be not too surprising considering that the residues in this region are different and are required to interact with distinct sets of effectors and regulators. Significant differences between Rho•GDP and Rho•GTPγS were found in the switch 1 and switch 2 regions, but not in the insert region, arguing that although the insert region may participate in the effector interactions, the nucleotide-derived specificity of the effector interactions is determined solely by the conformation of the switch regions. As in other tri-phosphate structures, a nucleophilic water molecule was also localized close (3.6 Å) to the γ-phosphate. One feature of Rho proteins is their sensitivity towards many bacterial toxins (see Chapter 10). *Clostridium botulinum* C3 ADP-ribosyltransferase modifies Asn41 in the effector region of Rho, and the structure shows this residue to be exposed. Likewise *Clostridium difficile* toxin glucosylates Rho on Thr 37, favouring the GDP-bound form. This is supported by the structures where Thr37 in the GDP-bound form is available for modification, but is not available in the GTPγS form, where it is coordinated by Mg^{2+}.

The solution structure of human Cdc42 both in the GDP and the $GppCH_2p$ form was solved by NMR, confirming most other observations on Rho family proteins and their differences to Ras (54). The major finding of this study was that the GDP/GTP-induced conformational change involves not only the canonical switch 1 and 2 regions, but also helix α3 and the α3-β4 loop, generating a much larger switch surface to interact with effectors. As in Rac and Rho, the insert region was not sensitive to the nucleotide state. The insert region is, however, involved in the mechanism of action of GDI; loss of the insert region does not reduce GDI binding affinity to Rho proteins, but it does destroy GDI-dependent inhibition of guanine nucleotide dissociation (55).

3.6   Rab

The structure of Rab7 has been solved and is very similar to that of Ras (P. Metcalf, personal communication).

4.   Heterotrimeric G proteins

Heterotrimeric G proteins consist of α, β and γ subunits, where the α subunit is the GTP-binding protein. More than 20 different mammalian G_α subunits have been described. Together with five β and 12 γ subunits, which form a non-separable βγ heterodimer, they are potentially able to form a large variety of different hetero-trimeric G proteins (see Chapters 1 and 2). Heterotrimeric G proteins are involved in transducing signals form the outside into the interior of cells thereby regulating a number of different cellular activities. The signal is initiated by the binding of a ligand to a heptahelical transmembrane receptor (56), which somehow activates the potential of the receptor to act as a guanine-nucleotide exchange factor (GEF) towards the heterotrimeric G protein (57, 58). After binding of GTP, $G_\alpha \bullet GTP$ dissociates from the activated receptor and from the $\beta\gamma$ subunit. The GTPase reaction in isolated G_α proteins is about 100 times faster than in Ras. Recently GAPs for G_α proteins termed RGS (for regulators of G-protein signalling) have also been described (59).

4.1 α subunit structures

G_α proteins are almost twice as large as the small, Ras-related GTP-binding protein (42–45 kDa). They were predicted to have a Ras-like G domain with four insertions (57), one of which is large enough to constitute an independent domain (Fig. 8). The first structure to be solved was that of the α subunit of transducin, $G_{\alpha t}$, in the active GTPγS form, and various other complexes of $G_{\alpha t}$, $G_{i\alpha 1}$, and $G_{s\alpha}$ were soon to follow (60–65), all of which were solved by X-ray crystallography. G_α subunits consist of two domains, one of which is very similar to that of Ras and to the G domain of the polypeptide elongation factors Tu and G. The second domain constitutes a helical domain which is inserted between helix α1 and β-strand β2 of the G-domain fold just before the invariant Thr (PM2) motif (Fig. 9). In addition G_α proteins contain three other insertions in the G domain, almost exactly as predicted through sequence alignments (57).

The helical domain is a six-helix bundle which can be expressed as an independent domain and was shown by NMR to have a solution structure similar to that which it assumes in the complete G_α protein (66). The G and the helical domain form a narrow crevice, which has been postulated to reduce the dissociation rate and increase the

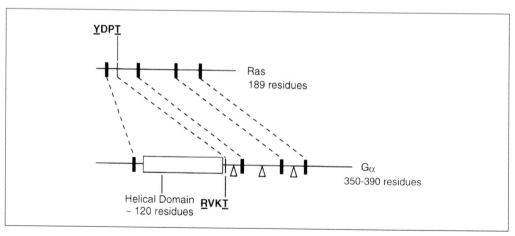

Fig. 8 Structural homology between small GTP-binding protein and the α subunits of heterotrimeric G proteins. G_α contains several insertions, one of which forms an independent helical domain and several small insertions indicated by the triangles. The C-terminal end of the helical insertion joins the G domain with the invariant threonine (PM2 motif), which is preceded in G_α by the invariant Arg and by Tyr in Ras (underlined), both of which have important functional roles.

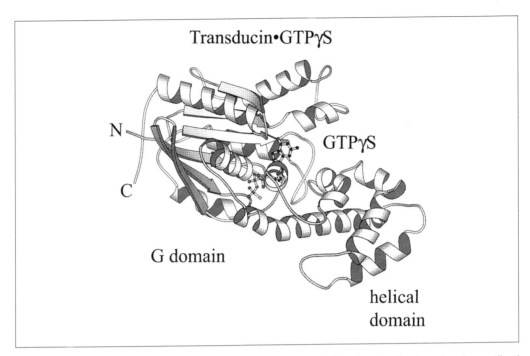

Fig. 9 Structure of the transducin subunit $G_t \cdot GTP\gamma S$ (pdb-file 1TND), showing the two domains outlined schematically in the previous figure.

binding affinity for nucleotide. However, judging from the dissociation rate constants of Ras proteins (in the order of 10^{-5} s^{-1}) versus that of isolated G_α proteins, which are usually faster except for $G_{\alpha t}$ (67), the dissociation of nucleotide is probably not generally decreased by the helical domain. Assuming similar association rates (which tend to be very similar, in the order of 10^5 M^{-1} s^{-1} (6, 68, 69)), which have not been measured for G_α so far, the affinity is also not likely to be increased by the helical domain. This is in line with the structure in which the helical domain does not appear to contribute any significant interacting residue. The helical domain has been assumed to constitute a tethered intrinsic GAP. Indeed, a tight link between the helical domain and the G domain via an ion pair is necessary for formation of a GTPase transition-state mimic (70) and the GTPase can in fact be partly reconstituted by combining the independently expressed domains (71). From the structural analysis, the helical domain appears to orient the crucial arginine, which is situated in the linker between the helical domain and the G domain, such that it can participate in the GTPase reaction.

The other feature that distinguishes G_α proteins from small GTP-binding proteins concerns the Mg^{2+} binding site. In Ras, Mg^{2+} stabilizes the binding of both GDP and GTP, and the rate of nucleotide release is stimulated several hundredfold in the presence of EDTA (72). Mg^{2+} binds to Ras·GDP with a μM dissociation constant which decreases to approximately 10 nM with Ras·GTP (19). Based on this observation, small GTP-binding proteins (in the GDP-bound form) can easily be loaded with any desired nucleotide (provided it binds tightly enough): addition of excess EDTA allows nucleotide exchange, while readdition of Mg^{2+} inhibits further exchange and stabilizes the binding. In contrast, monomeric G_α and heterotrimeric G proteins in the GDP-bound form do not bind Mg^{2+} appreciably, consequently nucleotide release is not inhibited by the presence of Mg^{2+} and Mg^{2+} is not required to maintain the GDP-bound state (73, 74). Correspondingly, in the structures of $G_{i\alpha1}$•GDP complexes, no Mg^{2+} could be located in the nucleotide-binding site. $G_{t\alpha}$•GDP did contain Mg^{2+} bound to the β-phosphate binding site, but this was most likely due to the presence of 200 mM Mg^{2+} in the crystallization medium. As will be shown below it appears that the βγ subunit binding to G_α•GDP acts to inhibit the release of nucleotide, thus taking over the role of the metal ion. In the triphosphate state, G_α proteins bind the metal ion with high affinity (10 nM K_D) as in Ras-like proteins, and the structures of G_α proteins in the triphosphate state unequivocally contain bound Mg^{2+} in the canonical binding site between the β- and γ-phosphate oxygens.

4.2   The conformational change

The trigger for the conformational change in all GTP-binding proteins is universal (20), but the details are different for each protein. Structural comparisons provide insights into these details. For both $G_{i\alpha1}$ and $G_{t\alpha}$, structural changes in the canonical switch 1 and 2 regions are found, while a third region, switch 3, located in the β4–α3 loop, also changes in conformation. Together the three switch regions are adjacent on one face of the protein.

The switch 1 region shows a similar change in both $G_{t\alpha}$ and $G_{i\alpha1}$, but the major difference involves the switch 2 and 3 regions, which simply become disordered in $G_{i\alpha1}$ (63, 64) but show a clearly defined structural change in $G_{t\alpha}$ (60, 61). Loss of the γ-phosphate releases the ordered α-helical conformation of switch 2 which relaxes into a very different conformation with part of the helix melted and its angle to the G domain widened. Since the switch 2 helix and the switch 3 helix pack tightly against each other through a number of salt bridges between invariant residues, helix α3 experiences a concomitant structural change. These structural details of the conformational change helped to explain two observations that have long been associated with the GTP–GDP transition. First, the decrease in fluorescence upon GTP hydrolysis which can be used to follow the reaction (74, 75) is due to a tryptophan residue (assigned to Trp207 in $G_{t\alpha}$ (76)) becoming exposed to the solvent. Secondly, trypsin digestion in switch 2 region is only possible in the GDP-bound form and has been used to monitor the conformational change (77); the structure shows that this effect is due to positively charged residues in switch 2 becoming exposed to the solvent after the conformational change.

Comparing the structures of $G_{s\alpha}$, $G_{t\alpha}$, and $G_{i\alpha1}$ in the triphosphate form (61, 63, 65) with the GDP forms of the latter two (60, 64), one can conclude that the triphosphate structures are very similar, with the nucleotide-binding site being organized in a similar way as in Ras-like proteins. Significant deviations are only seen in the GDP-bound conformations. This argues that for the G_α proteins, as for the Ras-related proteins, the GTP-bound conformation applies more constraints on the structure, most likely by hydrogen bonding the flexible switch elements to the γ-phosphate via the conserved Thr and Gly residues.

4.3 G_α subunits complexed to GDP and AlF_4^-

During the course of elucidating the action of hormones on cells, Sutherland discovered that cAMP production by adenylyl cyclase was stimulated by fluoride (78). Twenty years later it was shown that this stimulation was due to the formation of aluminium fluoride (79), which forms in a concentrated solution of fluoride in glass, and the subsequent activation of heterotrimeric G proteins (80). The hypothesis that aluminium fluoride forms a tetracoordinate AlF_4^- or $AlF_3(OH)^-$ and is trapped in the γ-phosphate binding site, thereby mimicking at least some aspects of the GTP-bound form, was confirmed by the crystal structures of corresponding G_α-complexes, $G_{\alpha t}$ and $G_{\alpha i1}$ (62, 63). These structures showed that AlF_4^- forms a planar square; with aluminium occupying the centre and fluorides the corners, the aluminium being further coordinated by the oxygen of GDP and the oxygen of a water molecule on the other side of the plane, complementing its octahedral coordination. This observation, together with biochemical data on the role of Gln204 and Arg174 of $G_{\alpha i1}$ in binding AlF_4^-, (63) supported the idea that GDP•AlF_4^- does not simply act as a GTP analogue, but is rather a mimic of the pentavalent phosphate in the transition state of the GTPase reaction. On the basis of numerous other studies on GTP- and ATP-converting enzymes, aluminium fluoride complexes are now considered to be general mimics of the phosphoryl group transferred during GTP or ATP hydrolysis and other phosphotransfer reactions, with aluminium fluoride being bound as either AlF_4^- or AlF_3 (81).

Since the GDP•AlF_4^- complex in G_α proteins is a mimic of the transition state of the GTPase reaction, the structure allows one to draw conclusions about the reaction mechanism. Although the overall structure of the α subunit does not change appreciably between the GTPγS and the GDP• AlF_4^- state, and although the nucleotide binds as expected from a triphosphate-like conformation, there are significant changes around the γ-phosphate binding site. The most important feature of the transition state mimic are the positions of the conserved glutamine (Gln61 in Ras), conserved in most GTP-binding proteins, and arginine, found only in G_α proteins. Gln had been anticipated to be involved in the GTPase reaction since its mutation in Ras or G_α blocks the GTPase reaction. It had been considered to be acting as a general base (12) but theoretical considerations and biochemical experiments suggested otherwise (82–84). The structure of $G_{i\alpha1}$•GDP•AlF_4^- and $G_{t\alpha}$•GDP•AlF_4^- show that the carboxamido side-chain of the corresponding glutamine contacts one of the fluorides of the planar AlF_4^- and the nucleophilic water molecule, suggesting that Gln is involved in stabilizing the transition state of the reaction, supporting an earlier suggestion by Prive *et al.* (14), although a role of abstracting a proton from water via its imino tautomeric form was not excluded (62). The arginine residue, Arg 178 in $G_{i\alpha1}$ and Arg174 in $G_{t\alpha}$, contacts two other fluorine atoms of AlF_4^-. Its role in catalysis is to neutralize negative charges that develop in the transition state. Depending on whether the transition state is dissociative or associative, this negative charge is more on the leaving group oxygen, which is the βγ bridging oxygen of GTP, or the transferred phosphoryl group, respectively (85). The fact that the arginine contacts AlF_4^-, a mimic of the transferred phosphoryl group, argues for a mostly associative character of the transition state, since such a contact would be anticatalytic in the case of a dissociative mechanism. Whereas in the structure of $G_{t\alpha}$ (62), both glutamine and arginine contact the nucleotide also in the ground state GTPγS complex, in $G_{i\alpha1}$, these two residues appear to be involved only in binding the GDP•AlF_4^- moiety, which is in line with biochemical experiments showing that the mutation of these residues has no influence on GTPγS binding but strongly influences GDP•AlF_4^- binding (63).

4.4   Mutants of $G_{i\alpha1}$ and the GDP•Pi state

The G42V mutant of $G_{i\alpha1}$ has a 30-fold reduced GTPase activity. It can only be slightly (twofold) stimulated by the $G_{i\alpha1}$-specific GAP, RGS4 (86), but no other biochemical properties are significantly affected. The structure of the mutant was solved in both the GDP-bound and the GTPγS forms. Whereas the GDP-bound structure was similar to that of the wild type, the triphosphate structure showed a significant deviation, since the bulky isopropyl side-chain in the P-loop pushes the catalytically important Gln204 into a position where it is unable to support GTP hydrolysis (see below), similar to the results obtained with oncogenic Ras (G12V) (25). The G42V mutant could also be crystallized as a $G_{i\alpha1}$•GDP•P_i complex, which is believed to represent an intermediate on the GTPase reaction pathway. The fact that such a complex could be trapped in the crystal was surprising since P_i release is fast and not rate-limiting in the intrinsic GTPase reaction (86), and complex formation between $G_{i\alpha1}$•GDP and inorganic phosphate could not be demonstrated in solution even at 50 mM phosphate. The structure showed that the positioning of P_i away from the β-phosphate causes a substantially altered conformation of the switch 1 and 2 regions as compared to the GTPγS structure.

A GDP•P_i intermediate has also been observed for the $G_{i\alpha1}$ mutant G203A (87). This mutation of the invariant Gly of the PM3 motif was first described as a G226A mutant of $G_{s\alpha}$ in an S49 mouse lymphoma cell line that lacked stimulatable adenylyl cyclase activity (88). The mutant G_α protein (G226A) has properties similar to wild type, but is unable to undergo the conformational change after GTP binding that is necessary to release the βγ subunits (77, 89). The homologous mutant $G_{i\alpha1}$ (G203A) is also unable to undergo the required structural change after GTP binding and to release βγ subunit. The mutant $G_{i\alpha1}$ was crystallized in the presence of GTPγS but, surprisingly, the structure showed that GTPγS had apparently hydrolysed and that the ternary product complex G_α•GDP•P_i was trapped in the active site (87).

The overall structure of $G_{i\alpha1}$ (203A) resembles that of other complexes of $G_{i\alpha1}$, and is much closer to the wild-type GTPγS structure than the GDP structure, except that no Mg^{2+} ion was found. From modelling Ala203, it had been anticipated that the methyl side-chain of Ala would interfere with Gly42 in the P-loop. The structure shows that, indeed, helix α2 adopts a different conformation, which avoids the steric clash and does not produce the tight packing interface between the α2 and α3 helices seen in the GTPγS structure. At the same time, this conformational change seems to create a binding site for the phosphate ion. Even though the GTPγS state was itself not determined, it has been argued that the inability of the βγ subunits to dissociate from the mutant after GTP binding is due to the inability to adopt the triphosphate-like structure. Furthermore the $G_{i\alpha}\bullet GDP\bullet P_i$ structure resembles very much that of the $G_{\alpha\beta\gamma}$ heterotrimer conformation (see below) indicating that $G_{\beta\gamma}$ can be bound to the mutant even in the presence of three phosphates. In the GDP•P_i state of the mutant protein, Arg178 still contacts the phosphate whereas Gln204 is moved out of the reaction centre. The structures of GTPγS, GDP•AlF_4^-, and GDP from wild-type or mutant versions of $G_{i\alpha 1}$, have provided snapshots of intermediates and transition states on the reaction pathway of the GTPase.

5. Guanine nucleotide dissociation inhibitors

The Rab and Rho families of small GTPases are regulated by so-called guanine nucleotide dissociation inhibitors (GDIs) (see Chapters 4 and 5) (90). RabGDIs are sequence-related to the component A subunit of Rab geranylgeranyl transferase II, also referred to as Rab escort protein (Rep), and also to the mammalian choroideraemia gene product CHM, which contributes to late-onset retinal degeneration and a related isoform CHML (90). Since the βγ subunits of G proteins appear to have a similar biochemical effect and physiological role as GDIs, they will also be considered in this section. In addition, NTF2, a nucleocytoplasmic transport protein involved in regulating the pool of RanGDP destined for nuclear import, will be discussed.

5.1 RabGDI structure

The first structural characterization of a GDI was that of the bovine α-isoform of RabGDI, a 50 kDa protein (91). GDI consists of two major domains, a large, cylindrical, complex folding module of ~340 residues and a smaller helical module of ~100 residues (Fig. 10). The larger domain consists of two major β-sheets (five parallel strands and seven mostly antiparallel strands nearly perpendicular to each other), an α-helical part consisting of four helices and three smaller sheets. The helical domain consists of five helices inserted between two β-sheets, where N- and C-terminal ends are close to each other.

Unexpectedly, the large domain is closely related in structure to FAD-containing mono-oxygenases and oxidases. All members of this family contain a GxGxxG motif involved in binding the phosphate moiety of FAD, and all members of the GDI family contain a GxG remnant of this motif in a structurally similar position. Because of the structural similarity between flavoproteins and GDI, it is speculated that FAD may be a cofactor for an unsuspected GDI hydroxylase or oxidase activity, which may participate in regulating membrane transport. Biochemically, however, no flavin binding to recombinant or bovine brain purified GDI has been detected.

Structure-based mutational analysis has indicated that the Rab binding site

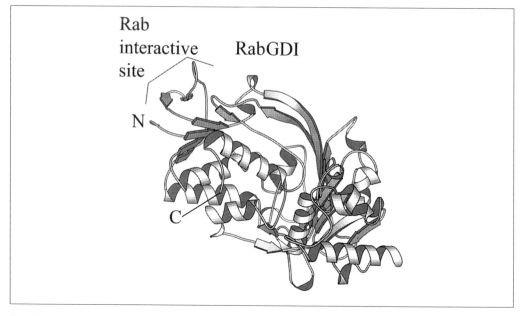

Fig. 10 Structure of RabGDI (pdb-file 1GND) shows a complex topology with several subdomains, not related to the fold of RhoGDI.

constitutes the two most invariant stretches of the molecule, SCR1 and SCR3A, located close to each other at the apex of the cylindrical larger domain, as analysed by the ability to bind to prenylated Rab and to extract Rab3A from membranes. The binding activities of the mutants correlated closely with their ability to inhibit transport through the early exocytic pathway. Given the exposed β-sheet nature of this binding area and the fact that the effector region of Rab is involved in the interaction with GDI, it was speculated that the interaction between Rab and RabGDI may involve an interprotein β-sheet, similar to that found in the Ras- and Rap-effector complexes.

5.2 RhoGDI structure

RhoGDI is a 204-residue protein, that binds C-terminally isoprenylated Rho family members and can solubilize them from membranes. It also inhibits nucleotide exchange. Although the isoprene moiety is important for binding, other protein–protein contacts are necessary for high-affinity interaction. The three-dimensional structure and mode of binding have been determined independently by a combined biochemical and NMR (92) and a combined X-ray and NMR (93) approach, leading to basically identical results (Fig. 11). The ^{15}N-^{1}H-HSQC spectrum of RhoGDI indicated that there are two groups of peaks with different line width, one of which corresponds to the N-terminal unstructured region. By trypsin digestion, the N-terminal region (59 residues) can be cleaved off, generating a 17 kDa fragment. Studies with

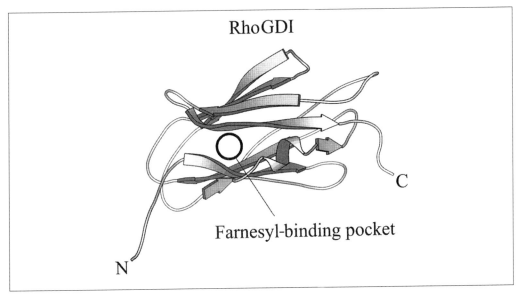

Fig. 11 Structure of the core domain of RhoGDI (pdb-file 1GDF). The N-terminal extension of the protein (not visible here), which is necessary for the guanine nucleotide dissociation inhibitory (GDI) effect, has been shown to be very flexible (92, 93).

this deletion construct had indicated that the Δ59-RhoGDI retains the ability to bind to the GTP-binding protein, but loses the ability to inhibit GDP dissociation (92), whereas a Δ22 construct appears to be fully functional in both binding and inhibition (92, 94). The structure of the stable domain was determined and shows an αβ-sandwich motif of two largely antiparallel β-sheets that pack against each other in a parallel manner very much resembling the Greek key topology of the immuno-globulin fold, with a small insertion of two β-strands and a 3_{10} helix. In contrast to the immunoglobulins, however, RhoGDI does not have a disulphide bond, indicating that the latter is not necessary for the stability of the fold. The structure of RhoGDI does not resemble the structure of RabGDI.

Using recombinant non-isoprenylated Rac1 (93) or Cdc42 (92) in the GDP-bound form and ^{15}N-labelled RhoGDI, the interface of the complex was determined by identifying shifts of resonances in the ^{15}N-^{1}H-HSQC NMR spectrum. It could be located towards the N-terminal side of the core RhoGDI domain, and the N-terminal flexible end was also found to be involved in binding, although it remains largely flexible even in the presence of saturating concentrations of the GTP-binding protein. The structure reveals that residues 23–58 are involved in binding to prenylated Rho proteins, presumably close to the effector region such that guanine nucleotide dissociation is inhibited, whereas the first 22 residues are unstructured even in the complex and are not involved in binding. The exact binding mode of Rho proteins to RhoGDI is, however, still largely unknown.

The two reports did, however, identify the binding pocket for the prenyl group, located as an open cavity between the two β-sheets (95). This binding site is lined by a number of conserved hydrophobic groups. The evidence for this cleft being the isoprenyl-binding site has been obtained by NMR titration experiments using a farnesylated peptide, acetyl-KKSRRC(S-farnesyl), and GDIΔ59. Large, saturable chemical shift changes were detected in residues in and around the binding pocket, whereas an unfarnesylated peptide did not induce these effects. Furthermore N-acetyl-S-farnesyl cysteine, which mimics the isoprenylated C terminus of Rho GTPases, causes a similar set of shift perturbations, albeit to a lesser degree, probably due to the lower affinity of this hydrophobic moiety. The combined structural and biochemical data suggest a model for the RhoGDI•Cdc42 complex in which the GTPase is on top of the core domain of GDI, with its attached isoprene buried in the hydrophobic pocket. This interaction anchors the inhibitory N-terminal peptide to the GTPase, allowing it to block exchange of GDP, and probably binding of GAPs and GEFs.

It has been observed that in T cells the N-terminus of RhoGDI-2 is cleaved by caspases such as ICE, following induction of apoptotic signalling pathways. The protease cleaves the N-terminal flexible end of RhoGDI at the position of tryptic digestion. The smaller of the resulting fragments will probably be unable to bind to farnesylated Rho with sufficiently high affinity to extract it from membranes. This proteolysis of RhoGDI might thus constitute an elegant way of regulating the activity of the protein in the cell during apoptosis.

5.3 NTF2 and its complex with Ran•GDP

Ran is the major regulator of nuclear transport. During an import–export cycle at the nuclear pore, it interacts with the downstream targets exportin and importins. *In vitro* import assays have shown that under limiting assay conditions, NTF2 (nuclear transport factor 2) is required for an efficient import reaction (see Chapter 7) (96, 97). NTF2 recognizes specifically the GDP-bound form of Ran, with a dissociation constant in the micromolar range, and is thus not an effector in the classical sense. Its role in nuclear transport appears to be to shuttle Ran•GDP from the cytoplasm to the nucleus so that it can function in another import/export cycle (98) (Chapter 7). In this sense the NTF2•Ran complex represents the cytosolic pool of Ran, just as the Rab•GDI and Rho•GDI complexes represent the cytosolic pools of Rab and Rho.

NTF2 is a small 10 kDa protein that forms a dimer in solution. The structure of NTF2 alone was solved by X-ray crystallography (99). It consists of a distinctly bent β-sheet with one long α-helix. The arrangement of secondary structure elements produces a cone-shaped molecule which generates a conspicuous hydrophobic cavity surrounded by negative charges.

The crystal structure of rat NTF2 bound to canine RanGDP was solved by X-ray crystallography to 2.5 Å resolution (Fig. 12), using the RanGDP and NTF2 models for molecular replacement (100). The asymmetric unit contains a NTF2 dimer and three molecules of Ran. Two of the Ran molecules interacted with NTF2 in a similar manner, whereas the third interacted differently with each NTF2 chain. The latter Ran and its interaction with NTF2 is believed to be a crystallographic artefact, since

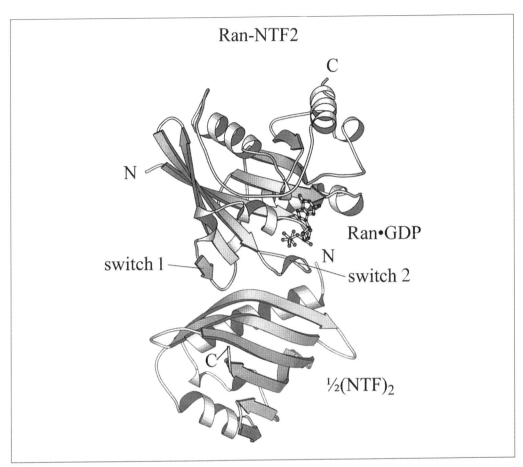

Fig. 12 The structure of the NTF2-Ran•GDP complex (pdb-file 1A2K) shows that NTF2, which is a dimer both in solution and in the crystal, uses the switch 1 and 2 regions for conformation-specific binding.

gel filtration chromatography shows only a dimer–dimer complex in solution (100). The structure of Ran and NTF2 in the crystal showed only minor differences to the structure of the isolated molecules (46, 99), the rms deviation ranging from 0.34 to 0.55 Å.

The interface of the complex involves mostly residues from switch 2 and to a minor extent also residues of switch 1 of Ran, which is consistent with the NTF2 interaction being dependent on nucleotide state. The most striking feature of the interface is the insertion of Phe72 from the switch 2 region into a hydrophobic cavity in NTF2 formed by a number of hydrophobic residues. This interaction alone buries 365 Å^2 (22%) of the total 1640 Å^2 solvent-accessible surface between each pair of chains (1 Ran-1 NTF2). Considering the structural changes of RanGDP versus RanGTP (see below), the structure explains nicely the nucleotide-state-dependent binding of NTF2 to Ran•GDP only. The structure supports studies with mutants of NTF2 from a genetic screen, which still bind to nucleoporins, but were unable to bind to Ran. In the complex, salt bridges are formed between K71 and R76 on Ran and D92, D94, and E42 on NTF2. In either the RanR76E and NTF2 (E42K) mutants, this salt bridge is abolished and the affinity is drastically reduced. These mutants were used to prove that a tight interaction of Ran and NTF2 is necessary for the transport of Ran into the nucleus (98).

5.4 The βγ subunit and the heterotrimeric G protein

The structures of complete G-protein heterotrimers were solved independently for $G_{t\alpha}$ and $G_{\alpha i1}$ (101, 102) (Fig. 13) and the uncomplexed βγ subunit was also solved for the transducin system (103). The β subunit forms a propeller-like structure where the blades of the propeller are formed by four-stranded β-sheets arranged around a water-filled central shaft. Such a propeller structure has been found in enzymes such as neuramidinase, galactose oxidase, and methanol dehydrogenase. The repeating

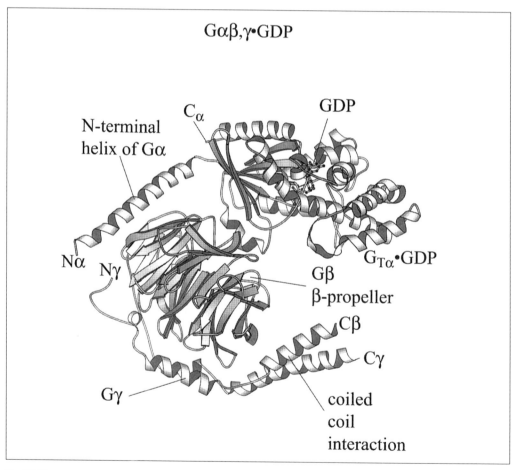

Fig. 13 The structure of the heterotrimeric G protein G_t (pdb-file 1GP2) shows G_β as a sevenfold repeated propeller which forms an interface with the switch 2 region and the N-terminal α-helical extension of G_α.

nature of the G_β sequence, the so-called WD40 repeat, is found in a number of regulatory proteins in higher eukaryotes, and is so named because the repeating core unit contains a conserved WD sequence motif (104). The number of blades in the propeller proteins is variable; seven in the case of the G_β protein. The structure shows that the borders of one blade of the propeller do not coincide with the order predicted from the consensus WD40 repeat. Thus, the last (outermost) β-strand of the each antiparallel sheet (blade) is the first element of the sequence repeat, while the rest of the sequence within one repeat actually specifies the three inner strands of the next blade of the propeller. It seems that this rearrangement ensures the stability of the circular structure. This structure predicts that the expression of a single WD40 repeat, as defined by the sequence comparison, would most likely not lead to a properly folded, stable protein.

The γ subunit is an extended molecule with two helices, but without any additional intrachain hydrogen bonds. The N-terminal helix of γ forms a coiled-coil with the N-terminal end of the β subunit, as previously predicted by sequence analysis (105, 106). The extensive hydrophobic interactions with β explain why the γ subunit cannot be isolated in the absence of β and why β is only moderately stable in the absence of γ.

The structures of the heterotrimeric G_t and G_{i1} are rather similar. Due to the independent structure determination of the uncomplexed βγ subunit alone (103) it was concluded that its structure does not change on binding to G_α•GDP, the rms deviation between free and bound βγ being 0.72 Å. It appears to behave as a rigid unit that is released when G_α binds GTP and rebinds G_α when the GTPase timer has run out. The interaction between G_α and βγ is formed by two distinct interfaces. One of these is the switch 2 region on α which changes conformation between GDP- and GTP-bound state, the other is the N-terminal region, whose conformation is probably independent of the nature of the nucleotide. The conformational change in the switch 2 region of G_α has been mapped very precisely and it is here that most of the contacts with β are made. Comparing the structure of G_αGTP with that of G_αGDP revealed that the triphosphate conformation could not be accomodated in the heterotrimer (101, 102), providing a structural explanation for why GTP binding disrupts the heterotrimer and releases G_α-GTP from $G_{\beta\gamma}$.

The second interface between α and βγ is formed between the N-terminus of G_α and the first blade of the propeller, assigning a function to the N-terminus of α that is in line with previous observations. In $G_{i\alpha1}$–GDP the N-terminus has been found to make a mini-domain with the C terminus, that could be due to crystal packing, since it is not seen in the heterotrimeric structure of either $G_{i\alpha1}$ or transducin (64). The interface of βγ with the N-terminus of α is 900 Å^2, as compared to the switch interface which is 1800 Å^2, and is therefore probably more important, although a correlation between size of interface and strength of interaction does strictly exist. The γ subunit does not contribute to the interface between α and βγ.

The structure of the mutant G203A of $G_{i\alpha1}$ in the heterotrimer form was solved in an attempt to explain why this mutant (and the homologous $G_{s\alpha}$ mutants) is defective in releasing the βγ subunit after binding of GTP (107). The mutation does not greatly perturb the structure of the switch 2 region in the heterotrimer, but on comparing the $G_\alpha \bullet GDP\beta\gamma$ structure with the active $G_\alpha \bullet GTP$ structure, it is postulated that steric effects of the methyl side-chain of Ala203 prevent the conformational change necessary for release of $G_{\beta\gamma}$.

6. Guanine nucleotide exchange factors—GEFs

Guanine nucleotide exchange factors (GEFs) for most families of small GTP-binding proteins have now been identified. They are related by sequence within, but not between, the families (90).

The concept of guanine nucleotide exchange stimulation on small GTP-binding proteins was first established for Ras after identification of a biochemical activity which stimulated guanine nucleotide release. GEFs for the Ras family (Ras/Rap/Ral/R-ras) contain a so-called Cdc25 homology domain, after the *S. cerevisiae* Cdc25 protein, which was shown genetically to be an upstream regulator of Ras. Sdc25, a homologue of Cdc25 in yeast, was the first purified recombinant protein for which *in vitro* activity for a Ras-related protein was demonstrated (108). The largest number of GEFs have been found for the Rho subfamily. All contain a DH (Dbl) homology domain, and its biochemical GEF activity has been verified *in vitro* (109). The gene for the Ran-GEF, RCC1, was cloned as a regulator of chromosome condensation (110) and later it was purified as a complex with Ran from nuclear extracts of HeLa cells shown to possess GEF activity towards Ran (111). Sec7 was identified in a screen for mutants defective in secretion in *S. cerevisisae* and the identification of the Sec7 domain as a GEF for Arf came with the cloning of the human protein ARNO (for Arf nucleotide-binding site opener) and the biochemical identification of its GEF activity (112). Several proteins containing the sec7 domain with GEF activity towards Arf have now been identified. The exchange factor for Sar1 has been identified in yeast as Sec12, a gene responsible for secretion (113). A Rab5-specific exchange factor, Rabex-5, has been found (114) and shows high homology to the *S. cerevisisae* Vps9p gene product, which is involved in vacuolar sorting (115) and most likely constitutes a GEF for another Rab protein. Rab GEFs have not been studied in any detail at the biochemical or structural level.

The largest class of GEFs (estimated to be several thousand in the human genome) are G-protein-coupled receptors (GPCRs) which act on heterotrimeric G proteins. They are seven transmembrane helical proteins, with the N-terminus located outside and the C-terminal end located inside the cell. The enzymatic GEF activity of most of these is activated by ligand binding, although some are activated by light or proteolysis (56).

6.1 Mechanism and kinetics of GEFs

Most small GTP-binding proteins bind guanine nucleotides very tightly and the intrinsic dissociation rate is very slow, typically of the order of 10^{-5} s^{-1} for Ras and

Ran, and about tenfold faster for Rho proteins. GEFs increase the dissociation rate dramatically; in the case of Ras and Ran, where it has been measured under saturating conditions, by several orders of magnitude such that the nucleotide comes off in less than one second (116–118). The general minimal model for the reaction is that the GEF approaches the Rax•GDP (Rax: any Ras-like protein) binary complex and forms a ternary complex (116–118). Thereafter the nucleotide binding pocket changes from a tight to a loose binding conformation, resulting in a fast release of nucleotide from the ternary Rax•GDP•GEF complex. By binding GTP, the dominant nucleotide in the cell, formation of a ternary complex and release of GEF, Rax•GTP is generated.

It should be stressed that in principle the guanine nucleotide exchange reaction can proceed in either direction, since all the partial reaction steps are very fast in each direction (116–118). In the case of EF-Tu or Ran (117, 118), the reaction is actually faster in the direction of formation of the GDP-bound state. Since GEF merely works as a catalyst, in the absence of other restraints that might drive the reaction towards the GTP-bound state (e.g. incorporation of the products into the membrane, as is the case with Arf•GTP), the outcome of the reaction is determined solely by the relative nucleotide affinities of the GTP-binding proteins, the concentrations of GDP versus GTP in the cell (or in the particular compartment of the cell where the reaction occurs), and the concentration of components that sequester either the GDP- or the GTP-bound state.

Very little information on the kinetic mechanism of the GEF reaction has been obtained so far. The existence of a binary complex between GEF and the GTP-binding protein, anticipated from the EF-Tu•EF-Ts complex described earlier, was demonstrated for Ras-Cdc25 (116, 119–121), and for Ran-RCC1 (111), Arf-Arno (44), and Mss4-Sec4 (122). For the Cdc25-mediated exchange on Ras it has been postulated that the rate-limiting step of the overall reaction is the conformational change from a tight to a loose binding conformation of the nucleotide on Ras (116). This step corresponds to the second step of the two-step binding reaction of nucleotide to Ras alone, where the initial encounter complex with loosely bound nucleotide (K_D ~10 μM) isomerizes to a tightly bound complex ($K_D \sim$ 10 pM) (68). The most thorough investigation of the GEF mechanism has been done on Ran, where the minimal set of individual rate and equilibrium constants has been described. It is also the only system for which the existence of the trimeric Ran•GDP•RCC1 complex has been documented spectroscopically. Here it was shown that the 10^5-fold increase in the rate of nucleotide dissociation is due to a similar 10^5-fold decrease in the affinity of GDP in the trimeric complex as compared to the binary Ran•GDP complex (117, 118). Recent structural studies on the binary complexes have supplied a great deal of information on the nature of structural changes that lead to the destabilization of the nucleotide-binding site (see below).

The structures of five types of guanine nucleotide exchange factor have been solved so far, ARNO (for Arf), RCC1 (for Ran), Sos (for Ras), DH (for Rho/Rac), and Mss4 (putative Rab-GEF). As anticipated from the sequence comparison, the structures of all five types are indeed different, where ARNO, DH, and Sos are purely α-helical and RCC1 and Mss4 are β-proteins.

6.2 Mss4

Mss4 is the human homologue of DSS4, a regulator of the yeast GTPase Sec4, and is presumed to be a GEF for Rab proteins (123, 124). Mss4 is not, however, a bona fide GEF, since it acts stoichiometrically to promote guanine nucleotide exchange, whereas other GEFs have been shown to act catalytically by increasing the rate of attainment of the equilibrium between the GDP- and GTP-bound state. Furthermore, the nucleotide-free Rab–Mss4 complex is very stable towards nucleotides (125), which disqualifies the protein as a proper GEF. However, since its function partially overlaps with that of other GEFs, and since it was the first interacting protein of the Ras-related proteins whose three-dimensional structure was solved, it is described briefly below (122).

The structure of human Mss4 has been determined by NMR spectroscopy (122). The 14 kDa protein can be described as a three-layered β-protein with a central seven-stranded antiparallel sheet flanked by a small three-stranded sheet on one face and a β-hairpin on the other. Mss4 also binds a Zn^{2+} ion, which is tetrahedrally coordinated by two CXXC motifs, arranged in the so-called rubredoxin 'knuckle' structure. The Zn^{2+}-binding region and a neighbouring loop define the active site of the protein. The Rab-interacting surface has been determined by NMR experiments using a nucleo-tide-free form of the GTPase, Sec4, a Rab family member from *S. cerevisiae*.

6.3 Arf GEFs, the sec7 domain

The cloning of the yeast Arf GEF genes, *Gea1* and *Gea2* (for guanine nucleotide exchange on Arf) (126) allowed the cloning of the human homologue ARNO (112). ARNO contains a domain, now called the sec7 domain, found as a module of about 200 amino acids in a number of proteins such as sec7 and cytohesin-1 (127) (see Chapter 6). This domain is necessary and sufficient to act as a GEF for Arf, although the presence of certain lipids is also required to allow the myristoylated N-terminus of Arf to become exposed (45, 128). Subsequent membrane insertion of the N-myristoyl group is believed to be part of the driving force for nucleotide exchange and might, therefore, ensure that the exchange reaction *in vivo* is always in the Arf-GDP to Arf-GTP direction. Using N-terminally truncated Arf, where the first 17 residues have been removed, the sec7 domain is sufficient for a fast and catalytic GEF exchange reaction in both directions (129).

The structure of the sec7 domain of ARNO was solved independently, by Cherfils *et al.* (130) and Mossesssova *et al.* (131), by X-ray crystallography (Fig. 14), and that of the sec7 domain of cytohesin by NMR (132). The structure is an elongated domain consisting of ten α-helices, which form a right-handed superhelix somewhat reminiscent of the armadillo repeat region of catenin (133). Overall, sec7 is a compact rod-shaped molecule with a single, deep, surface groove. The canyon shape of this groove and the localization of most of the conserved residues of the sec7 family identified it as the active binding site around the totally invariant motif FRLPGE (the FG loop) connecting helices F and G (or a6, a7). The identification of the binding site allowed the computer-simulated docking of Arf with ARNO and suggested a number

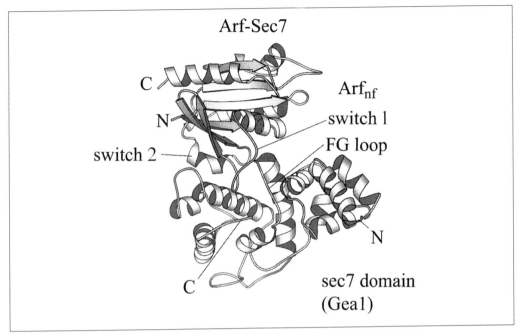

Fig. 14 The sec7 domain of Gea1, a guanine nucleotide exchange factor for Arf, is shown in complex with Arf (coordinates from J. Goldberg). The action of the GEF displaces switch 1 and 2 and their nucleotide-binding interactions, the most important element being the F-G loop on the sec7 domain.

of residues for mutational analysis which confirmed the original docking proposal.

In particular, charge reversal of the invariant glutamic acid (Glu156) of the FG loop to lysine, a mutation originally found as an embryonic differentiation mutant in *Arabidopsis thaliana*, turned out to be critical for the enzymatic activity of ARNO (131, 132, 134). Surprisingly it was shown that this mutation, although eliminating catalytic activity, led to a very tight ternary complex between Arf, ARNO, and nucleotide (134). Such a complex is normally very transient, since nucleotide is released rapidly from the ternary complex in the course of the exchange reaction. These observations led to the proposal that the glutamic acid residue acts like a finger, analogous to the arginine finger of GAP, that is inserted into the phosphate-binding site. Since both the phosphate group and the finger are negatively charged, this would destabilize binding and induce the rapid release of nucleotide. Other mutations of invariant residues in the proposed active site also showed more or less drastic effects on exchange activity, the most prominent being Met194, an exposed hydrophobic residue whose mutation to Lys also completely abolishes GEF activity.

6.4 Ran GEF, RCC1

The best kinetically characterized GEF is RCC1, the GEF for Ran (117, 118). When RCC1 was cloned, an internal sevenfold sequence repeat was immediately recognized that was highly conserved among RCC1 homologues/isologues (110). X-ray crystallography showed a propeller structure, consisting of a sevenfold repeated β-sheet each of four antiparallel β-strands, arranged in a circular fashion around a water-filled channel (Fig. 15) (135). Ring closure and thus stabilization is apparently achieved such that two strands from the N terminus and two from the C terminus form the first β-sheet, to constitute a kind of a molecular 'Velcro'. A topologically similar fold has been found for the β subunits of G proteins (101–103). However the β subunit WD40 repeat is very different from the RCC1 type of repeat; the latter has a different set of residues conserved between the blades, mostly glycines, and prolines, instead of the conserved W and D of the WD40 repeat. A number of different genetic screens had identified temperature-sensitive mutants of RCC1 in both *S. cerevisisae* and *S. pombe*. In the light of the structure, these mutants could be rationalized as having a mutation in the residues invariant between the repeats, usually glycines, which are thus necessary for the structural integrity of the fold rather than being directly involved in function.

Not much is known about the structural details of the interaction between RCC1 and Ran. However, due to a number of mutations obtained from an alanine scan of conserved, hydrophilic residues, the side of the propeller interacting with Ran has been identified (136). These and more extended mutational studies show that two invariant aspartic acid residues and one histidine residue, all of which are located in the tight β-turn between strands A and B of different blades, are the most crucial residues for catalysis. In a computer-generated docking model of the Ran–RCC1 complex, the same residues are found in the interface of the complex, positioned such that they could interfere with guanine nucleotide binding. An experimentally determined structure of a binary or ternary complex is not available at the moment to verify the proposed model and to draw conclusions about the structural mechanism

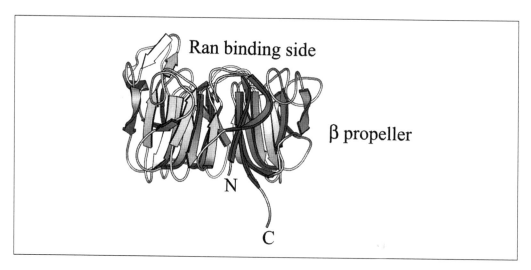

Fig. 15 RCC1 (pdb-file 1A12), the GEF for Ran, is a sevenfold β-propeller with a highly conserved internal sequence repeat different from that of the WD40 repeat motif of G$_\beta$. The interaction face of the propeller, as deduced from mutagenesis studies, is indicated.

of nucleotide release. The fact that both the β subunits of G proteins and the GEF of Ran have a sevenfold propeller structure might be an evolutionary accident, but it might also give some hints about common mechanistic details. The β subunit, although not the actual exchange factor of G proteins, is nevertheless involved in both the inhibition of nucleotide release (107) and supposedly also in the exchange mechanism via heptahelical receptors (58), as outlined below.

6.5 Rho GEFs, the DH domain

Guanine nucleotide exchange factors for the Rho family of proteins contain the DH domain (137), first identified in a transforming gene, *Dbl*, and later shown to be a GEF for Cdc42 (109). Many more Rho GEFs have been identified subsequently, the majority of which have been isolated as oncogenes, making the DH domain proteins the largest group of biochemically related oncogenes (137). The biochemical properties and the kinetic mechanism of Rho GEFs have not been studied in any detail.

All DH domains thus far identified are connected at their C-terminal end to a pleckstrin homology (PH) domain, which, at least *in vivo*, is necessary for GEF function. *In vitro* the DH domain alone appears to be sufficient to promote nucleotide exchange. PH domains are found in a number of proteins involved in signal transduction and cytoskeletal reorganization and have been shown to be involved in binding phospholipids. In the context of Rho GEF proteins, it is believed that the main purpose of the attached PH domain is to localize the DH domain to phospolipid-containing membranes.

The structure of the DH domain of Trio was solved by NMR (138) and the structure of the DH–PH fragment from Sos (Fig. 16) was solved by X-ray crystallography (139). Trio is unusual in that it contains two DH–PH domains, one specific for Rac and the other for Rho. Since Sos also contains a Cdc25 domain and is a GEF for Ras, it may provide a potential biochemical link between the Ras and Rho family of GTPases. The structures of the two solved DH domains are similar and are all α-helical. The Sos DH domain consists of 11 helical segments and the Trio-DH domain of nine, three of which are longer and form the core of the elongated molecule with a length of approximately 70Å and a width of 40 Å. The differences between the Trio and Sos DH domains have not been analysed in detail, but appear to be located outside the central fold. The structural database shows that the DH domain is a new and unique fold. The three major helices constitute the conserved sequence elements, identified before as the conserved regions CR1, 2 and 3 (90, 140). Since these conserved regions constitute the central core of the structure and many of the residues are apparently conserved for structural reasons, it is to be expected that all DH domains will have a similar structure.

Several conserved residues, located in the central portion of the DH domain on two helices, H1α and H8 in Sos, α1 and α9 in Trio, are exposed to the solvent. Mutations have been described in other DH proteins which map to this region, although their GEF activity has not been determined. Furthermore, the same patch of residues is shown to be shifted in the ^{1}H-^{15}N-HSQC spectrum of Trio after addition

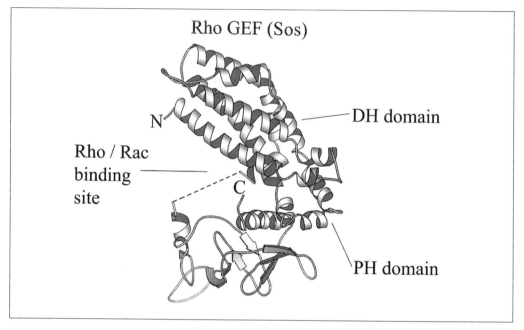

Fig. 16 The DH-PH motif (coordinates from J. Kuriyan) from human Sos contains the all-helical DH (Dbl homology) domain, responsible for the GEF activity on Rho proteins, and the mostly β-sheet PH (pleckstrin homology) domain. The Rho/Rac binding site is located close to the interface between these two domains.

of Rac1 (138). Mutation of these residues by Liu *et al.* shows a variable effect on the GEF activity of Trio-DH towards Rac1, some for example completely eliminate GEF activity without disturbing the structure, confirming that this region is important for catalytic activity.

In the X-ray structure of the DH–PH tandem unit, the two domains form a tight interface, burying 1100 Å² of water-accessible surface. Surprisingly, residues in the interface are not conserved either in the DH or the PH domains, arguing for a variable interface in different Rho GEFs. In comparison to canonical PH domains, the structure of the PH domain in the DH–PH unit contains an N-terminal extension called the linker, that is partly ordered, otherwise the structure corresponds to that of the isolated PH domain of Sos determined earlier by NMR (141, 142). The PH domain of Sos has been shown to bind phospholipids and chemical shift perturbations by NMR have shown that the location of the binding site is very similar to that for IP₃ in the PH domain of phospholipase C-δ (143). The binding site of the GTP-binding protein on DH, identified from mutational studies, would locate the Rho protein such that it would be in contact with the side ofn the PH domain distal to the phospholipid-binding site. From the structural comparison between the unliganded PH domain structure of Sos and that of PLC-β in complex with IP₃, it has been suggested that binding would induce structural changes in the GTPase binding site of DH. That the PH domain does indeed biochemically modify the GEF reaction (independent of any membrane targetting role) has been demonstrated by the mutational studies of Liu *et al.* (138) on Trio. It was shown that the DH–PH domain unit has a 100-fold higher GEF activity on Rac1 compared to the DH domain alone and that the effect of mutating interface residues in the DH domain is different in the DH- versus the DH–PH-domain-mediated GEF reaction *in vitro*. Thus, from the structural and biochemical studies, it appears that the PH domain does not serve merely as an anchor or translocation signal for the DH domain, but is rather actively involved in controlling catalysis.

7. Complexes and structural mechanism

7.1 Ras–Sos

The structure of the Ras–Sos complex, solved by Borriak-Sjodin *et al.* (144), has given important insights into the mechanism of Ras activation. The minimal domain of Cdc25, and presumably of Sos, responsible and sufficient for efficient catalytic activity, contains 248 residues (Cool and Wittinghofer, unpublished) and most of the detailed kinetic studies have been performed with fragments of a similar size (116, 120, 121). The fragment of Sos used for crystallization was 485 residues long, corresponding to residues 564–1049 from human Sos. This fragment contains an extra, small domain of approximately 180 residues, also called REM (for Ras exchanger motif (145)) or CR0 (conserved region 0, (144)). This region had been shown by sequence analysis to be conserved in other Cdc25-like Ras exchange factors, but is found at different distances relative to the catalytic domain. In the structure, it apparently supports a protruding helical hairpin from the catalytic domain and is probably responsible for its stability, since smaller fragments of Sos, and probably of Cdc25, tend to be labile. The catalytic domain, including the sequence-conserved regions SCR1, 2 3, is almost completely α-helical (Fig. 17).

The catalytic domain of Sos has been described as an oblong bowl where Ras binds at the centre of the bowl. The Ras–Sos complex structure shows that the P-loop, the switch 1 and 2 regions, and the region around α3 of Ras are involved in the interaction, as predicted from a number of mutational studies in many laboratories (1, 146). The most prominent, although not necessarily most important, interaction between Ras and Sos in terms of binding energy for the binary complex involves switch 1 and a protruding helical hairpin. The switch 1 region is completely flipped out of its normal position such that residues contacting the base and the phosphates (e.g. Phe28 and Thr35) are removed from their binding site position and are stabilized by the hairpin of Sos. The other important feature of binary complex formation involves almost all the residues from the switch 2 region, which form direct contacts with helices B, D, E, and G from Sos. The switch 2 region is very well structured in the Ras–Sos complex and involves a number of features that disfavour nucleotide binding. These features are the intrusion of a glutamic acid (Glu942) and a hydrophobic leucine (L938) into the Mg^{2+}- and phosphate-binding area, the flipping of a peptide bond such that the carbonyl oxygen from Ala59 occupies the Mg^{2+}-binding site, and the complete rearrangement of the β3-α2 loop

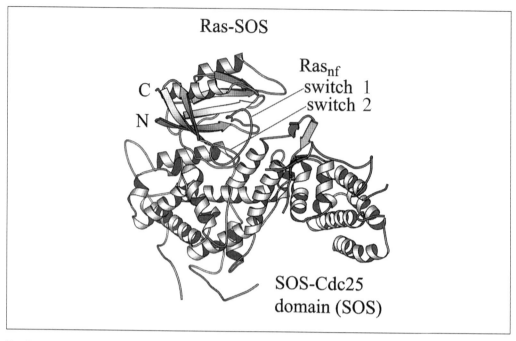

Fig. 17 The Ras-Sos binary complex (coordinates from J. Kuriyan) shows the Cdc25-like domain from Sos in complex with nucleotide-free Ras. The switch 1 and 2 regions in Ras have been displaced by Sos to allow nucleotide release.

such that Gly60 and, more importantly, Lys16 are involved in a tight interaction with Glu62 of Ras, a residue that has been shown to be crucial for the exchange reaction with Cdc25 and Sdc25 (147). Lys16 is particularly important for the binding of phosphates, as it contacts the β-phosphate oxygen in the GDP- and the β,γ-phosphate oxygen in GTP-complex. It is totally invariant in all P-loop proteins (148), and when mutated in Ras, it reduces affinity to guanine nucleotides drastically (149). Reorganization of the switch 2 region of Ras also involves Arg68 which forms internal hydrogen bonds in Ras-GTP, but reorients completely in the complex to make contact with Glu1002 of Sos.

7.2 The Sec7–Arf complex

The structure of the Arf–ARNO complex was solved (Fig. 14) using the Sec7 domain of the yeast protein Gea2 and a truncated Arf where the 17 N-terminal residues that form the extra helix containing the myristoyl group have been deleted (44). As was observed for the Ras–Sos complex, Arf in the Sec7–Arf complex more closely resembles the Arf–GTP than the Arf–GDP structure. Structural changes in Arf, compared to the binary nucleotide complex, are localized in the switch 1 and 2 regions, which are the only contact sites. This is different from the Ras–Sos complex, which shows a more extensive interface on Ras. The Sec7 domain in the complex is likely to be similar to the unligated molecule except that the two subdomains of the Gea–Sec7 domain have rotated by 12° relative to each other following (or allowing) interaction with Arf. The interface of the complex buries 2680 Å^2 of solvent-accessible surface area upon complex formation, compared to 3600 Å^2 in the Ras–Sos complex.

The recognition surface on Sec7 is formed by helices F, G, and H, the C terminus and the hydrophilic FG loop, ^{93}RLPGESQ^{99}, which is the most conserved sequence (also called block 1) element in the Sec7 domain. Residues of Sec7 forming the interface of the complex form a continuous, mostly hydrophobic surface except for the FG loop. The most relevant residues in Arf involved in the binding are the hydrophobic residues Ile49 and Phe51 of switch 1, which penetrate deeply into the hydrophobic pocket of Sec7, and Leu77, Trp78, and Tyr81 of switch 2.

The mechanism leading to the reduction of affinity and eventually release of the nucleotide from Arf includes binding of residues Ile49/Phe51 of Sec7, which induces a conformational change of switch 1 such that the phosphate-binding area becomes modified. The most dramatic effect of Sec7 on nucleotide release, however, is probably due to the FG loop, which wedges into the phosphate- and nucleotide-binding area. Particularly important is the totally invariant Glu97, mutation of which causes a drastic defect in catalytic activity (131, 132, 134). As suggested from previous experiments (134), Glu97 occupies the γ-phosphate/Mg^{2+}-binding site of Arf, and the carboxylate oxygens accept a hydrogen bond from the P-loop lysine. Although lysine itself and the following Thr residue, which normally binds the metal ion, are unperturbed, the preceding residues which direct their peptide NH groups towards βγ-phosphate (148), have altered conformations and are no longer able to interact with phosphate. In addition, the switch 2 Gly70 and the following Glu71 are also moved out of their binding position in Arf•GppNHp. The mutation of Glu97 to Lys has been shown to reduce the catalytic activity 1200-fold and to allow the formation of a stable trimeric Arf•ARNO•GDP complex. This, together with the structural data, now gives a coherent view of the ARNO interaction, whereby Glu97 in the FG loop of Sec7 approaches the β-phosphate binding site from the site of the Mg^{2+} ion and creates an unfavourable electrostatic repulsion of phosphate and glutamic acid. Together with the other interactions involving the switch 2 region this leads to a weakening of nucleotide affinity and an increased dissociation rate.

As in the Ras–Sos complex the guanine and ribose binding sites of Arf are more or less unperturbed. This would indicate that in the reverse reaction, when GTP binds to induce dissociation of the binary Arf–ARNO complex, the base enters first, to form a ternary weak binding complex that then rearranges into the binary Arf–nucleotide complex, similar to the conclusions reached for the Ras–Sos structure (144, 150).

7.3 A universal mechanism of GEF-catalysed exchange?

The minimal model for GEF-catalysed nucleotide exchange involves a series of fast kinetic steps via transient ternary and stable binary complexes and conformational transitions (150). The presence of a ternary complex and its properties have been documented only for the Ran–RCC1 system (117, 118) where the affinities of the ternary complex are much lower than for the Ras–Cdc25 system (116) and thus amenable for biochemical analysis. Both the structure of the Ras–Sos (144) and of the Arf–Sec7 complexes (44) give a very detailed view of the structure of the binary complex, and can be compared to the structure of the EF-Tu•EF-Ts solved earlier (151, 152). Incidentally, the effect of disturbing the Mg^{2+}-binding site is not alone sufficient to achieve fast nucleotide release, as suggested by Kawashima *et al.* (151), since it has been shown for Ras (116) that even in the absence of Mg^{2+} (presence of EDTA), Cdc25 still has a strong stimulating effect. Although the way in which these GEFs bind to the GTP-binding protein and disrupt the nucleotide-binding site look similar in principle, the details vary considerably. All three GEFs directly approach and perturb the binding site by removing and interfering with the switch 1 region, the P-loop, and the Mg^{2+}-binding site. In addition, all GEFs interact with and dismantle the nucleotide-binding interactions of the N-terminal part of the switch 2 region. However, different GEFs use different residues and mechanisms to achieve their goal of destabilizing the binding site. In the case of EF-Ts, the most crucial residue is a hydrophobic phenylalanine inserted into the Mg^{2+}-binding site, whereas in the case of ARNO it is a glutamic acid occupying a similar position. Thus, in contrast to the GTPase stimulation by GAPs, which appear to use a similar principle to achieve stimulation of GTP hydrolysis, GEFs use an array of hydrophilic and hydrophobic residues inserted into the nucleotide-binding site to stimulate a similar magnitude of guanine nucleotide release.

As shown above, the binary complex is only an intermediate in the overall reaction; it is quickly dissociated under physiological concentrations of nucleotide. At present we know little about the intermediary ternary complexes, about conformational changes from tight to loose binding complexes, and about rate-determining steps in the kinetic mechanism. Nevertheless the three-dimensional structures tell us why the affinity of the nucleotide is reduced so dramatically.

In addition the structures suggest how the nucleotide might re-enter its binding site and reverse the reaction. Since the base and ribose are open, the nucleotide most likely comes in with its base moiety first to form what may be the weak-binding ternary complex. In order to achieve full strength binding, a conformational change(s) break(s) the interactions with GEF in the phosphate binding area, releasing GEF and allowing tight binding of nucleotide. By the principle of microreversibility one can thus also deduce the structural changes in the forward reaction where the action of GEF first flips open the base and ribose binding area and a subsequent conformational change which releases the nucleotide and forms a tight binary complex

7.4 Structural models for the receptor–G protein exchange mechanism

The α subunits of many G proteins are myristoylated and most are reversibly palmitoylated near the N terminus (see Chapters 1 and 2). The G_γ subunits contain a CaaX box at the C terminus, which is the signal for either farnesylation (C15 thioether) or geranylgeranylation (C20), dependent on the nature of residue X. These post-translational modifications are responsible for membrane attachment of the heterotrimer. Although the lipophilic moieties are missing from the recombinant G proteins whose structures were determined, the presumed localization of the N- and C-terminal ends on the same face of the heterotrimer allows conclusions to be drawn about membrane attachment and receptor interaction. The side of the heterotrimer believed to interact with the membrane is remarkably flat and its electrostatic surface is predominantly neutral with some dispersed positive charge. This agrees with a number of studies using ADP-ribosylation by pertussis toxin and peptide inhibition which implicate the N and C termini of G_α and residues 311–329 as important for interaction with the receptor (56).

Using the available evidence, a consistent proposal for receptor-mediated GEF activity and the mechanism of nucleotide exchange has recently been put forward (58). Since the intracellular loops of the activated receptor are presumably too short to reach the guanine nucleotide binding pocket, which is estimated to be located approximately 30 Å away from the membrane surface, it is proposed that $G_{\beta\gamma}$ is actively involved in destabilizing the nucleotide-binding site on G_α. This concept is based on a number of site-directed and disease mutations in both G_α and $G_{\beta\gamma}$. and on the structure of the EF-Tu•EF-Ts complex (151, 152). The binding site of EF-Ts on EF-Tu actually overlap with the contact site between G_α and $G_{\beta\gamma}$. A number of mutations of G_β have been tested which apparently do not change change the affinity between α and $\beta\gamma$ and yet have a drastic effect on receptor-mediated nucleotide release (153, 154). Another interesting mutation is a polymorphism of $G_{\beta3}$ associated with hypertension, where, due to a splicing defect, a short isoform $G_{\beta3-s}$ is produced that lacks one complete blade of the propeller and leads to a gain-of-function subunit, which is presumed to increase $G_{\beta\gamma}$-mediated GDP dissociation (155). The proposed model of the GEF mechanism of heptahelical receptors needs to be verified by a three-dimensional model of such a receptor, hopefully caught in the act of the switch on, in complex with $G_{\alpha\gamma\beta}$. That the β-propeller of G_β is a mediator of GEF action is interesting because RCC1 is also a seven-bladed propeller. Therefore clues to the mechanism of heptahelical receptor action might also come from solving the structure of the Ran–RCC1 complex.

8. The GTPase reaction and GTPase-activating proteins—GAPs

GTP hydrolysis is a key process in intracellular signal transduction and is the common timing mechanism for returning the molecular GTP-binding switches to the GDP-bound 'off'-state. Intrinsic GTP hydrolysis by Ras-related proteins is very slow (of the order of $0.001–0.1$ min^{-1}), ten- to 100-fold slower than G_α subunits. Hydrolysis rates can be accelerated by several orders of magnitude by interaction with GTPase-activating proteins (GAPs), such that the reaction has a half-life of less than a second.

A large number of studies have shown that GTPase activation by GAPs is a general principle of regulation for all GTP-binding proteins. GAPs have been described for the Ras, Rho, Rab, Ran, and Arf families of Ras-related GTP-binding proteins, and within families these are sequence-related to each other (90, 157, 158). Accordingly, they are termed Ras GAPs, Rho GAPs, Ran GAPs, etc. Modular architecture in GAPs is commonly used to combine the downregulatory GAP activity with various other functions, including signalling. A large number of GAPs for the αsubunits of hetero-trimeric G proteins (G$_\alpha$) have also been described. These were first detected genetic-ally in yeast (159) and are now commonly referred to as regulators of G-protein signalling (RGS) (59). The variability of GAP function is demonstrated by elongation factor Tu (EF-Tu): its almost unmeasurable intrinsic GTPase activity is stimulated dramatically by the mRNA-primed ribosome, with the C-terminal domain of the L7/L12 protein of the large subunit presumably acting as an EF-Tu GAP (160).

GAPs are primarily downregulators of the GTP-bound form, but some can also be active signal transduction molecules (157, 161, 162). The importance of proper functioning of GAPs *in vivo* is demonstrated by the occurrence of diseases associated with the loss of function of the Ras GAP, neurofibromin in type 1 neurofibromatosis patients (163), or the loss of function of Rho GAP responsible for X-linked mental retardation (MRX) (164) or a skin-defect syndrome, MLS (165).

Biochemical experiments have been done on a number of GAPs, but most of our knowledge is derived from Ras GAP, Rho GAPs, and Ran GAP. For the interaction between Ras and Ras GAPs (166–168) and for Ran and Ran GAPs (117, 118), it has been shown that the intrinsic GTPase reaction rates are stimulated by five orders of magnitude; in both cases this is the same order of magnitude as the stimulation of the exchange reaction by the respective exchange factor (116–118). It had long been argued that the rate-limiting step of the intrinsic GTPase reaction of Ras is a con-formational change and that GAP acts by accelerating this step. Experiments favour-ing or disfavouring this mechanism have been put forward (168–171). However, it was found that Ras, unlike G$_\alpha$ proteins (80, 172), binds aluminium and beryllium fluoride complexes only in the presence of stoichiometric amounts of GAP and that an arginine residue of GAP, Arg789 from p120GAP and Arg1276 from neuro-fibromin, is crucial for this effect (173, 174). Binding of aluminium fluoride was not observed with an oncogenic GTPase-deficient mutant of Ras. This suggested that Ras proteins are ineffective GTPase catalysts that need active participation from GAPs to achieve fast GTP hydrolysis.This argues that the actual chemical step of GTP hydrolysis is stimulated by Ras GAP, which participates in the reaction directly. This has now been shown, by biochemical (174) and structural (see below) experiments for Ras and for the Rho family proteins and their corresponding GAPs (175–177).

8.1 Structures of Ras GAP and Rho GAP

Numerous Rho GAPs active on Rho/Rac/Cdc42 have been identified, which are homologous to the C-terminal domain of Bcr, the break-point cluster region gene product (178). The structures of the GAP-like domain derived from the p85 subunit of PI3 kinase, comprising a 200-residue helical protein (179) and that of the corresponding domain of p50Rho GAP (180) were solved and shown to be highly similar. Its core contains a four-helix bundle, one face of which carries most of the conserved residues and has been proposed as the binding site for the Rho proteins.

Of the five mammalian Ras GAPs described to date p120 GAP and neurofibromin are the best studied examples (161). The structure of a catalytic fragment of p120 GAP, GAP-334, was solved by crystallography and found to be an α-helical elongated protein (181). The structure defined a central domain of 218 amino acids that contains all residues conserved among Ras GAPs, and corresponds to a minimal catalytic domain of neurofibromin that retains full GAP activity (182). The structure of a similar catalytic fragment from neurofibromin, NF1–333, was found to be similar to the GAP-334 structure (183). By solving this structure it was possible to locate residues in the catalytic domain of neurofibromin which were mutated in patients with neurofibromatosis; the mutation of Arg1276, which affects the most crucial residue involved in GTPase stimulation on Ras, was particularly interesting (184).

The catalytic domains of p120 GAP and p50Rho GAP are not strikingly similar. However, alignment of the models based on the knowledge of how they communicate with their GTP-binding partners reveals at least distant structural relationships. In an overlay, the extra domain and the C-terminal part of the central domain of Ras GAP are missing in Rho GAP, leaving the helical core described for Rho GAP as a possible evolutionary module conserved between Ras GAP and Rho GAP (157, 185, 186). This has been called the cradle fold (187). It is likely, however, that other GAPs will have a different fold. The Ran GAP catalytic domain, for example, consists of leucine-rich repeats, and analysis of an RNA inhibitor with leucine-rich repeats revealed repeating α-β units arranged in a horseshoe-like structure (188, 189). Arf GAP, on the other hand, is predicted to constitute a Zn-binding domain (190).

8.2 The Ras–Ras GAP complex

In the study of phosphoryl transfer enzymes, such as G_α, myosin, F1-ATPase, or nucleotide diphosphate (NDP) kinase, beryllium fluoride, aluminium fluoride, and vanadate have been used as analogues of either the γ-phosphate in the ground state (beryllium fluoride) or the transition state (aluminium fluoride, vanadate) of the reaction. The structure of the Ras–Ras GAP complex, formed by H-Ras•GDP and GAP-334 in the presence of aluminium fluoride shows GAP-334 (Fig. 18) interacting predominantly with the switch 1 and 2 regions and the P-loop. This is similar to the proposed docking model (181) and in line with mutational analysis (1, 146), and involves both hydrophobic and hydrophilic contacts stabilizing the interface (41). The interaction between Ras and Ras GAP places an exposed loop of Ras GAP in proximity to the nucleotide, with the guanidinium group of Arg789 interacting with the β-phosphate of GDP and a planar AlF_3. The same arginine mediates a hydrogen bond between its main-chain carbonyl oxygen and the side-chain amide group of the catalytically important Gln61 of Ras. Since Arg789 and the loop point into the active site, they have been called the 'arginine finger' and the 'finger loop', respectively. Glutamine 61 in turn contacts AlF_3 and a water molecule representing the attacking nucleophile. As in unligated GAP-334 (181), Arg903 of the most highly conserved FLR motif stabilizes the finger loop by side-chain/main-chain interactions.

Mutations of Gly12 and Gln61, which are common in Ras-activated tumours, usually lock GTP-binding proteins in their active conformation (191, 192) and

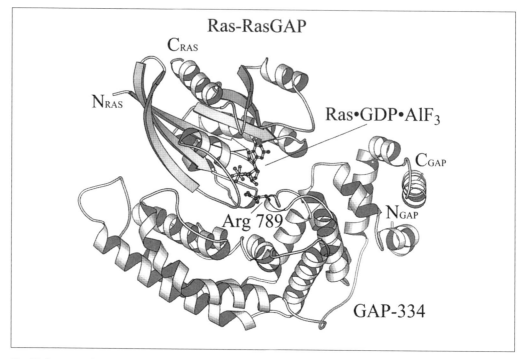

Fig. 18 Complex of Ras with Ras GAP (pdb-file 1WQ1) in the presence of aluminium fluoride shows the all-helical structure of Ras GAP with Arg789 on an exposed loop penetrating the active site of Ras. Switch 2 of Ras is the major site of interaction.

homologous mutations in other GTP-binding proteins show a similar phenotype. The structure provides a simple explanation why these mutants are insensitive to GAP: Gly12 is in close proximity to the finger loop such that even the smallest possible change to alanine would sterically interfere with the geometry of the transition state. Interestingly, Gly12-mutants bind to GAP with almost wild-type affinity, arguing that larger side-chains in position 12 can apparently be tolerated in the Ras•Ras GAP ground-state complex, but not in the transition state. The apparent involvement of Gln61 in the stabilization of the transition state, along with biochemical (173) data, confirms the notion that Gln61 has a vital role in catalysis, which cannot be played by other residues.

8.3 Rho–Rho GAP complexes

The situation observed in the active site of Ras•GDP•AlF$_3$•GAP-334 was basically confirmed by the structure of the corresponding complex between the catalytic domain of p50Rho GAP and RhoA (Fig. 19) (193). One noticeable difference was the presence of an AlF$_4^-$ moiety in the γ-phosphate binding site, whereas the Ras structure showed AlF$_3$, albeit at lower resolution. AlF$_4^-$ has been found in the complexes of G$_\alpha$ proteins with aluminium fluoride and myosin, whereas AlF$_3$ has

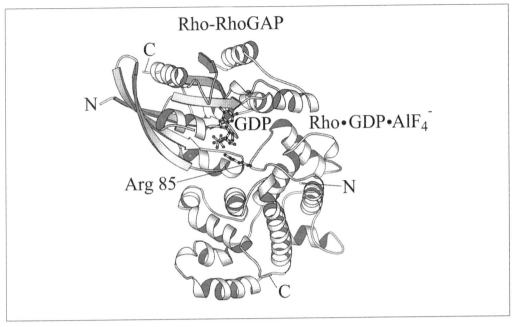

Fig. 19 The Rho-Rho GAP complex (pdb-file 1TX4) in the presence of aluminium fluoride shows the all-helical conformation and Arg85 interacting with the active site of Rho. Rho GAP and Ras GAP are believed to be evolutionarily related (199).

been found in transition state structures of NDP kinase (194) and UMP kinase (195) and Cdc42•Rho GAP (see below). The reason for the difference is not clear, but AlF_3 would certainly be a better mimic of the pentavalent transition state of phosphoryl transfer. An invariant arginine in Rho GAP (Arg85, equivalent to Arg 282 in full-length Rho GAP) contacts the nucleotide and a fluoride ligand of the square-planar AlF_4^-. As in the Ras–Ras GAP complex, the carbonyl oxygen of this arginine forms a hydrogen bond with the amide group of the critical glutamine in Rho (Gln63), which in turn contacts AlF_4^- and the nucleophilic water molecule. The loop carrying Arg85 is stabilized by an invariant lysine (Lys122), apparently in a similar way as the finger loop of Ras GAP is stabilized by another invariant arginine.

Comparison of the AlF_4^--bound complex with the ground-state complex, as represented by p50Rho GAP•Cdc42•GppNHp (196) revealed a major structural change upon formation of the transition state, involving a 20° rigid body rotation of the two proteins relative to each other (193). In the ground-state complex Arg85 contacts the P-loop of Cdc42 and is not in a position to support catalysis. Phosphorus NMR experiments with Ras have shown that the addition of Ras GAP to Ras•GppNHp does not induce a chemical shift change on any of the phosphate resonances, which would be expected if they were contacted by a positively charged side-chain of arginine (33). Together with binding studies of oncogenic mutants and with mutants of GAP, this suggests that for both the Cdc42/Rho•Rho GAP and the Ras•Ras GAP interaction, the finger arginine of GAP in the ground state is not in a position where it could accelerate GTP hydrolysis, but only gets there in the transition state. In the critical Gly12 position of Cdc42/RhoA, larger side-chains could be accommodated in the ground state, but would cause steric hindrance upon transition-state formation, as observed in the Ras•Ras GAP complex (41).

The structure of Rho GAP was also solved in complex with Cdc42•GDP in the presence of aluminium fluoride, and showed AlF_3 rather than AlF_4^- in the active site. The crucial Arg205 finger contacts a fluoride of AlF_3 and the bridging oxygen and an α-phosphate oxygen of the nucleotide, a feature not seen in the other structures. Other than that, the active site features are very similar to the other transition-state complexes. In an interesting twist to the story, the structure of the complex between a mutant Rho GAP and Cdc42 was solved. In the mutant, the arginine finger R205 was substituted, resulting in only a fortyfold loss of GAP activity (197), surprising considering that the same mutation causes a 2000-fold loss of activity in Ras GAPs (174). In the structure, Tyr32 of Cdc42 has moved into the active site to fill the void created by the R205A mutation, and its hydroxyl group makes a strong hydrogen bond with one of the fluorines, thus apparently explaining the partial stabilization of the transition state and the relatively small effect on catalytic activity.

8.4   Mechanism of GTPase activation: arginine finger crosses the GAP

Arginines or lysines have been proposed to play critical roles in phosphoryl transfer reactions. Positively charged under physiological conditions, they are, in principle, able to neutralize negative charges developing on the transferred phosphoryl group or on the leaving-group oxygen, depending on whether the mechanism is associative or dissociative (85). In addition to being charged, their side-chains are comparatively long, enabling them to bridge larger distances within the interfaces of protein–protein complexes. The structures of Rho and Ras in complexes with their respective GAPs did not reveal residues that could act as a general base for the activation of the nucleophilic water molecule. This is consistent with the notion that the proposed mechanism of substrate-assisted catalysis (84) may also hold for the GAP-catalysed reaction (198). This is different from the situation with kinases and phosphatases, where invariant general bases have been identified as additional elements of catalytic rate enhancement.

Both Ras GAPs and Rho GAPs have invariant arginines and lysines (180, 198). In Ras GAP, mutation of the crucial arginine (Arg789) to lysine or alanine reduces the GTPase activity 2000-fold. Another invariant arginine (Arg903) is less crucial for catalysis as it can be replaced by lysine without serious effects on GAP catalysis (174). This is in perfect agreement with the complex structure:

1. Arg789 points into the active site to neutralize negative charges on the phosphates (or the fluorides mimicking the phosphate oxygens) and stabilizes the position of Gln61 of Ras.

2. The finger loop itself is stabilized by another invariant arginine, Arg903 in p120 GAP.

3. Gln61 apparently positions the water molecule for nucleophilic attack and contacts a γ-phosphate oxygen, thus also stabilizing the transition state.

In summary, GAP supplies a catalytic residue into the active site, the scaffold of which is provided by the GTP-binding protein. In Rho GAP a very similarly located arginine also contacts the active site, although its mutation to Ala does not have such a dramatic effect (177, 197, P. Lowe, pers. Communication) and it is not stabilized by another arginine, but rather by a lysine. The higher resolution of the Rho–Rho GAP or Cdc42–Rho GAP complex structures shows that the ε-imido group of the arginine side-chain contacts the fluoride and the terminal negative charge contacts the βγ bridging oxygen. Considering that the participation of arginine residues is crucial for GTPase stimulation, it is interesting to note that other GAPs such as Ran GAP, Arf GAP, RapGAPs, and Rab GAPs, also contain apparently invariant positively charged residues (199). This indicates a possible common mechanism of GAP-mediated GTP hydrolysis by small GTPases, in contrast to GEF-mediated nucleotide exchange, which is more variable.

8.5 A heterodimer enzyme

Transition-state stabilization is the basic principle of enzyme catalysis. In GTP-binding proteins, the substrate-binding site has a catalytic machinery that is, in principle, able to perform GTP hydrolysis at a significantly increased rate compared to spontaneous hydrolysis in water. However, this rate is still relatively slow and can be increased substantially upon interaction with GAP. These two components form a highly efficient heterodimeric enzymatic catalyst, which seems to represent a new principle in evolution.

Separation of the components of the enzymatic machinery is most certainly of major physiological importance. p120 GAP is a cytosolic protein and becomes localized, via the SH2 domain, to the plasma membrane by binding to activated receptor tyrosine kinases such as the PDGF receptor. Considering the low affinity of p120 GAP to Ras•GTP, which is in the 10–20 μM range, and considering the concentrations of the reaction partners, it seems reasonable to assume that GAP acts on Ras only when localized to the plasma membrane where the local concentration, in two-dimensional space, can become very high. This is analogous to the guanine nucleotide exchange factors, which also act on Ras only when localized to the plasma membrane, since their K_M for Ras is in the 200–300 μM range (116).

8.6 RGS proteins and the $G_{i\alpha1}$–RGS4 complex

G_α subunits of heterotrimeric G proteins hydrolyse GTP with a rate constant of 2–5 min^{-1}, on the average 100 times faster than the small GTPases. However, from physiological studies it appears that downregulation of G proteins *in vivo* is much faster. It was originally thought that effectors were stimulators of the GTPase reaction and, indeed, both phospholipase C-β (200) and the γ subunit of GMP phosphodiesterase (201) can stimulate the intrinsic GTPase reaction rate of their respective G_α proteins, $G_{q\alpha}$ and $G_{t\alpha}$. G_α proteins differ from small GTP-binding proteins in that they contain an additional helical domain and an arginine is crucial for catalysis (202).

RGS proteins constitute a large family of proteins with varying length, containing a conserved catalytic core of about 120 amino acids. They act as GAPs for G protein αsubunits. Biochemically they have been shown to stimulate the GTPase activity of G_α subunits about 100-fold, such that the maximally stimulated GTPase reaction rate is in the order of s^{-1}, similar to the GAP-catalysed rate of small GTPases (59). Since GAPs for Ras-related proteins supply the missing Arg residue into the active site and thereby stabilize the transition state, it was of special interest to find out how RGS proteins increase the GTPase rate of heterotrimeric G proteins, which already have a catalytic arginine.

The first clue to the mechanism came from studies on $G_{i\alpha1}$ and RGS4, where it was shown that the latter binds very tightly to $G_{i\alpha1}$ in the $G_{i\alpha1} \bullet GDP \bullet AlF_4^-$ transition-state complex and much weaker to the $G_{i\alpha1} \bullet GTP\gamma S$ ground state (203). The crystal structure of the complex between the catalytic core of RGS4 and $G_{\alpha i} \bullet GDP \bullet AlF_4^-$ has now been solved by X-ray crystallography (204). RGS4 contains 205 amino acids, but only residues 58–178 comprising the actual RGS box were visible in the structure, supporting the view that only the conserved box is involved in contacting G_α. The RGS box consists of nine α-helices, four of which form an helical bundle-like subdomain. Three helixes α7,α,8,α9, can actually be considered a single curved α-helix interrupted by two bends. The RGS structure, although purely α-helical, does not resemble the structure of either Rho GAP or Ras GAP.

The bottom of this four-helix bundle forms a surface that interacts with the G_α protein. Only the three switch regions and almost no residues from the helical domain of G_α are involved in the contact with RGS. The interface buries 1100 Å^2 of solvent-accessible surface. It is rich in electrostatic and hydrogen-bonding interactions, with only few hydrophobic side-chains making a contribution. Conserved residues in switch 2 appear to be the most important for complex formation as they form a large number of contacts with RGS. Since RGS covers the switch surface of G_α, it is expected (and has been confirmed biochemically) that binding of RGS and effector proteins is mutually exclusive. The conformation of $G_{i\alpha}$ in the complex structure resembles most closely that of free $G_\alpha \bullet GDP \bullet AlF_4^-$, with Gln 204 and R178 in the same position, but different from that in the GTPγS state, confirming the expectation that the structure resembles that of the transition state and that RGS is sensitive to the nucleotide state of G_α. None of the residues of RGS is close to the active site, except the conserved Asn126 which contacts Q204 of $G_{i\alpha1}$.

Comparing the role of GAPs for small GTP-binding proteins with that of RGSs we can conclude that nature has developed at least two concepts to realize efficient GTP hydrolysis:

(1) use of arginine residues for stabilizing developing negative charges; and

(2) stabilization of the switch regions for optimum orientation of the catalytic machinery in the GTP-binding protein, the most important element of which is a glutamine.

In heterotrimeric G proteins the arginine is part of the GTP-binding protein itself, being supplied 'in *cis*' with an extra domain necessary for orientation and an extra protein (RGS) for proper alignment of the whole machinery. For small GTP-binding proteins, the arginine finger is supplied 'in *trans*', which along with stabilization of the catalytic machinery, is achieved by GAPs. Although the intrinsic arginine of G_α subunits and the external arginine from GAPs are approaching the active site from a different angle, the catalytically important head group of the side-chain is in the same position. Stabilization of the switch regions is best documented in Ras, where the Gln61 region is highly mobile in the isolated protein (13, 20, 30), but becomes stabilized in the complex with GAP (41).

9. Interactions with effectors

9.1 Ras effectors

Numerous bona fide or putative effectors of Ras have been identified, including Raf, (three isoforms c-Raf-1, B-Raf, and A-Raf) and RalGEF (is a guanine nucleotide exchange factor for Ral with three isoforms, RalGDS, Rgl, and Rlf). In addition to its Cdc25-like GEF domain, RalGEF proteins have a domain at the C terminus involved in Ras binding. Further Ras effectors include PI3 kinase (three isoforms), Rin, Nore-1, and AF6, all of which have been shown to contain a domain that binds to Ras at the effector region in a nucleotide-dependent manner (see Chapter 3) (205, 206). The region in Raf responsible for Ras binding has been mapped to a fragment comprising residues 51–131 from human c-Raf-1 (207) and was subsequently called the Ras-binding domain. The affinity of this domain for the GTP- and GDP-bound forms of Ras differs by a factor of 1000, as would be anticipated for an effector (208). Ras-binding domains have been identified in RalGEFs, PI3 kinases, AF6, and Nore-1, all of which (except PI3 kinases) seem to have a stable fold.

9.2 Structure of the Ras–Raf RBD complex

The structure of the Ras-binding domain (RBD) of c-Raf-1 was solved by NMR spectroscopy (209). It showed a ubiquitin-like fold with a five-stranded mixed β-sheet and one major helix on one face of the sheet. The structure of RafRBD in complex with the Ras homologue Rap1A and the non-hydrolysable triphosphate analogue GppNHp was solved by X-ray crystallography (38) and represented, at the time, the first effector complex structure (Fig. 20). Rap1A is a close homologue of Ras and has the same core effector region (residues 32–40) as Ras. The structure showed that the main interactions between Rap and Raf are mediated by an inter-protein β-sheet, which is formed by two antiparallel strands from the two proteins, and by residues on the helix. Most of the residues of Ras that had been implicated in effector

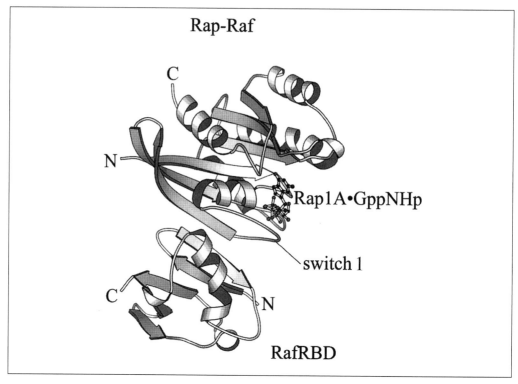

Fig. 20 The Ras homologous protein Rap in its GppNHp form interacting with the Ras-binding domain of the Ras (and possibly Rap) effector c-Raf-1 (pdb-file 1GUA). The interface between the proteins is formed by an inter-protein β-sheet.

binding were indeed found to be located at the interface and involved in binding. This type of interaction appears to be a common type of protein–protein interaction, and has also been found for the complex between immunoglobulin and an immunoglobulin-binding domain (210, 211).

The puzzle is that Rap and Ras have different, if not opposing biological roles, yet both bind to the effector Raf kinase in a similar way, as the residues found in the Rap–Raf interface are conserved between Ras and Rap. The role of Rap1 is unknown at present (see Chapter 3), but this does raise the problem of the specificity of the Rap–Raf versus the Ras–Raf interaction. Ras binds approximately 100-fold tighter than Rap to Raf (212), and the residue responsible for this specificity is residue 31, which has an opposite charge in Rap and Ras (Lys versus Glu, respectively). Mutation of Lys31 to Glu in Rap clearly demonstrates that this is indeed the main residue responsible for specificity and the three-dimensional structure of this Rap mutant, called Raps, in complex with RafRBD has also been solved (40). It showed, due to the interaction of Lys84 from Raf and Glu31 from Raps, a more favourable interaction (compared to the Rap–Raf interaction) between the proteins. The structure of the Raps–Raf complex is believed to mimic that of the real Ras–Raf interaction very closely.

^{31}P-NMR experiments have shown that Tyr32 is a very mobile residue which flips in and out of two conformations in the triphosphate-bound state of Ras (33). These conformations are in rapid equilibrium, but only one is able to bind to RafRBD. In the Rap–Raf complex, the tyrosine hydroxyl is coordinated to the γ-phosphate either directly or via a water molecule (38, 40). It is therefore likely that the binding of the effector stabilizes the interaction between Tyr32 and the nucleotide, and produces a GDI-like effect, whereby the binding of an effector inhibits nucleotide dissociation (208). Furthermore, the release of the terminal phosphate after GTP hydrolysis and the subsequent release of the mobile Tyr32 may represent an additional trigger for the changes leading to the GDP-bound conformation (24). A residue corresponding to Tyr32 of Ras is conserved between Ras, Rho, and Ran subfamily members, and may have a similar function in these other proteins. Indeed, conformational flexibility is also observed for Ran, where the γ-phosphate has two rapidly interconverting states, which may be affected by an aromatic residue inducing a chemical shift when close to γ-phosphate.

9.3 Ras and RalGEF

All three isoforms of RalGEF (Ral GDS, Rgl, and Rlf) have a C-terminal Ras-binding domain (RGF–RBD). Rap binds 100-fold tighter to RGF–RBD than Ras, and again the charge reversal in residue 31 is mostly responsible for the effect (212). The structure of the Ras-binding domain of RalGEF (RGFRBD)has been determined by NMR (34) and X-ray crystallography (213), and that of the homologous Rlf-RBD has also been solved by NMR (214). They show that their Ras-binding domain has the same ubiquitin-like topology as RafRBD. The structure of the complex of RalGEF with either a tighter binding Ras (E31K) mutant (215) or real Ras (216) followed soon after. The structures showed, not unexpectedly, that complex formation is very similar to that of Raps–Raf and uses a similar inter-protein β-sheet interaction. The major difference between the two structures is, apart from using a Ras mutant instead of wild-type Ras, that in the former, RGF forms a tetrameric complex with two Ras molecules such that Ras uses both its effector region and the switch 2 region for the interaction. Biochemical studies show however that the Ras-RGFRBD complex is a monomer in solution and that dimerization is due to disulphide formation in the crystallization batch—tetramerization is a crystallographic artefact (216).

Comparison between the Ras–Raf and the Ras–RGF complexes show that although in both complexes the effector/switch 1 region is used and a inter-protein β-sheet interaction occurs, the individual residues involved in the contact are partly over-lapping but also partly different. This gives a structural basis for the observations of partial loss-of-function mutations, where certain residues in the effector region, such as D38A, inhibit the interaction of Ras with all effectors, whereas mutations such as E37G or Y40C inhibit only a subset of effector interactions (see Chapter 3) (206). Indeed, Asp38 is involved in both Raf and RGF complexes, whereas Glu37 is only relevant for the Ras–Raf interaction. In both complexes Thr35 is not involved in forming the interface, which is surprising considering that the T35S mutation blocks interaction with RalGEF but not with Raf. The possible solution to this puzzle came from NMR studies of the conformational flexibility of Ras(T35S)•GppNHp. Whereas the wild-type protein shows two conformations in the P-NMR spectrum (33), the mutant shows only one conformation (Wittinghofer, unpublished). The reason for this is unclear, but might explain the finding that this threonine residue is totally invariant in all GTP-binding proteins, although as a ligand of Mg^{2+} or phosphate, Ser should be a suitable substitute. It appears that the threonine is also responsible for the flexibility of the effector region, which in turn is necessary for the switch function of Ras.

The structure of the complexes is also different around residue 31. In wild-type Ras, the highly negatively charged surface side of RGF is seen to make long-range unfavourable interactions with Glu31 of Ras (216). In a mutant complex (E31K) which mimics the Rap–RGFRBD complex, Lys31 can now make favourable interactions with negatively charged residues in its vicinity (215).

9.4   Ran effectors

Ran has been shown to bind in a GTP-dependent manner to a large number of proteins. They can be grouped into two major classes. The first is a superfamily of proteins related to importin-β (or karyopherin-β). They are proteins of approximately 100 kDa molecular mass which are involved in shuttling cargo either into or out of the nucleus (see Chapter 7). They contain a small domain that binds to Ran•GTP and, in the case of importin, with such high affinity that the dissociation of the complex is unmeasurably slow. Exportins have a somewhat lower affinity. In the case of importins, the cargo–importin complex is dissociated by Ran•GTP, whereas exportin–cargo complexes need Ran•GTP for high-affinity binding. The second class of effector proteins is related to the 25 kDa proteins RanBP1 (Ran-binding protein 1) which contains a Ran-binding domain (RanBD), a conserved sequence motif found in several proteins involved in nucleocytoplasmic transport, such as RanBP1, RanBP2, RanBP3, the yeast proteins Yrb1p, Yrb2p, and Nup2p of *S. cerevisiae* (217, 218). They bind Ran•GTP with high affinity, using a binding site that is non-overlapping with the importin-binding site, such that ternary complexes between Ran•RanBP•importin can be generated. These are stable even in the presence of Ran•GDP, conditions under which the binary complex would be unstable. RanBP-like effectors are co-activators of Ran•GAP since they stimulate the Ran•GAP-mediated GTPase reaction on Ran (219). The are also described as essential release factors for complexes of Ran with transport factors (220).

9.5   Structure of the Ran•GppNHp•RanBD1 complex

RanBP2 is a large protein, found in long fibrils on the cytoplasmic side of the nuclear pore complex (NPC) and is believed to play an essential role in protein import and export in higher eukaryotic cells (221–224). The 3224 residues of RanBP2 comprise an

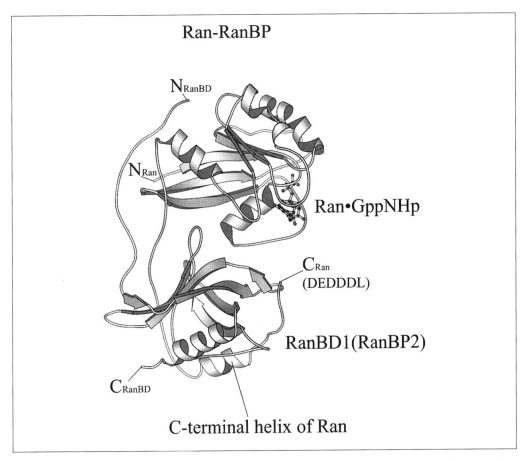

Fig. 21 Ran•GppNHp in complex with a new type of effector motif, RanBP (coordinates from I. Vetter), derived as the Ran-binding domain from RanBP2, shows a molecular embrace whereby Ran wraps its C-terminal end around RanBD1.

amino-terminal, 700-residue, leucine-rich region; four Ran-binding domains (homologous to RanBP1); eight zinc-finger motifs; and a carboxy terminus with high homology to cyclophilin, a protein with unknown function that can act potentially as peptidyl-prolyl-isomerase (221, 222). Electron microscopy has shown RanBP2 to be a filamentous molecule of about 36 nm length and it has been proposed that it forms a major part of the cytoplasmic fibrils of the NPC (224). RanBP2 has been identified as the only prominent protein of rat-liver nuclear envelopes that binds Ran–GTP, and a model has been developed where Ran targets members of the importin-β/karyopherin-β superfamily to the NPC via binding of a RanBP2 domain.

The crystal structure of Ran•GppNHp in complex with the first Ran-binding domain (RanBD1) of human RanBP2 was recently solved (Fig. 21). Ran–GppNHp shows an extensive conformational change as compared to Ran–GDP, involving the switch 1 region and the long Ran-specific C terminus. Secondary-structure predictions had indicated RanBD1 to consist almost entirely of α-helices. Unexpectedly, the RanBD1 domain turned out to have the topology of a pleckstrin domain (PH) and the related phosphotyrosine binding (PTB) domain, although the sequence shows no detectable homology (225). This fold features two, almost orthogonal, β-sheets consisting of four and three strands, with a helix sitting on top of the sheet like a lid. PH domains have been shown to bind to phosphatidylinositiol lipids but may also constitute protein–protein interaction domains. The direct interaction of a PH-like domain with a small GTP-binding protein (Ran) indicates that the PH-domain fold is a stable domain useful for many purposes. In addition to the PH-domain fold, the N- and C-terminal ends of RanBD1 have no secondary structure. The C terminus is not visible in the structure, whereas the N terminus is an important part of the Ran–RanBD interface.

The interface between Ran and RanBD1 has several main contact sites:

1. The effector region of Ran around residues 32–35, and β2 interact with the core of RanBD1.

2. The linker and the helical part of the C terminus of Ran make extensive contact.

3. The unstructured N-terminal part of RanBD1 reaches across the Ran molecule and makes contact mostly with the region around β2–β3 loop of Ran.

The large extended regions in both Ran and RanBD1 give the complex the impression not only of a molecular handshake between the two proteins, but that of a molecular embrace, because the Ran wraps its C-terminal end around RanBD1 and forms an almost complete turn. By doing so, the highly conserved C-terminal DEDDDL motif of Ran is situated on a large positively charged patch on RanBD1. The DEDDDL motif of the Ran C terminus has been shown biochemically to contribute significantly to the affinity of Ran for RanBPs and, when deleted, it reduces the affinity 8000 fold.

The Ran–RanBD1 structure supplies a structural explanation for the role of RanBP-like molecules and that of the DEDDDL motif of Ran, in nucleocytoplasmic transport. The latter is inhibitory to the Ran GAP-mediated GTPase activation of Ran. It also negatively regulates the interaction between Ran and importin-β-like molecules, since the association rate constant between Ran–GTP and importins is higher in the absence of the motif. Furthermore, this motif is accessible to DEDDDL-specific antibody in the Ran–importin complex, but not in free Ran–GTP. Thus it is envisioned that the role of RanBD1 is to recognize the loose C-terminal end of Ran–exportin complexes, to cause the C-terminal end to wrap around itself and thereby release the importin-like protein from the complex with Ran. This in turn renders the active site of Ran accessible to the action of Ran GAP, because in the Ran–RanBD1 complex the active site can be approached by Ran GAP in a similar way as Ras is approached by Ras GAP. Since four RanBD domains, together with the covalently attached Ran GAP, are located on RanBP2 on the exit site of the nuclear pore (226), this combined action of the five domains dissociates all export–cargo complexes after leaving the nuclear pore (see Chapter 7).

9.6 G_α protein–adenylyl cyclase interaction

Adenylyl cyclase is responsible for converting ATP to cAMP, a ubiquitous second messenger. cAMP in turn activates diverse cellular responses by activating cAMP-dependent protein kinases and cAMP-gated ion channels, and recently it was found that a GEF for Rap is also activated by cAMP (227). Nine isoforms of adenylyl cyclases have been identified, AC I to AC IX, all of which are regulated by G_α subunits (see Chapter 1). The protein is a twofold repeat unit each consisting of a transmembrane module, modelled as six transmembrane helices, followed by a 40 kDa cytosolic region. The two cytosolic regions each contain a homologous domain called C1 and C2 . The cytosolic C1 and C2 domains can be expressed in *E. coli* as soluble proteins and can be combined to form a highly active enzyme that retains the sensitivity to be stimulated by $G_{s\alpha}$ and by forskolin (228, 229). Both C1 and C2 can also form homodimers, but these have only low activity.

The structure of the C2 homodimer of type II adenylyl cyclase in the presence of two molecules of forskolin was solved. It showed a three-layered $\alpha\beta$-sandwich fold (230). Although no structural homology was noticed, it later turned out that the central part of the structure has a high homology to the palm domain of prokaryotic DNA polymerases (231). The end of the central β-sheet is bent back towards itself such that the monomers form a wreath-like dimer with a central cleft. The two forskolin molecules bind to a hydrophobic pocket at one end of this cleft, which is also the presumed binding site for adenine nucleotides.

The crystal structure of a catalytically active C1/C2 heterodimer in complex with $G_{s\alpha}\bullet GTP\gamma S$ (Fig. 22) was determined by Tesmer *et al.* (204). The heterodimer was constructed from the C1 domain of AC V and the C2 domain of AC II. This VC1–IIC2 heterodimer is stabilized by, and could be crystallized in the presence of, one molecule each of forskolin and $G_{s\alpha}\bullet GTP\gamma S$ complex. The overall structure of the VC1–IIC2 heterodimer is basically similar to that of the IIC2 homodimer. This may not be so surprising, since the latter was used as a search model in the molecular replacement solution, (except for a 7° rotation of the two domains relative to each other, which appears to be crucial) together with a $G_{s\alpha}$ model. As in the homodimer, there is a large amount of surface (3300 Å^2) buried in the interface between the two monomers. The VC1 and IIC2 monomers superimpose with an rms of 1.5 Å for 153 structural equivalent Ca atoms, in addition to some notable differences between the two domains. The importance of the heterogeneity of the interface, found to a lesser extent also for the homodimer and most likely very relevant for the catalytic activity, is that, of the 28 interface residues in VC1 and 33 in IIC2, only 17 are found at structurally equivalent positions and only 12 are conserved. The asymmetry of the interface is responsible for binding of one molecule of forskolin, $G_{s\alpha}$, and ATP to the heterodimer.

In contrast to the homodimer, only one molecule of forskolin is bound to the heterodimer. The reasons for the inability to bind a second molecule are apparent from the structure and are mostly due to steric constraints generated by the asymmetry of the heterodimer. It had been proposed that forskolin induces adenylyl

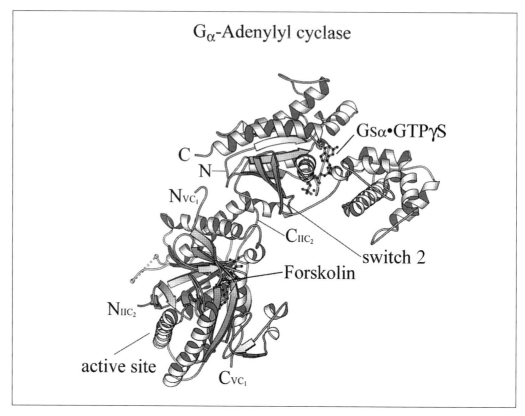

Fig. 22 The complex between $G_{s\alpha}$ in the GTPγS conformation and its effector adenylyl cyclase (pdb-file 1AZS), using the C1 domain from adenylyl cyclase type V and the C2 domain from type II, together with the diterpene stimulator forskolin. The active site has been deduced from inhibitor-binding sites determined crystallographically. The switch 2 region from $G_{s\alpha}$ is interacting with the C2 domain such that residues in the active site are allosterically adjusted to be in a better position for catalysis.

cyclase activity by stabilizing dimer formation. Although this is certainly measurable, using the isolated C2 domain, its influence on activity seems to be more complex, since its activating effect is larger than its effect on dimerization, which may be less relevant in intact adenylyl cyclase where the C1 and C2 catalytic domains are part of the same molecule.

The interaction of $G_{s\alpha}$ is mostly with the IIC2 protein and the most important interaction is the insertion of the switch 2 loop of $G_{s\alpha}$ into a groove of the IIC2 molecule, confirming the primary role of switch elements for effector interactions for G_α proteins. The residues contacted by $G_{s\alpha}$ are conserved between all adenylyl cyclases and some had been predicted as contact sites by mutagenesis. It is predicted that the inhibitory $G_{i\alpha}$, whose inhibition of adenylyl cyclase is not competitive with respect to $G_{s\alpha}$, binds to the VC1 subunit, opposite to the binding site of $G_{s\alpha}$.

Cyclase crystals have also been soaked with P-site inhibitors (2′,d3′-AMP or d2′,3′-

AMP and Mg^{2+}) which, in the presence of pyrophosphate, bind with high affinity and act as dead-end inhibitors of product (pyrophosphate) release. The structure of this complex was used as a mimic of the product complex in order to map the ATP-binding site and to probe the catalytic mechanism. It has been shown that the attack of the 3′-hydroxyl group on the α-phosphate is by a direct in-line attack. A general base activating the nucleophile could not be identified, but an arginine is located close to the α- phosphate and the αβ bridging oxygen and this could stabilize the pentavalent transition state.

The major question posed by the structure is how $G_{s\alpha}$ stimulates the activity of adenylyl cyclase. As mentioned above, if the IIC2 domain of the homodimer and the heterodimer are superimposed, a relative movement of the VC1 domain has been observed which is believed to be the result of the IIC2–$G_{\alpha s}$ interaction. Although the structure of the VC1•IIC2 in the absence of $G_{s\alpha}$ is not known, the movement of the interface between the heterodimers created by $G_{s\alpha}$ binding on one side of the cyclase is believed to change the orientation of residues in the ATP-binding site on the other side. A major effect, therefore, could be the reorientation of Arg1029 such that it can now stabilize the transition state of the reaction. This would be similar to the mechanism of small GTPase GAPs, where the proper placement of an Arg residue in phosphoryl transfer can stimulate the reaction by several orders of magnitude (41, 193, 196).

10. Concluding remarks

Within the past 10 years a large number of structural studies on GTP-binding proteins have supplied enormous insights into the function and mechanism of these switch molecules. They have demonstrated, on the one hand, a universal mechanism of structural transition from the GTP-bound to the GDP-bound state of the G domain proper which, on the other hand, shows spectacular differences in structural details. Whereas the triphosphate form of these proteins shows a universal way of attaching the structural switch regions to the γ-phosphate, the GDP-bound forms relax in many different ways. The structures of complexes with regulators and effectors show that the switch regions are the primary sites for forming the interface. Complexes with GAPs show an arginine residue from GAP stabilizing the transition state, a mechanism of GTPase acceleration that may be universal. Complexes with GEF show a basic mechanism of disturbance of the nucleotide-binding site via the switch 1 and 2 regions, leading to fast nucleotide dissociation, but the details of this interference are vastly different in the three examples so far reported. The few available structures of effector complexes show that different strategies are used to achieve specific and tight interactions with the GTP-bound state of small GTP-binding proteins and α subunits of G proteins.

This is clearly just a beginning and much more remains to be done to understand the full spectrum of conformational transitions and protein–protein interactions: a few dishes have been served and tasted, but it has stimulated our appetite for a complete menu and a lot more.

Acknowledgements

I thank Rita Schebaum for a lot of secretarial assistance, Michael Hess for preparing the figures, Thilo Brinkmann for carefully reading the manuscript, and John Kuriyan and Jonathan Goldberg for coordinates.

References

1. Schweins, T. and Wittinghofer, A. (1994) Structures, interactions and relationships. *Curr. Biol.*, **4**, 547.
2. Bourne, H. R., Sanders, D. A., and McCormick, F. (1990) The GTPase superfamily: a conserved switch for diverse cell functions. *Nature*, **348**, 125.
3. Bourne, H. R., Sanders, D. A., and McCormick, F. (1991) The GTPase superfamily: conserved structure and molecular mechanism. *Nature*, **349**, 117.
4. Sprang, S. R. (1997) G protein mechanisms: insights from structural analysis. *Annu. Rev. Biochem.*, **66**, 639.
5. Kjeldgaard, M., Nyborg, J., and Clark, B. F. (1996) The GTP binding motif: variations on a theme. *FASEB J.*, **10**, 1347.
6. Rensland, H., John, J., Linke, R., Simon, I., Schlichting, I., Wittinghofer, A., and Goody, R. S. (1995) Substrate and product structural requirements for binding of nucleotides to H-ras p21: the mechanism of discrimination between guanosine and adenosine nucleotides. *Biochemistry*, **34**, 593.
7. Kabsch, W., Gast, W. H., Schulz, G. E., and Leberman, R. (1977) Low resolution structure of partially trypsin-degraded polypeptide elongation factor, EF-TU, from *Escherichia coli*. *J. Mol. Biol.*, **117**, 999.
8. Jurnak, F. (1985). Structure of the GDP domain of EF-Tu and location of amino acids homologous to ras oncogene proteins. *Science*, **230**, 32.
9. McCormick, F., Clark, B. F. C., la Cour, T. F. M., Kjelgaard, M., Norskov-Lauritsen, L., and Nyborg, J. (1985) A model for the tertiary structure of p21, the product of the *ras* oncogene. *Science*, **230**, 78.
10. LaCour, T. F. M., Nyborg, J., Thirup, S., and Clark, B. F. C. (1985). Structural details of the binding of guanosine diphosphate to elongation factor tu from *E.coli* as studied by X-ray crystallography. *EMBO J.*, **4**, 2385.
11. Pai, E .F., Kabsch, W., Krengel, U., Holmes, K. C., John, J., and Wittinghofer, A. (1989) Structure of the guanine-nucleotide-binding domain of the Ha-ras oncogene product p21 in the triphosphate conformation. *Nature*, **341**, 209.
12. Pai, E. F., Krengel, U., Petsko, G. A., Goody, R. S., Kabsch, W., and Wittinghofer, A. (1990) Refined crystal structure of the triphosphate conformation of H-ras p21 at 1.35 Å resolution: Implications for the mechanism of GTP hydrolysis. *EMBO J.*, **9**, 2351.
13. Milburn, M. V., Tong, L., DeVos, A. M., Brünger, A., Yamaizumi, Z., Nishimura, S., and Kim S.-H. (1990) Molecular switch for signal transduction: structural differences between active and inactive forms of protooncogenic ras proteins. *Science*, **247**, 939.
14. Privé, G. G., Milburn, M. V., Tong, L., DeVos, A. M., Yamaizumi, Z., Nishimura, S., and Kim, S. H. (1992) X-ray crystal structures of transforming p21 ras mutants suggest a transition-state stabilization mechanism for GTP hydrolysis. *Proc. Natl Acad. Sci. USA* , **80**, 3649.
15. Yamasaki, K., Kawai, G., Ito, Y., Muto, Y., Fujita, J., Miyazawa, T., Nishimura, S., and

Yokoyama, S. (1989) Conformation change of effector-region residues in antiparallel, β-sheet of human c-Ha-ras protein on GDP-GTPgS excange: a two-dimensional NMR study. *Biochem. Biophys. Res. Commun.*, **162**, 1054,

16. Halliday, K. R. (1984) Regional homology in GTP-binding proto-oncogene products and elongation factors. *J. Cyclic Nucl. Prot. Phosphor. Res.*, **9**, 435.

17. Valencia, A., Chardin, P., Wittinghofer, A., and Sander, C. (1991) The ras protein family: Evolutionary tree and role of conserved amino acids. *Biochemistry*, **30**, 4637.

18. Farnsworth, C. L. and Feig, L. A. (1991) Dominant inhibitory mutations in the Mg^{2+}-binding site of ras[H] prevent its activation by GTP. *Mol. Cell. Biol.*, **11**, 4822.

19. John, J., Rensland, H., Schlichting, I., Vetter, I., Borasio, G. D., Goody, R. S., and Wittinghofer, A. (1993) Kinetic and structural analysis of the Mg^{2+}-binding site of the Guanine Nucleotide-binding protein p21[H-ras]. *J. Biol. Chem.*, **268**, 923.

20. Wittinghofer, A. and Pai, E. (1991) The structure of ras protein: a model for a universal molecular switch. *TIBS*, **16**, 383.

21. Schmidt, G., Lenzen, C., Simon, I., Deuter, R., Cool, R. H., Goody, R. S., and Wittinghofer, A. (1996) Biochemical and biological consequences of changing the specificity of p21[ras] from guanosine to xanthosine nucleotides. *Oncogene*, **12**, 87.

22. Scheidig, A. J., Franken, S. M., Corrie, J. E. T., Reid, G. P., Wittinghofer, A., Pai, E. F., and Goody, R. S. (1995) X-ray crystal structure analysis of the catalytic domain of the oncogene product p21[H-ras] complexed with caged GTP and mant dGppNHp. *J. Mol. Biol.*, **253**, 132.

23. Schlichting, I., Almo, S. C., Rapp, G., Wilson, K., Petratos, K., Lentfer, A., Wittinghofer, A., Kabsch, W., Pai, E. F., Petsko, G. A., and Goody, R. S. (1990) Time-resolved x-ray crystallographic study of the conformational change in Ha-Ras p21 protein on GTP hydrolysis. *Nature*, **345**, 309.

24. Wittinghofer, A. and Nassar, N. (1996) How Ras-related proteins talk to their effectors. *Trends Biochem. Sci.*, **21**, 488.

25. Krengel, U., Schlichting, I., Scherer, A., Schumann, R., Frech, M., John, J., Kabsch, W., Pai, E. F., and Wittinghofer, A. (1990) Three-dimensional structures of H-ras p21-mutants: Molecular basis for their inability to function as signal switch molecules. *Cell*, **62**, 539.

26. Franken, S. M., Scheidig, A. J., Krengel, U., Rensland, H., Lautwein, A., Geyer, M., Scheffzek, K., Goody, R. S., Kalbitzer, H. R., Pai, E. F. and Wittinghofer, A. (1993) Three-dimensional structures and properties of a transforming and a nontransforming glycine-12 mutant of p21[H-ras]. *Biochemistry*, **32**, 8411.

27. Campbell-Burk, S. (1989) Structural and dynamic differences between normal and transforming N-ras gene products: A ^{31}P and isotope-edited ^{1}H NMR study. *Biochemistry*, **28**, 9478.

28. Miller, A.-F., Papastavros, M. Z., and Redfield, A. G. (1992) NMR studies of the conformational change in human N-p21[ras] produced by replacement of bound GDP with the GTP analog GTPγS. *Biochemistry*, **31**, 10208.

29. Miller, A.-F., Halkides, C. J., and Redfield, A. G. (1993) An NMR comparison of the changes produced by different guanosine 5'-triphosphate analogs in wild-type and oncogenic mutant p21[ras]. *Biochemistry*, **32**, 7367.

30. Kraulis, P. J., Domaille, P. J., Campbell-Burk, S. L., Van Aken, T., and Laue, E. D. (1994) Solution structure and dynamics of Ras p21•GDP determined by heteronuclear three- and four-dimensional NMR spectroscopy. *Biochemistry*, **33**, 3515.

31. Ito, Y., Yamasaki, K., Iwahara, J., Terada, T., Kamiya, A., Shirouzu, M., Muto, Y., Kawai, G., Yokoyama, S., Laue, E.D., *et al.* (1997) Regional polysterism in the GTP-bound form of the human c-Ha-Ras protein. *Biochemistry*, **36**, 9109.

32. Hu, J. S. and Redfield, A. G. (1997) Conformational and dynamic differences between N-Ras p21 bound to GTP-γ-S and to GMPPNP as studied by NMR. *Biochemistry*, **36**, 5045.

33. Geyer, M., Schweins, T., Herrmann, C., Prisner, T., Wittinghofer, A., and Kalbitzer, H. R. (1996) Conformational transitions in p21ras and in its complexes with the effector protein Raf-RBD and the GTPase activating protein GAP. *Biochemistry*, **35**, 10308.

34. Geyer, M., Herrmann, C., Wohlgemuth, S., Wittinghofer, A., and Kalbitzer, H. R. (1997) Structure of the Ras-binding domain of RalGEF and implications for Ras binding and signalling. *Nature, Struct. Biol.*, **4**, 694.

35. Halkides, C. J., Bellew, B. F., Gerfen, G. J., Farrar, C. T., Carter, P. H., Ruo, B., Evans, D. A., Griffin, R. G., and Singel, D. J. (1996) High frequency (139.5 GHz) electron paramagnetic resonance spectroscopy of the GTP form of p21 ras with selective ^{17}O labeling of threonine. *Biochemistry*, **35**, 12194.

36. Latwesen, D. G., Poe, M., Leigh, J. S., and Reed, G. H. (1992) Electron paramagnetic resonance studies of a ras p21-Mn^{II}GDP complex in solution. *Biochemistry*, **31**, 4946.

37. Bellew, B. F., Halkides, C. J., Gerfen, G. J., Griffin, R. G., and Singel, D. J. (1996) High frequency (139.5 GHz) electron paramagnetic resonance characterization of Mn(II)-H$_2$ ^{17}O interactions in GDP and GTP forms of p21 *ras*. *Biochemistry*, **35**, 12186.

38. Nassar, N., Horn, G., Herrmann, C., Scherer, A., McCormick, F., and Wittinghofer, A. (1995). The 2.2 Å crystal structure of the Ras-binding domain of the serine/threonine kinase c-Raf1 in complex with Rap1A and a GTP analogue. *Nature*, **375**, 554.

39. Cherfils, J., Menetrey, J., Lebras, G., Janoueixlerosey, I., Degunzburg, J., Garel, J. R., and Auzat, I. (1997) Crystal structures of the small G protein Rap 2A in complex with its substrate GTP, with GDP and with GTP-Gamma-S. *EMBO J.*, **16**, 5582.

40. Nassar, N., Horn, G., Herrmann, C., Block, C., Janknecht, R., and Wittinghofer, A. (1996) Ras/Rap effector specificity determined by charge reversal. *Nature Struct. Biol.*, **3**, 723.

41. Scheffzek, K., Ahmadian, M. R., Kabsch, W., Wiesmüller, L., Lautwein, A., Schmitz, F., and Wittinghofer, A. (1997) The Ras–RasGAP complex: Structural basis for GTPase activation and its loss in oncogenic Ras mutants. *Science*, **277**, 333.

42. Amor, J. C., Harrison, D. H., Kahn, R. A., and Ringe, D. (1994). Structure of the human ADP-ribosylation factor 1 complexed with GDP. *Nature*, **372**, 704.

43. Greasley, S. E., Jhoti, H., Teahan, C., Solari, R., Fensome, A., Thomas, G. M. H., Cockcroft, S., and Bax, B. (1995) The structure of rat ADP-ribosylation factor-1 (ARF-1) complexed to GDP determined from two different crystal forms. *Nature Struct. Biol.*, **2**, 797.

44. Goldberg, J. (1998) Structural basis for activation of ARF GTPase: Mechanisms of guanine nucleotide exchange and GTP-myristoyl switching. *Cell*, **95**, 237.

45. Antonny, B., Berauddufour, S., Chardin, P., and Chabre, M. (1997) N-terminal hydrophobic residues of the G-protein ADP-ribosylation factor-1 insert into membrane phospholipids upon GDP to GTP exchange. *Biochemistry*, **36**, 4675.

46. Scheffzek, K., Klebe, C., Fritz-Wolf, K., Kabsch, W., and Wittinghofer, A. (1995) Crystal structure of the nuclear Ras-related protein Ran in its GDP-bound form. *Nature*, **374**, 378.

47. Abel, K., Yoder, M. D., Hilgenfeld, R., and Jurnak, F. (1996) An α to β conformational switch in EF-Tu. *Structure*, **4**, 1153.

48. Polekhina, G., Thirup, S., Kjeldgaard, M., Nissen, P., Lippmann, C., and Nyborg, J. (1996) Helix unwinding in the effector region of elongation factor EF-Tu-GDP. *Structure*, **4**, 1141.

49. Vetter, I. R., Nowak, C., Nishimoto, T., Kuhlmann, J., and Wittinghofer, A. (1999) Structure of the Ran-GppNHp-RanBD1 complex: A molecular embrace and ist implication for nuclear transport. *Nature*, **398**, 39.

50. Richards, S. A., Lounsbury, K. M., and Macara, I. G. (1995) The C terminus of the nuclear RAN/TC4 GTPase stabilizes the GDP-bound state and mediates interactions with RCC1, RAN-GAP, and HTF9A/RANBP1. *J. Biol. Chem.*, **270**, 14405.

51. Hirshberg, M., Stockley, R. W., Dodson, G., and Webb, M. R. (1997) The crystal structure of human rac1, a member of the rho-family complexed with a GTP analogue. *Nature Struc. Biol.*, **4**, 147.

52. Wei, Y., Zhang, Y., Derewenda, U., Liu, X., Minor, W., Nakamoto, R. K., Somlyo, A. V., Somlyo, A. P., and Derewenda, Z. S. (1997) Crystal structure of RhoA-DGP and its functional implications. *Nature Struct. Biol.*, **4**, 699.

53. Ihara, K., Muraguchi, S., Kato, M., Shimizu, T., Shirakawa, M., Kuroda, S., Kaibuchi, K., and Hakoshima, T. (1998) Crystal structure of human RhoA in a dominantly active form complexed with a GTP analogue. *J. Biol. Chem.*, **273**, 9656.

54. Feltham, J. L., Dötsch, V., Raza, S., Manor, D., Cerione, R. A., Sutcliffe, M. J., Wagner, G., and Oswald, R. E. (1997) Definition of the switch surface in the solution structure of Cdc42Hs. *Biochemistry*, **36**, 8755.

55. Leonhard, D., Hart, M. J. Platko, J. V., Eva, A., Henzel, W., Evans, T., and Cerione, R. A. (1992) The identification and characterization of a GDP-dissociation inhibitor (GDI) for the Cdc42Hs protein. *J. Biol. Chem.*, **267**, 22860.

56. Bourne, H. R. (1997) How receptors talk to trimeric G proteins. *Curr. Opinion Cell Biol.*, **9**, 134.

57. Conklin, B. R., and Bourne, H. R. (1993). Structural elements of Ga subunits that interact with Gbg, receptors, and effectors. *Cell*, **73**, 631.

58. Iiri, T., Farfel, Z., and Bourne, H. R. (1998) G-protein diseases furnish a model for the turn-on switch. *Nature*, **394**, 35.

59. Berman, D. M. and Gilman, A. G. (1998) Mammalian RGS proteins – barbarians at the gate. *J. Biol. Chem.*, **273**, 1269.

60. Noel, J. P., Hamm, H. E., and Sigler, P. B. (1993). The 2.2 Å crystal structure of transducin-α complexed with GTPγS. *Nature*, **366**, 654.

61. Lambright, D. G., Noel, J. P., Hamm, H. E., and Sigler, P. B. (1994). Structural determinants for activation of the a-subunit of a heterotrimeric G protein. *Nature,* **369**, 621.

62. Sondek, J., Lambright, D. G., Noel, J. P., Hamm, H. E., and Sigler, P. B. (1994). GTPase mechanism of G proteins from the 1.7-Å crystal structure of transducin α·GDP·AlF$_4$. *Nature*, **372**, 276.

63. Coleman, D. E., Berghuis, A. M., Lee, E., Linder, M. E., Gilman, A. G., and Sprang, S. R. (1994). Structures of active conformations of $G_{i\alpha 1}$ and the mechanism of GTP hydrolysis. *Science*, **265**, 1405.

64. Mixon, M. B., Lee, E., Coleman, D. E., Berghuis, A. M., Gilman, A. G., and Sprang, D. E. (1995) Tertiary and quaternary structural changes in $G_{i\alpha 1}$ induced by GTP hydrolysis. *Science*, **270**, 954.

65. Sunahara, R. K., Tesmer, J. J. G., Gilman, A. G., and Sprang, S. R. (1997) Crystal structure of the adenylyl cylase activator G(s-alpha). *Science*, **278**, 1943.

66. Benjamin, D. R., Markby, D. W., Bourne, H. R., and Kuntz, I. A. (1995) Complete ^{1}H, ^{13}C, and ^{15}N assignments and secondary structure of the GTPase activating domain of G$_s$. *Biochemistry,* **34**, 155.

67. Fawzi, A. B. and Northup, J. K. (1990) Guanine nucleotide binding characteristics of transducin: esential role of rhodopsin for rapid exchange of guanine nucleotides. *Biochemistry*, **29**, 3804.

68. John, J., Sohmen, R., Feuerstein, J., Linke, R., Wittinghofer, A., and Goody, R. S. (1990)

Kinetics of interaction of nucleotides with nucleotide-free H-ras p21. *Biochemistry*, **29**, 6059.

69. Simon, I., Zerial, M., and Goody, R. S. (1996) Kinetics of interaction of Rab5 and Rab7 with nucleotides and magnesium ions. *J. Biol. Chem.*, **271**, 20470.

70. Codina, J. and Birnbaumer, L. (1994) Requirement for intramolecular domain interaction in activation of G protein α subunit by aluminum fluoride and GDP but not by GTPβ,γS*. *J. Biol. Chem.*, **269**, 29339.

71. Markby, D. W., Onrust, R., and Bourne, H. R. (1993) Separate GTP binding and GTPase activating domains of a Gα subunit. *Science*, **262**, 1895.

72. John, J., Frech, M., and Wittinghofer, A. (1988) Biochemical properties of Ha-ras encoded p21 mutants and mechanism of the autophosphorylation reaction. *J. Biol. Chem.*, **263**, 11792.

73. Higashijima, T., Ferguson, K. M., Sternweis, P. C., Smigel, M. D., and Gilman, A. G. (1987). Effects of Mg^{2+} and the βγ subunit complex on the interaction of guanine nucleotides with G proteins. *J. Biol. Chem.*, **262**, 762.

74. Phillips, W. J. and Cerione, R. A. (1988) The intrinsic fluorescence of the α subunit of transducin. *J. Biol. Chem.*, **263**, 15498.

75. Higashijima, T., Ferguson, K. M., Smigel, M. D., and Gilman, A. G. (1987) The effect of GTP and Mg^{2+} on the GTPase activity and the fluorescent properties of Go. *J. Biol. Chem.*, **262**, 757.

76. Faurobert, E., Otto-Bruc, A., Chardin, P., and Chabre, M. (1993) Tryptophan W207 in transducin Tα is the fluorescence sensor of the G protein activation switch and is involved in the efector binding. *EMBO J.*, **12**, 4191.

77. Miller, R. T., Masters, S. B., Sullivan, K. A., Beiderman, B., and Bourne, H. R. (1988) A mutation that prevents GTP-dependent activation of the a-chain of G_s. *Nature*, **334**, 712.

78. Rall, T. W. and Sutherland, E. W. (1958) Formation of a cyclic adenine ribonucleotide by tissue particles. *J. Biol. Chem.*, **232**, 1065.

79. Sternweis, P. C. and Gilman, A. G. (1982) Aluminum: a requirement for activation of the regulatory component of adenylate cyclase by fluoride. *Proc. Natl Acad. Sci., USA*, **79**, 4888.

80. Chabre, M. (1990) Aluminofluoride and beryllofluoride complexes: new phosphate analogs in enzymology. *TIBS*, **15**, 6.

81. Wittinghofer, A. (1997) Signaling mechanistics: Aluminum fluoride for molecule of the year. *Curr. Biol.*, **7**, R682.

82. Schweins, T., Langen, R., and Warshel, A. (1994) Why have mutagenesis studies not located the general base in the GTP-hydrolysis of the ras-p21 oncoprotein? *Nature Struct. Biol.*, **1**, 476.

83. Chung, H. H., Kim, R., and Kim, S. H. (1992) Biochemical and biological activity of phosphorylated and non-phosphorylated ras p21 mutants. *Biochim. Biophys. Acta*, **1129**, 278.

84. Schweins, T., Geyer, M., Scheffzek, K., Warshel, A., Kalbitzer, H. R., and Wittinghofer A. (1995) Substrate-assisted catalysis as a mechanism for GTP hydrolysis of p21^{ras} and other GTP-binding proteins. *Nature Struct. Biol.*, **2**, 36.

85. Maegley, K. A., Admiraal, S. J., and Herschlag, D. (1996) Ras-catalyzed hydrolysis of GTP: A new perspective from model studies. *Proc. Natl Acad. Sci. USA*, **93**, 8160.

86. Raw, A. S., Coleman, D. E., Gilman, A. G., and Sprang, S. R. (1997) Structural and biochemical characterization of the GTP-γ-S-, GDP-center-dot-P-I-, and GDP-bound forms of a GTPase-deficient Gly(42)-]Val mutant of G(I-α-1). *Biochemistry*, **36**, 15660.

87. Berghuis, A. M., Lee, E., Raw, A. S., Gilman, A. G., and Sprang, S. R. (1996) Structure of the GDP-Pi complex of Gly203 →Ala gialpha: a mimic of the ternary product complex of galpha-catalyzed GTP hydrolysis. *Structure*, **4**, 1277.

88. Salomon, M. R. and Bourne, H. R. (1981) Novel S49 lymphoma variants with aberrant cyclic AMP metabolism. *Mol. Pharmacol.*, **19**, 109.

89. Lee, E., Taussig, R., and Gilman, A. G. (1992) The G226A mutant of $G_{s\times}$ highlights the requirement for dissociation of G protein subunits. *J. Biol. Chem.*, **267**, 1212.

90. Boguski, M. S. and McCormick, F. (1993) Proteins regulating Ras and its relatives. *Nature*, **366**, 643.

91. Schalk, I., Zeng, K., Wu, S. K., Stura, E. A., Matteson, J., Huang, M., Tandon, A., Wilson, I. A., and Balch, W. E. (1996) Structure and mutational analysis of Rab GDP-dissociation inhibitor. *Nature*, **381**, 42.

92. Gosser, Y. Q., Nomanbhoy, T. K., Aghazadeh, B., Manor, D., Combs, C., Cerione, R. A., and Rosen, M. K. (1997) C-terminal binding domain of Rho GDP-dissociation inhibitor directs N-terminal inhibitory peptide to GTPases. *Nature*, **387**, 814.

93. Keep, N. H., Barnes, M., Barsukov, I., Badii, R., Lian, L.-Y., Segal, A. W., Moody, P. C. E., and Roberts, G. C. K. (1997) A modulator of rho family G proteins, rhoGDI, binds these G proteins via an immunoglobulin-like domain and a flexible N-terminal arm. *Structure*, **5**, 623.

94. Platko, J. V. *et al.* (1995) A single residue can modify target-binding affinity and activity of the functional domain of the Rho-subfamily GDP dissociation inhibitors. *Proc. Natl Acad. Sci. USA*, **92**, 2974.

95. Flower, D. R. (1996) The lipocalin protein family – structure and function. *Biochem. J.*, **318**, 1–14.

96. Moore, M. S. and Blobel, G. (1994) Purification of a Ran-interacting protein that is required for protein import into the nucleus. *Proc. Natl Acad. Sci., USA*, **91**, 10212.

97. Paschal, B. M. and Gerace, L. (1995) Identification of NTF2, a cytosolic factor for nuclear import that interacts with nuclear pore complex protein P62. *J. Cell Biol.*, **129**, 925.

98. Ribbeck, K., Lipowsky, G., Kent, H. M., Stewart, M., and Görlich, D. (1998) NFT2 mediates nuclear import of Ran. *EMBO J.*, **17**, 6587.

99. Bullock, T. L., Clarkson, W. D., Kent, H. M., and Stewart, M. (1996) The 1.6 Å resolution crystal structure of nuclear transport factor 2 (NTF2). *J. Mol. Biol.*, **260**, 422.

100. Stewart, M., Kent, H. M., and McCoy, A. J. (1998) Structural basis for molecular recognition between nuclear transport factor 2 (NFT2) and the GDP-bound form of the Ras-family GTPase Ran. *J. Mol. Biol.*, **277**, 635.

101. Wall, M. A., Coleman, D. W., Lee, E., Iñiguez-Lluhi, J. A., Posner, B. A., Gilman, A. G., and Sprang, S. T. (1995) The structure of the G protein heterotrimer $G_{i\alpha1}$ $\beta_{1\gamma2}$. *Cell*, **83**, 1047.

102. Lambright, D. G., Sondek, J., Bohm, A., Skiba, N. P., Hamm, H. E., and Sigler, P. B. (1996). The 2.0Å crystal structure of a heterotrimeric G protein. *Nature*, **379**, 311.

103. Sondek, J., Bohm, A., Lambright, D. G., Hamm, H. E., and Sigler, P. B. (1996) Crystal structure of a G_A protein β,γ dimer at 2.1Å resolution. *Nature*, **379**, 369.

104. Neer, E. J., Schmidt, C. J., Nambudripad, R., and Smith, T. F. (1994). The ancient regulatory-protein family of WD-repeat proteins. *Nature*, **371**, 297.

105. Lupas, A., Van Dyke, M., and Stock, J. (1991). Predicting coiled coils from protein sequences. *Science*, **252**, 1162.

106. Garristen, A., Van Galen, P. J. M., and Simonds W. F. (1993). The N-terminal coiled-coil domain of β is essential for β,γ association: a model for G-protein βγ subunit interaction. *Proc. Natl Acad. Sci., USA*, **90**, 7706.

107. Wall, M. A., Posner, B. A., and Sprang, S. R. (1998) Structural basis of activity and subunit recognition in G protein heterotrimers. *Structure*, **6**, 1169.

108. Créchet, J.-B., Poulet, P., Mistou, M.-Y., Parmeggiani, A., Camonis, J., Boy-Marcotte, E., Damak, F., and Jacquet, M. (1990) Enhancement of the GDP–GTP exchange of ras proteins with the carboxyl-terminal domain of SCD25. *Science,* **248**, 866.

109. Hart, M. J., Eva, A., Evans, T., Aaronson, S. A., and Cerione, R. A. 81991) Catalysis of guanine nucleotide exchange on the CDC42hs protein by the dbl oncogene product. *Nature*, **354**, 311.

110. Ohtsubo, M., Kai, R., Furuno, N., Sekiguchi, T., Sekiguchi, M., Hayashida, H., Kuma, K.-I., Miyata, T., Fukushige, S., Murotsu, T., *et al.* (1987) Isolation and characterisation of the human cell cycle gene (RCC1) involved in rge regulation of onset of chromosome condensation. *Genes Devel.*, **1**, 585.

111. Bischoff, F. R. and Ponstingl, H. (1991) Catalysis of guanine nucleotide exchange on Ran by the mitotic regulator RCC1. *Nature*, **354**, 80.

112. Chardin, P., Paris, S., Antonny, B., Robineau, S., Berauddufour, S., Jackson, C. L., and Chabre, M. (1996) A human exchange factor for arf contains sec7- and pleckstrin-homology domains. *Nature*, **384**, 481.

113. Barlowe, C. and Schekman, R. (1993) SEC12 encodes a guanine-nucleotide-exchange factor essential for transport vesicle budding from the ER. *Nature*, **365**, 347.

114. Horiuchi, H., Lippe, R., Mcbride, H. M., Rubino, M., Woodman, P., Stenmark, H., Rybin, V., Wilm, M., Ashman, K., Mann, M., and Zerial, M. (1997) A novel Rab5 GDP GTP exchange factor complexed to rabaptin-5 links nucleotide exchange to effector recruitment and function. *Cell*, **90**, 1149.

115. Burd, C. G., Mustol, P. A., Schu, P. V., and Emr, S. D. (1996) A yeast protein related to a mammalian ras-binding protein, vps9p, is required for localization of vacuolar proteins. *Mol. Cell. Biol.*, **16**, 2369.

116. Lenzen, C., Cool, R. H., Prinz, H., Kuhlmann, J., and Wittinghofer, A. (1998) Kinetic analysis by fluorescence of the interaction between Ras and the catalytic domain of the guanine nucleotide exchange factor Cdc25[Mm]. *Biochemistry*, **37**, 7420.

117. Klebe, C., Bischoff, F. R., Ponstingl, H., and Wittinghofer, A. (1995) Interaction of the nuclear GTP-binding protein Ran with its regulatory proteins RCC1 and RanGAP1. *Biochemistry*, **34**, 639.

118. Klebe, C., Prinz, H., Wittinghofer, A., and Goody, R. S. (1995) The kinetic mechanism of Ran—nucleotide exchange catalyzed by RCC1. *Biochemistry*, **34**, 12543.

119. Powers, S., O'Neill, K., and Wigler, M. (1989) Dominant yeast and mammalian *RAS* mutants that interfere with the *CDC25*-dependent activation of wild-type *RAS* in *Saccharomyces cerevisiae*. *Mol. Cell. Biol.*, **9**, 390.

120. Haney, S. A. and Broach, J. R. (1994) Cdc25p, the guanine nucleotide exchange factor for the Ras proteins of *Saccharomyces cerevisiae*, promotes exchange by stabilizing Ras in a nucleotide-free state. *J. Biol. Chem.* **269**, 16541.

121. Jacquet, E., Baouz, S., and Parmeggiani, A. (1995) Characterization of mammalian C-DCD25[Mm] exchange factor and kinetic properties of the exchange reaction intermediate p21•C-CDC25[Mm]. *Biochemistry*, **34**, 12347.

122. Yu, H. and Schreiber, S. L. (1995) Structure of guanine-nucleotide-exchange factor human Mss4 and identification of its Rab-interacting surface. *Nature*, **376**, 788.

123. Moya, M., Roberts, D., and Novick, P. (1993) DSS4–1 is a dominant suppressor of sec4–8 that encodes a nucleotide exchange protein that aids Sec4p function. *Nature*, **361**, 460.

124. Burton, J., Roberts, D., Montaldi, M., Novick, P., and DeCamilli, P. (1993) A mammalian guanine-nucleotide-releasing protein enhances function of yeast secretory protein Sec4. *Nature*, **361**, 464.

125. Nuoffer, C., Wu, S.-K., Dascher, C., and Balch, W. E. (1997) MSS4 does not function as an exchange factor for Rab in endoplasmic reticulum to golgi transport. *Mol. Biol. Cell*, **8**, 1305.

126. Peyroche, A., Paris, S., and Jackson, C. L. (1996) Nucleotide exchange on arf mediated by yeast gea1 protein. *Nature*, **384**, 479.

127. Meacci, E., Tsai, S. C., Adamik, R., Moss, J., and Vaughan, M. (1997) Cytohesin-1, a cytosolic guanine nucleotide-exchange protein for ADP-ribosylation factor. *Proc. Natl Acad. Sci., USA*, **94**, 1745.

128. Walker, M. W., Bobak, D. A., Tsai, S.-C., Moss, J., and Vaughan, M. (1992) GTP but not GDP analogues promote association of ADP-ribosylation factors, 20-kDa protein activators of cholera toxin, with phospholipids and PC-12 cell membranes. *J. Biol. Chem.*, **267**, 3230.

129. Paris, S., Beraud-Dufour, S., Robineau, S., Bigay, J., Antonny, B., Chabre, M., and Chardin, P. (1997) Role of protein–phospholipid interactions in the activation of ARF1 by the guanine nucleotide exchange factor Arno. *J. Biol. Chem.*, **272**, 22221.

130. Cherfils, J., Menetrey, J., Mathieu, M., Labras, G., Robineau, S., Berauddufour, S., Antonny, B., and Chardin, P. (1998) Structure of the sec7 domain of the arf exchange factor arno. *Nature*, **392**, 101.

131. Mossessova, E., Gulbis, J. M., and Goldberg, J. (1998) Structure of the guanine nucleotide exchange factor Sec7 domain of human Arno and analysis of the interaction with Arf GTPase. *Cell*, **92**, 415.

132. Betz, S. F., Schnuchel, A., Wang, H., Olejniczak, E. T., Meadows, R. P., Lipsky, B. P., Harris, E. A. S., Staunton, D. E., and Fesik, S. W. (1998) Solution structure of the cytohesin-1 (B2–1) Sec7 domain and its interaction with the GTPase ADP ribosylation factor 1. *Proc. Natl Acad. Sci.*, USA, **95**, 7909.

133. Huber, A. H., Nelson, W. J., and Weis, W. I. (1997) Three-dimensional structure of the armadillo repeat region of beta-catenin. *Cell*, **90**, 871.

134. Beraud-Dufour, S., Robineau, S., Chardin, P., Paris, S., Chabre, M., Cherfils, J., and Antonny, S. (1998) A glutamic finger in the guanine nucleotide exchange factor ARNO displaces Mg^{2+} and the β-phosphate to destabilize GDP on ARF1. *EMBO J.*, **17**, 3651.

135. Renault, L., Nassar, N., Vetter, I., Becker, J., Roth. M., and Wittinghofer, A. (1998) The 1.7 Å crystal structure of the regulator of chromosome condensation (RCC1) reveals a seven-bladed propeller. *Nature*, **392**, 97.

136. Azuma, Y., Seino, H., Seki, T., Uzawa, S., Klebe, C., Ohba, T., Wittinghofer, A., Hayashi, N., and Nishimoto, T. (1996) Conserved histidine residues of RCC1 are essential for nucleotide exchange on Ran. *J. Biochem.*, **120**, 82.

137. Cerione, R. A. and Zheng, Y. (1996) The DBL family of oncogenes. *Curr. Opinion Cell Biol.*, **8**, 216.

138. Liu, X., Wang, H., Eberstadt, M., Schnuchel, A., Olejniczak, E. T., Meadows, R. P., Schkeryantz, J. M., Janowick, D. A., Harlan, J. E., Harris, E. A. S., *et al.* (1998) NMR structure and mutagenesis of the N-terminal Dbl homology domain of the nucleotide exchange factor trio. *Cell*, **95**, 269.

139. Soisson, S. M., Nimnuao, A. S., Uy, M., Bar-Sagi, D., and Kuriyan, J. (1998) Crystal structure of the Dbl and pleckstrin homology domains from the human Son of sevenless protein. *Cell*, **95**, 259.

140. Whitehead, I. P., Campbell, S., Rossman, K. L., and Der, C. J. (1997) Dbl family proteins. *Biochem. Biophys. Acta*, **1332**, F1–F23.

141. Koshiba, S., Kigawa, T., Kim, J.-H., Shirouzu, M., Bowtell, D., and Yokoyama, S. (1997)

The solution structure of the pleckstrin homology domain of mouse son-of-sevenless 1 (mSos1). *J. Mol. Biol.*, **269**, 579.

142. Zheng, J., Chen, R.-H., Corblan-Garcia, S., Cahill, S. M., Bar-Sagi, D., and Cowburn, D. (1997) The solution structure of the pleckstrin homology domain of human SOS1. *J. Biol. Chem.*, **272**, 30340.

143. Ferguson, K. M., Lemmon, M. A., Schlessinger, J., and Sigler, P. B. (1995) Structure of the high affinity complex of inositol trisphosphate with a phospholipase C pleckstrin homology domain. *Cell*, **83**, 1037.

144. Boriack-Sjodin, P. A., Margarit, S. M., Bar-Sagi, D., and Kuriyan, J. (1998) The structural basis of the activation of ras by sos. *Nature*, **394**, 337.

145. Fam, N. P., Fan, W-T., Wang, Z., Zhang,L.-J., Chen, H., and Moran, M. (1997) Cloning and characterization of Ras-GRF2, a novel guanine nucleotide exchange factor for Ras. *Mol. Cell. Biol.*, **17**, 1396.

146. Polakis, P. and McCormick, F. (1993) Structural requirements for the interaction of p21[ras] with GAP, exchange factors, and its biological effector target. *J. Biol. Chem.*, **268**, 9157.

147. Mistou, M.-Y., Jacquet, E., Poullet, P., Rensland, H., Gideon, P., Schlichting, I., Wittinghofer, A. and Parmeggiani, A. (1992) Mutations of Ha-ras p21, that define important regions for the molecular mechanism of the SDC25 C-domain, a guanine nucleotide dissociation stimulator. *EMBO J.*, **11**, 2391.

148. Saraste, M., Sibbald, P. R., and Wittinghofer, A. (1990) The P-loop—a common motif in ATP- and GTP-binding proteins. *TIBS*, **15**, 430.

149. Sigal, I. S., Gibbs, J. B., D'Alonzo, J. S., Temeles, G. L., Wolanski, B. D., Socher, S. H., and Scolnick, E. M. (1986) Mutant ras-encoded proteins with altered nucleotide binding exert dominant biological effects. *Proc. Natl Acad. Sci., USA*, **83**, 952.

150. Wittinghofer, F. (1998) Ras signalling: Caught in the act of the switch-on. *Nature*, **394**, 317.

151. Kawashima, T., Berthet-Colominas, C., Wulff, M., Cusack, S., and Leberman, R. (1996) The structure of the *Escherichia coli* EF-Tu•EF-Ts complex at 1.5Å resolution. *Nature*, **379**, 511.

152. Wang, Y., Jiang, Y., Meyering-Voss, M., Sprinzl, M., and Sigler, P. B. (1997) Crystal structure of the EF-Tu.EF-Ts complex form *Thermus thermophilus*. *Nature Struct. Biol.*, **4**, 650.

153. Onrust, R., Herzmark, P., Chi, P., Garcia, P. D., Lichtarge, O., Kingsley, C., and Bourne, H. R. (1997) Receptor and beta-gamma binding sites in the alpha subunit of the retinal G protein transducin. *Science*, **275**, 381.

154. Ford, C. W., Skiba, N. P., Bae, H. S., Daaka, Y. H., Reuveny, E., Shekter, L. R., Rosal, R., Wenig, G. Z., Yang, C. S., Iyengar, R., *et al.* (1998) Molecular basis for interactions of G protein β,γ subunits with effectors. *Science*, **280**, 1271.

155. Siffert, W., Rosskopf, D., Siffert, G., Busch, S., Moritz, A., Erel, R., Sharma, A. M., Ritz, E., Wichmann, H. E., Jakobs, K. H., and Horsthemke, B. (1998) Association of a human G-protein beta-3 subunit variant with hypertension. *Nature Genetics*, **18**, 45.

156. Trahey, M. and McCormick, F. (1987) A cytoplasmic protein stimulates normal N-ras p21 GTPase, but does not affect oncogenic mutants. *Science*, **238**, 542.

157. Scheffzek, K., Ahmadian, M. R., and Wittinghofer, A. (1998) GTPase activating proteins: helping hands to complement an active site. *TIBS*, **23**, 257.

158. Gamblin, S. J. and Smerdon, S. J. (1998) GTPase-activating proteins and their complexes. *Curr. Opinion Struct. Biol.*, **8**, 195.

159. Chan, R. K. and Otte, C. A. (1982) Isolation and genetic analysis of Saccharomyces cerevisiae mutants supersensitive to G1 arrest by a factor and alpha factor pheromones. *Mol. Cell. Biol.*, **2**, 21.

160. Liljas, A. and Alkaradaghi, S. (1997) Structural aspects of protein snythesis. *Nature Struct. Biol.*, **4**, 767.

161. Wittinghofer, A., Scheffzek, K., and Ahmadian, M. R. (1997) The interaction of Ras with GTPase-activating proteins. *FEBS Lett.*, **410**, 63.

162. Tocque, B., Delumeau, I., Parker, F., Maurier, F., Multon, M. C., and Schweighoffer, F. (1997) Ras-GTPase activating protein (GAP) – a putative effector for Ras. *Cell. Signal.*, **9**, 153.

163. Viskochil, D., White, R., and Cawthon, R. (1993) The neurogibromatosis type 1 gene. *Annu. Rev. Neurosci.*, **16**, 183.

164. Billuart, P., Bienvenu, T., Ronce, N., Desportes, V., Vinet, M. C., Zemni, R., Crollius, H. R., Carrie, A., Fauchereau, F., Cherry, M., *et al.* (1998) Oligophrenin-1 encodes a rhogap protein involved in X-linked mental retardation. *Nature*, **392**, 923.

165. Schaefer, L., Prakash, S., and Zoghbi, H. Y. (1997) Cloning and characterization of a novel RHO-type GTPase-activating protein gene (ARHGAP6) from the critical region for microphthalmia with linear skin defects. *Genomics*, **46**, 268.

166. Gideon, P., John, J., Frech, M., Lautwein, A., Clark, R., Scheffler, J. E., and Wittinghofer, A. (1992) Mutational and kinetic analysis of the GTPase-activating protein (GAP)-p21 interaction: The C-terminal domain of GAP is not sufficient for full activity. *Mol. Cell Biol.*, **12**, 2050.

167. Eccleston, J. F., Moore, K. J. M., Morgan, L., Skinner, R. H., and Lowe, P. N. (1993) Kinetics of interaction between normal and proline 12 Ras and the GTPase-activating proteins, p120-GAP and neurofibromin. *J. Biol. Chem.*, **268**, 27012.

168. Ahmadian, M. R., Hoffmann, U., Goody, R. S., and Wittinghofer, A. (1997) Individual rate constants for the interaction of Ras proteins with GTPase-activating proteins determined by fluorescence spectroscopy. *Biochemistry*, **36**, 4535.

169. Neal, S. E., Eccleston, J. F., and Webb, M. R. (1990) Hydrolysis of GTP by p21*NRAS*, the *NRAS* protooncogene product, is accompanied by a conformational change in the wild type protein: Use of a single fluorescent probe at the catalytic site. *Proc. Natl Acad. Sci., USA*, **87**, 3652.

170. Moore, K. J. M., Webb, M. R., and Eccleston, J. F. (1993) Mechanism of GTP hydrolysis by p21N-ras catalyzed by GAP: Studies with a fluorescent GTP analogue. *Biochemistry*, **32**, 7451.

171. Rensland, H., Lautwein, A., Wittinghofer, A., and Goody, R. S. (1991) Is there a rate limiting step before GTP cleavage by H-ras p21? *Biochemistry*, **30**, 11181.

172. Kahn, A. (1991) Fluoride is not an activator of the smaller (20–25 kDa) GTP-binding proteins. *J. Biol. Chem.*, **266**, 15595.

173. Mittal, R., Ahmadian, M. R., Goody, R. S., and Wittinghofer, A. (1996) Formation of a transition state analog of the Ras GTPase reaction by Ras•GDP, tetrafluoroaluminate and GTPase-activating proteins. *Science*, **273**, 115.

174. Ahmadian, M. R., Stege, P., Scheffzek, K. , and Wittinghofer, A. (1997) Confirmation of the arginine-finger hypothesis for the GAP-stimulated GTP-hydrolysis reaction of Ras. *Nature Struct. Biol.*, **4**, 686.

175. Ahmadian, M. R., Mittal, R., Hall, A., and Wittinghofer, A. (1997) Aluminum fluoride associates with the small guanine nucleotide binding proteins. *FEBS Lett.*, **408**, 315.

176. Hoffman, G. R., Nassar, N., Oswald, R. E., and Cerione, R. A. (1998) Fluoride activation of the Rho family GTP-binding protein Cdc42Hs. *J. Biol. Chem.*, **273**, 4392.

177. Leonard, D. A., Lin, R., Cerione, R. A., and Manor, D. (1998) Biochemical studies of the mechanism of action of the Cdc42-GTPase-activating protein. *J. Biol. Chem.*, **273**, 16210.

178. Lamarche, N. and Hall, A. (1994) GAPs for rho-related GTPases. *Trends Genet.*, **10**, 436.

179. Musacchio, A., Cantley, L. C., and Harrison, S. C. (1996) Crystal structure of the break-point cluster region-homology domain from phosphoinositide 3-kinase p85α subunit. *Proc. Natl Acad. Sci., USA*, **93**, 14373.

180. Barrett, T., Xiao, B., Dodson, E. J., Dodson, G., Ludbrook, S. B., Nurmahomed, K., Gamblin, S. J., Musacchio, A., Smerdon, S. J., and Eccleston, J. F. (1997) The structure of the GTPase-activating domain from p50rhoGAP. *Nature*, **385**, 458.

181. Scheffzek, K., Lautwein, A., Kabsch, W., Ahmadian, M. R., and Wittinghofer, A. (1996) Crystal structure of the GTPase-activating domain of human p120GAP and implications for the interaction with Ras. *Nature*, **384**, 591.

182. Ahmadian, M. R., Wiesmüller, L., Lautwein, A., Bischoff, F. R., and Wittinghofer, A. (1996) Structural differences in the minimal catalytic domains of the GTPase-activating proteins p120^{GAP} and neurofibromin. *J. Biol. Chem.*, **271**, 16409.

183. Scheffzek, K., Ahmadian, M. R., Wiesmüller, L., Kabsch, W., Stege, P., Schmitz, F., and Wittinghofer, A. (1998) Structural analysis of the GAP related domain from neurofibromin and its implications. *EMBO J.*, **17**, 4313.

184. Klose, A., Ahmadian, M. R., Schuelke, M., Scheffzek, K., Hoffmeyer, S., Gewies, A., Schmitz, F., Kaufmann, D., Peters, H., Wittinghofer, A., and Nürnberg, P. (1998) Selective disactivation of neurofibromin GAP activity in neurofibromatosis type 1 (NF1). *Human Mol. Genet.*, **7**, 1261.

185. Bax, B. (1998) Domains of RasGAP and RhoGAP are related. *Nature*, **392**, 447.

186. Rittinger, K., Taylor, W. R., Smerdon, S. J., and Gamblin, S. J. (1998) Support for shared ancestry of GAPs. *Nature*, **392**, 448.

187. Calmels, T. P. G., Callebaut, I., Léger, I., Durand, P., Bril, A., Mornon, J.-P., and Souchet, M. (1998) Sequence and 3D structural relationships between mammalian Ras- and Rho-specific GTPase-activating proteins (GAPs): the cradle fold. *FEBS Lett.*, **426**, 205.

188. Kobe, B. and Deisenhofer, J. (1994) The leucine-rich repeat: a versatile binding motif. *TIBS*, **19**, 415.

189. Kobe, B. and Deisenhofer, J. (1993) Crystal structure of porcine ribonuclease inhibitor, a protein with leucine-rich repeats. *Nature*, **366**, 751.

190. Cukierman, E., Huber, I., Rotman, M., and Cassel, D. (1995) The arf1 GTPase-activating protein – zinc finger motif and golgi complex localization. *Science*, **270**, 1999.

191. Seeburg, P. H., Colby, W. W., Capon, D. J., Goedel, D. V., and Levinson, A. D. (1984) Biological properties of human c-Ha-ras 1 genes mutated at codon 12. *Nature*, **312**, 71.

192. Der, C. J., Weissmann, B., and MacDonald, M. J. (1988) Altered guanine nucleotide binding and H-ras transforming and differentiating activities. *Oncogene*, **3**, 105.

193. Rittinger, K., Walker, P. A., Eccleston, J. F., Smerdon, S. J., and Gamblin, S. J. (1997) Structure at 1.65 Å of RHOA and its GTPase-activating protein in complex with a transition-state analogue. *Nature*, **389**, 758.

194. Xu, Y. W., Morera, S., Janin, J., and Cherfils, J. (1997) Alf$_3$ mimics the transition state of protein phosphorylation in the crystal structure of nucleoside diphosphate kinase and mGADP. *Proc. Natl Acad. Sci., USA*, **94**, 3579.

195. Schlichting, I. and Reinstein, J. (1997) Structures of active conformations of UMP kinase from *Dictyostelium discoideum* suggest phosphoryl transfer is associative. *Biochemistry*, **36**, 9290.

196. Rittinger, K., Walker, P. A., Eccleston, J. F., Nurmahomed, K., Owen, D., Laue, E., and Smerdon, S. J. (1997) Crystal structure of a small G protein in complex with the GTPase-activating protein RHOGAP. *Nature*, **388**, 693.

197. Nassar, N., Hoffmann, G. R., Manor, D., Clardy, J. C., and Cerione, R. A. (1998)Structures of Cdc42 bound to the active and catalytically compromised forms of Cdc42GAP. *Nature Struct. Biol.*, **5**, 1047.

198. Schweins, T., Geyer, M., Kalbitzer, H. R., Wittinghofer, A., and Warshel, A. (1996) Linear free energy relationships in the intrinsic and GTPase activating protein-stimulated guanosine 5′-triphosphate hydrolysis of p21ras. *Biochemistry*, **35**, 14225.

199. Ahmadian, M. R., Scheffzek, K., and Wittinghofer, A. (1998) Are all GTP-binding proteins switched off by arginine finger? *TIBS*, **23**, 260.

200. Berstein, G., Blank, J. L., Jhon, D. Y., Exton, J. H., Rhee, S. G., and Ross, E. M. (1992) Phospholipase C-beta 1 is a GTPase-activating protein for Gq/11, its physiologic regulator. *Cell*, **70**, 411.

201. Arshavsky, V. Y., Dumke, C. L., Zhu, Y., Artemyev, N. O., Skiba, N. P., Hamm, H. E., and Bownds, M. D. (1994) Regulation of transducin GTPase activity in bovine rod outer segments. *J. Biol. Chem.*, **269**, 19882.

202. Van Dop, C., Tsubokawa, M., Bourne, H. R., and Ramachandran, J. (1984) Amino acid sequence of retinal transducin at the site ADP-ribosylated by cholera toxin. *J. Biol. Chem.*, **25**, 696.

203. Berman, D. M., Kozasa, T., and Gilman, A. G. (1996) The gtpase-activating protein rgs4 stabilizes the transition state for nucleotide hydrolysis. *J. Biol. Chem.*, **271**, 27209.

204. Tesmer, J. J. G., Berman, D. M., Gilman, A. G., and Sprang, S. R. (1997) Structure of RGS4 bound to AlF$_4^-$-activated G$_{i\alpha1}$: Stabilization of the transition state for GTP hydrolysis. *Cell*, **89**, 251.

205. Marshall, M. S. (1995) Ras target proteins in eukaryotic cells. *FASEB J.* **9**, 1311.

206. McCormick, F. and Wittinghofer, A. (1996) Interactions between Ras proteins and their effectors. *Curr. Opinion Biotech.*, **7**, 449.

207. Vojtek, A. B., Hollenberg, S. M., and Cooper, J. A. (1993) Mammalian Ras interacts directly with the serine/threonine kinase Raf. *Cell*, 74, 205.

208. Herrmann, C., Martin, G. A., and Wittinghofer, A. (1995) Quantitative analysis of the complex between p21*ras* and the Ras-binding domain of the human Raf-1 protein kinase. *J. Biol. Chem.*, **270**, 2901.

209. Emerson, S. D., Madison, V. S., Palermo, R. E., Waugh, D. S., Scheffler, J. E., Tsao, K. L., Kiefer, S. E., Liu, S. P., and Fry, D. C. (1995) Solution structure of the Ras-binding domain of c-Raf-1 and identification of its Ras interaction surface. *Biochemistry*, **34**, 6911.

210. Derrick, J. P. and Wigley, D. B. (1994) The third IgG-binding domain from streptococcal protein G. An analysis by X-ray crystallography of the structure alone and in a complex with Fab. *J. Mol. Biol.*, **243**, 906.

211. Lian, L. Y., Barsukov, I. L., Derrick, J. P., and Roberts, G. C. (1994) Mapping the interactions between streptococcal protein G and the Fab fragment of IgG in solution. *Nature Struct. Biol.*, **1**, 355.

212. Herrmann, C., Horn, G., Spaargaren, M., and Wittinghofer, A. (1996) Differential interaction of the Ras family GTP-binding proteins H-Ras, Rap1A, and R-Ras with the putative effector molecules Raf kinase and Ral-guanine nucleotide exchange factor. *J. Biol. Chem.*, **271**, 6794.

213. Huang, L., Wenig, X., Hofer, F., Martin, G. S., and Kim, S.-H. (1997) Three-dimensional structure of the Ras-interacting domain of RalGDS. *Nature Struct. Biol.*, **4**, 609.

214. Esser, D., Bauer, B., Wolthuis, R. M. F., Wittinghofer, A., Cool, R. H., and Bayer, P. (1998) Structure determination of the Ras-binding domain of the Ral-specific guanine nucleotide exchange factor Rlf. *Biochemistry*, **37**, 13453.

215. Huang, L., Hofer, F., Martin, G. S., and Kim, S.-H. (1998) Structural basis for the interaction of Ras with RalGDS. *Nature Struct. Biol.*, **5**, 422.

216. Vetter, I. R. Linnemann, T., Wohlgemuth, S., Geyer, M., Kalbitzer, H. R., Herrmann, C., and Wittinghofer, A. (1999) Structural and biochemical analysis of Ras-effector signaling via RalGDS. *FEBS Letters*, **451**, 175.

217. Hartmann, E. and Görlich, D. (1995) A Ran-binding motif in nuclear pore proteins. *Trends Cell Biol.*, **5**, 192.

218. Dingwall, C., Kandels-Lewis, S., and Séraphin, B. (1995) A family of Ran binding proteins that includes nucleoporins. *Proc. Natl Acad. Sci., USA*, **92**, 7525.

219. Bischoff, F. R., Krebber, H., Smirnova, E., Dong, W., and Ponstingl, H. (1995) Co-activation of RanGTPase and inhibition of GTP dissociation by Ran-GTP binding protein RanBP1. *EMBO J.*, **14**, 705.

220. Bischoff, F. R. and Görlich, D. (1997) Ran BP1 is crucial for the release of RanGTP from importin beta-related nuclear transport factors. *FEBS Lett.*, **419**, 249.

221. Wu, J., Matunis, M. J., Kraemer, D., Blobel, G., and Coutavas, E. (1995) Nup358, a cyto-plasmically exposed nucleoporin with peptide repeats, Ran-GTP binding sites, zinc fingers, a cyclophilin A homologous domain, and a leucine-rich region. *J. Biol. Chem.*, **270**, 14209.

222. Yokoyama, N., Hayashi, N., Seki, T., Panté, N., Ohba, T., Nishii, K., Kuma, K., Hayashida, T., Miyata, T., Aebi, U., *et al.* (1995) A giant nucleopore protein that binds Ran/TC4. *Nature*, **376**, 184.

223. Wilken, N., Senécal, J.-L., Scheer, U., and Dabauvalle, M.-C. (1995) Localization of the Ran-GTP binding protein RanBP2 at the cytoplasmic side of the nuclear pore complex. *Europ. J. Cell Biol.*, **68**, 211.

224. Delphin, C., Guan, T., Melchior, F., and Gerace, L. (1997) RanGTP tarets p97 to RanBP2, a filamentous protein localized at the cytoplasmic periphery of the nuclear pore complex. *Mol. Biol.Cell*, **8**, 2379.

225. Hyvönen, M. and Saraste, M. (1997) Structure of the PH domain and BTK motif from Bruton's tyrosine kinase: Molecular explanations for X-linked a gammaglobulinaemia. *EMBO J.*, **16**, 3396.

226. Melchior, F. and Gerace, L. (1998) Two way trafficking with ran. *Trends Cell Biol.*, **8**, 175.

227. de Rooij, J., Zwartkruis, F. J. T., Verheijen, M. H. G., Cool, R. H., Nijman, S. M. B., Wittinghofer, A. and Bos, J. L. (1998) Epac is a Rap1 guanine-nucleotide-exchange factor directly activated by cyclic AMP. *Nature*, **396**, 474.

228. Whisnant, R. E., Gilman, A. G., and Dessauer, C. W. (1996) Interaction of the two cytosolic domains of mammalian adenylyl cyclase. *Proc. Natl Acad. Sci., USA*, **93**, 6621.

229. Tang, W.-J. and Gilman, A. G. (1995) Construction of a soluble adenylyl cyclase activated by Gs alpha and forskolin. *Science*, **268**, 1769.

230. Zhang, G., Liu, Y., Ruoho, A. E., and Hurley, J. H. (1997) Structure of the adenylyl cyclase catalytic core. *Nature*, **386**, 247.

231. Artymiuk, P. J., Poirrette, A. R., Rice, D. W., and Willett, P. (1997) A polymerase I palm in adenylyl cyclase? *Nature*, **388**, 33.

10 | GTPases targetted by bacterial toxins

KLAUS AKTORIES, GUDULA SCHMIDT, AND FRED HOFMANN

1. Introduction

Members of the GTPase superfamily operate as molecular switches in numerous signal-transduction pathways to control such diverse cellular processes as differentiation, proliferation, and the organization of the actin cytoskeleton. These master regulators are the specific targets of a large array of bacterial protein toxins (Table 1)(1). In an extremely efficient manner, the toxins cause the covalent modification of eukaryotic GTPases, resulting in drastic changes in their biological and biochemical properties. It has been known for many years that eukaryotic GTPases are substrates for numerous distinct bacterial ADP-ribosyltransferases. Diphtheria toxin was the first example of such a bacterial toxin and was shown to elicit its biological activity by ADP-ribosylation of the GTPase elongation factor 2 (2). Other well-known ADP-ribosylating bacterial toxins are cholera toxin (3, 4) the group of heat-labile *E. coli* enterotoxins (5), and pertussis toxin (6), all of which ADP-ribosylate heterotrimeric G proteins of the G_s and G_i families. More recently, small GTPases belonging to the Rho and Ras subfamilies have also been shown to be targets of bacterial toxins. The group of C3-like transferases, including the prototype *Clostridium botulinum* exoenzyme C3 (7–9), selectively ADP-ribosylates Rho, rendering it biologically inactive, while Ras appears to be an *in vivo* target of the ADP-ribosyltransferase exoenzyme S from *Pseudomonas aeruginosa*. Small GTPases are not only substrates for bacterial ADP-ribosyltransferases, but also serve as targets for toxins that act by monoglucosylation (e.g. the large clostridial toxins) (10) or by deamidation (e.g. cytotoxic necrotizing factors) (11).

2. Diphtheria toxin and *Pseudomonas aeruginosa* exotoxin A: ADP-ribosylation of elongation factor 2

Diphtheria toxin and *Pseudomonas* exotoxin A ADP-ribosylate elongation factor 2 (EF-2), a GTPase that participates in polypeptide chain elongation on eukaryotic ribosomes (12) (Table 1). Diphtheria toxin is the prototype for intracellularly acting

Table 1 Toxins acting on GTPases

Toxin/exoenzyme	Enzyme activity	Target GTPase
Diphtheria toxin	ADP-ribosyltransferase	Elongation factor 2
Pseudomonas exotoxin A	ADP-ribosyltransferase	Elongation factor 2
Cholera toxin	ADP-ribosyltransferase	$G\alpha_s$ family
Heat-labile *Escherichia coli* enterotoxins	ADP-ribosyltransferase	$G\alpha_s$ family
Pertussis toxin	ADP-ribosyltransferase	$G\alpha_i$ family
Clostridium botulinum C3 exoenzyme	ADP-ribosyltransferase	Rho
Clostridium limosum transferase	ADP-ribosyltransferase	Rho
Staphylococcus aureus transferase	ADP-ribosyltransferase	Rho
Bacillus cereus transferase	ADP-ribosyltransferase	Rho
Pseudomonas aeruginosa exotoxin S	ADP-ribosyltransferase	Ras (in intact cells)
Clostridium difficile toxin A and B	Glucosyltransferases	Rho, Rac, Cdc42
Clostridium sordellii lethal toxin	Glucosyltransferase	Rac, Cdc42, Ras, Ral, Rap
Clostridium sordellii haemorrhagic toxin	Glucosyltransferase	Rho, Rac, Cdc42
Clostridium novyi α-toxin	N-Acetylglucosaminyltransferase	Rho, Rac, Cdc42
Cytotoxic necrotizing factor 1, 2 (CNF)	Deamidase	Rho, Rac, Cdc42
Dermonecrotic toxin (DNT)	Deamidase	Rho, Rac, Cdc42

protein toxins and its analysis has been of general importance for understanding mechanisms of binding, internalization, translocation, and target modification. Diphtheria toxin is produced by lysogenic *Corynebacteria diphtheria*, the causative agent of diphtheria. The toxin is released as a single-chain toxin, but is readily 'nicked' to give an enzymatically active N-terminal fragment A ($M_r \sim 21\,000$) and a C-terminal binding fragment B ($M_r \sim 37000$) which remain associated via a single disulphide bond. Crystallographic analysis of diphtheria toxin has revealed a Y-shaped, three-domain structure; one arm of the 'Y' is occupied by the enzyme domain (fragment A), while the binding fragment B forms the other arm and the base of the 'Y' (13). Fragment B is involved in binding to receptors on the target cell and in translocation of the toxin into the cytosol.

Pseudomonas aeruginosa exotoxin A is a single-chain toxin (M_r 66 580). Although the regions responsible for receptor binding and enzymatic activity of exotoxin A and diphtheria toxin are located at opposite ends of the molecules, structural analysis reveals that both toxins have a very similar shape, i.e. a Y-shaped, three-domain organization. However, the two toxins bind to different cell-surface receptors: the heparin-binding, EGF-like growth factor precursor in the case of diphtheria toxin (14) and the α_2-macroglobulin receptor in the case of exotoxin A (15). Only restricted sequence similarities can be found in the enzymatic domains; nevertheless, both toxins modify the same amino-acid residue in EF-2 and appear to have the same mechanisms of internalization and translocation.

Elongation factor 2 (M_r 95700) is related to other GTPases, such as bacterial elongation factor Tu, elongation factor G, bacterial initiation factor 2α, and, to a lesser extent, heterotrimeric G proteins and small GTPases of the Ras superfamily (16). The regions of similarity are restricted to residues involved in GTP binding and GTP hydrolysis located in the N-terminal 160 amino-acid residues of EF-2. However,

ADP-ribosylation of EF-2 by diphtheria toxin and exotoxin A occurs at diphthamide, a post-translationally modified histidine residue (His715) (17) in the C terminus. ADP-ribosylation of diphthamide reduces the affinity of EF-2 for ribosomes (18) and the modified elongation factor is unable to promote translocation within the ribosome (19), resulting in inhibition of polypeptide synthesis and death of the target cell (2, 20).

3. Toxins modifying heterotrimeric G proteins

Cholera toxin, the related heat-labile *E. coli* enterotoxins, and pertussis toxin ADP-ribosylate heterotrimeric G proteins. Cholera toxin is produced by *Vibrio cholerae*, the causative agent of cholera diarrhoea. It consists of one A subunit and five B subunits (21), although the A component is cleaved into two further components—A1 (~21 kDa) and A2 (~6 kDa). A1 possesses ADP-ribosyltransferase activity (22) while A2 interacts with the B subunits. The B protomer consists of five identical subunits of M_r 11 677 arranged in a pentameric ring-like structure that includes the A2 subunit. Cholera toxin binds, via its B subunits, to monosialoganglioside GM1 on target cells (23). The heat-labile *E. coli* enterotoxins, which are responsible for the syndrome of 'traveller's diarrhoea', are very similar to cholera toxins in structure and in biological activity. All these toxins ADP-ribosylate the GTP-binding α subunit of G_s (24), G_t (25), and G_{olf} (the olfactory G_s protein)—ribosylation occurs at arginine-174 of G_t (26) and the corresponding residue, Arg201, of G_s. At least *in vitro*, ADP-ribosylation of G proteins by cholera toxin is greatly enhanced by various membrane-bound and cytosolic cellular factors, some of which are members of the Arf (ADP-ribosylation factor) small GTPase subfamily. ADP-ribosylation of α_s results in constitutive activation by inhibiting its intrinsic GTPase activity (27, 28), by enhancing GDP release (29), and by promoting the dissociation of the α and βγ subunits (30). Thus, signalling pathways controlled by G_s are dramatically affected (see Chapter 1).

Pertussis toxin is one of the exotoxins produced by *Bordetella pertussis*, the causative agent of whooping cough. The toxin (M_r 105 000) consists of the enzyme component S1 (M_r ~26 000) and a binding protomer B, comprised of subunits S2 (M_r ~22 000), S3 (M_r ~21 000), two S4 (M_r ~12 000), and S5 (M_r ~11 000) (31). Pertussis toxin catalyses the ADP-ribosylation of most α subunits belonging to the G_i family ($G_{i1,2,3}$; $G_{o1,2}$; $G_{tr,c}$), but members of the G_s, G_q, and G_{12} subfamilies are not modified. The toxin ADP-ribosylates a cysteine residue at the carboxy-terminal end of the α subunits of sensitive G proteins (32). Modification depends on the presence of βγ subunits, but has no effect on basal guanine nucleotide binding or intrinsic GTPase activity. However, modification blocks the interaction of the G protein with receptors, thereby inhibiting G_i-mediated signalling pathways (33, 34).

4. C3-like ADP-ribosyltransferases

Clostridium botulinum C3 ADP-ribosyltransferase is the prototype of the family of C3-like transferases (35). Other members of the protein family are *Clostridium limosum*

ADP-ribosyltransferase (36), *Bacillus cereus* transferase (37), and an ADP-ribosyl-transferase of *Staphylococcus aureus* called EDIN (epidermal differentiation inhibitor) (38). Exoenzyme C3 was serendipitously detected during screening for high producer strains of *Clostridium botulinum* C2 toxin, a transferase that ADP-ribosylates actin (39, 40). C3 is encoded by a bacteriophage, which also carries the genes for neurotoxins C1 or D1, which act as specific metalloproteases and target eukaryotic synaptic peptides (41). However, C3 is neither structurally nor functionally related to neurotoxins. Various isoforms of the C3 transferase have been described; for example, C3 from *Clostridium botulinum* strain C468 is a protein of 251 amino acids (M_r 27 823) (42), while C3 from the strain 003–9 has 244 amino-acid residues (M_r 27 362) with an overall identity of 65% (43). The mature proteins consist of 211 (M_r 23 546) and 204 (M_r 23 119) amino-acid residues, respectively; in both cases, the N-terminal 40 amino-acid residues function as a signal peptide. The C3-related trans-ferases produced by *Clostridium limosum* (36), *Staphylococcus aureus* (38), and *Bacillus cereus* (37) are about 63 and 30% identical with the enzymes from *Clostridium botulinum* strains. All C3-like transferases are basic proteins (pI 9–10) of 23–28 kDa.

4.1   ADP-ribosylation of Rho

All C3-like transferases modify the three closely related isoforms of Rho (RhoA, RhoB, and RhoC), while other members of the Rho family are poor substrates (36) (Fig. 1). The C3-like transferases ADP-ribosylate Rho at asparagine 41 (44). Aspara-gine is a unique acceptor residue for ADP-ribosylation; the ADP–ribose bond formed is stable towards neutral hydroxylamine (0.5 M, 2 hours) and mercury (2 mM, 1 hour), unlike ADP-ribosylation of arginine (e.g. by cholera toxin) or cysteine (e.g. by pertussis toxin) residues (45). Like other ADP-ribosyltransferases, C3-like exoenzymes also show NAD glycohydrolase activity, although this is around 100-fold lower than its transferase activity.

The acceptor amino acid, Asn41, is located within or near the effector region of Rho (Chapter 4) and ADP-ribosylation of this site renders the protein biologically in-active. So far the precise mechanism of inactivation is not clear since ADP-ribosylated Rho is still able to interact with at least some of the known Rho effectors, e.g. protein kinase N (PKN) (46). ADP-ribosylation of Rho does not have a major effect on nucleotide binding (GDP or GTP) or on the intrinsic or GAP-stimulated GTPase activity. It has been suggested that modified Rho might bind more strongly to effectors causing sequestration, although this does not appear to be the case for PKN. Interestingly, activation of Rho by the guanine nucleotide exchange factors such as Lbc is significantly decreased after ADP-ribosylation (Peter Sehr and Klaus Aktories, unpublished observation) (Fig. 2). Whether this is responsible for biological inactivation of Rho after ADP-ribosylation remains to be clarified.

In mammalian cell culture, the action of C3-like transferases is characterized by dramatic alterations in the actin cytoskeleton (47–49), leading in most cell types to rounding up. More precise studies have revealed that the primary effect of C3 is to induce the loss of actin stress fibres and integrin adhesion plaques, while actin-

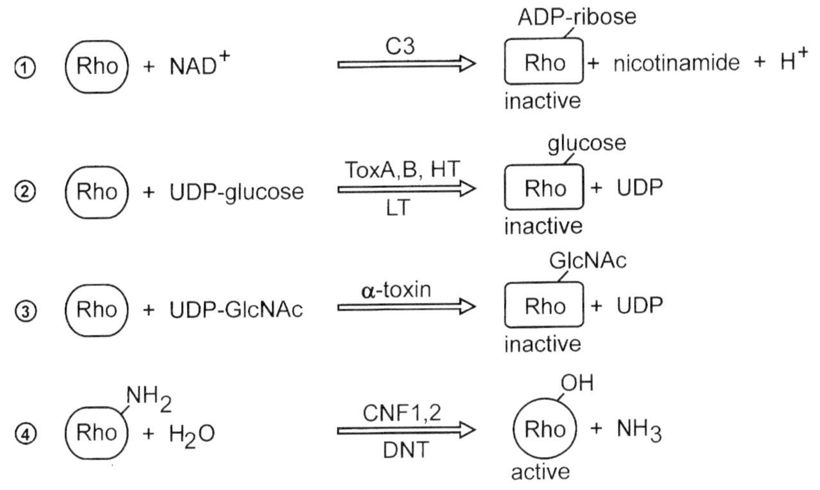

Fig. 1 Molecular mechanisms of Rho- and Ras-targetting toxins. (1) C3-like transferases ADP-ribosylate RhoA, B, and C by using NAD+ as a co-substrate. *Pseudomonas* exoenzyme S (not shown) modifies Ras by the same reaction. (2) UDP-glucose is the co-substrate for glucosylation by *C. difficile* toxins A and B (ToxA, B) and the *C. sordellii* haemorrhagic (HT) and lethal (LT) toxins. Whereas toxins A and B and HT glucosylate Rho, Rac, and Cdc42, LT modifies Rac and Cdc42 but not Rho. In addition, some Ras subfamily proteins are also substrates for LT. (3) The α-toxin from *C. novyi* causes the N-acetylglucosaminylation of all Rho subfamily proteins by using UDP-GlcNAc as a co-substrate. (4) The cytotoxic necrotizing factors (CNF) 1 and 2 from *E. coli* and the dermonecrotizing toxin (DNT) from *Bordetella* catalyse the deamidation of glutamine 63 of Rho or glutamine 61 of Rac and Cdc42.

dependent membrane ruffling and filopodia, which are regulated by Rac and Cdc42 (50, 51), respectively (see Chapter 4), are unaffected. However, rounding up of cells is not always observed upon C3 transferase treatment. In neuronal cells, for example, C3 induces flattening of cells and the expression of neurite-like cell processes, while activation of Rho (e.g. by lysophosphatidic acid) appears to enhance actin–myosin interactions (most likely via Rho kinase (52)) leading to contraction of the cell and collapse of cell processes (53).

Because C3 is a highly specific inhibitor of RhoA, B, and C, this exoenzyme is widely used as a tool to study Rho function. However, its use is hampered by its poor cell accessibility. In contrast to other bacterial ADP-ribosylating toxins, C3-like transferases consist only of a single enzymic component. Other toxins, for example the large clostridial cytotoxins and CNF (see Fig. 3), and the previously described diphtheria, cholera, and pertussis toxins, consist of at least three different components that mediate:

(1) binding to the surface of target cells;

(2) translocation of the biologically active enzyme component into the cytosol; and

(3) the enzymatic modification of the eukaryotic target proteins.

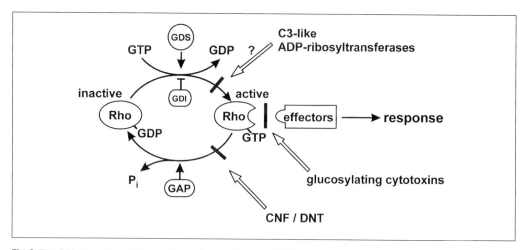

Fig. 2 The GTPase cycle of Rho and the effects of toxins. C3-like transferases induce ADP-ribosylation of Rho at asparagine 41 thereby blocking its biological activity. A possible mechanism is the inhibition of Rho activation. Large clostridial cytotoxins cause the glucosylation of Rho GTPases at Thr37 (Rho) or Thr35 (Rac, Cdc42), inhibiting their interaction with effectors. CNFs and DNT cause deamidation of Rho GTPase at Gln63 (Gln61 in Rac and Cdc42), thereby inhibiting the intrinsic and GAP-stimulated GTP hydrolysis and blocking the turn-off mechanisms.

Although C3 can be taken up by mammalian cells through non-specific mechanisms, high concentrations (>10 μg ml^{-1}) of C3 and long incubation times (24–48 h) are required to observe cellular effects. Alternatively, C3 can be introduced directly into cells by microinjection (48, 54), permeabilization with detergents or electrical discharge (55), and by transfection of DNA expression constructs (56). A chimeric toxin has been constructed consisting of the binding and translocation domain of diphtheria toxin (57) and the enzymatic domain of C3, while more recently, C3 has been engineered as a fusion toxin in a construct consisting of the active component of the binary C2 toxin (58). With these methods, C3 is active on cells within 2–3 hours at concentrations below 1 μg ml^{-1}.

5. *Pseudomonas aeruginosa* exoenzyme S

Besides exotoxin A, many *Pseudomonas aeruginosa* strains secrete another ADP-ribo-syltransferase termed exoenzyme S. Although the role of exoenzyme S as a virulence factor is not entirely clear, the enzyme has been associated with the establishment of infection and tissue damage. Recent data indicate that exoenzyme S inhibits eukaryotic cell proliferation (59) and causes alterations of the actin cytoskeleton (60). The transferase appears to be promiscuous compared to the previously described ADP-ribosylating toxins and various *in vitro* substrates have been described, including vimentin (61), the extracellular proteins IgG3 and apolipoprotein A1 (62), and the small GTPases Ras, Ral, Rap1A, Rab3, and Rab4 (63, 64). Recently, Ras was shown to be a substrate in intact cells (65) leading to inhibition of Ras-mediated

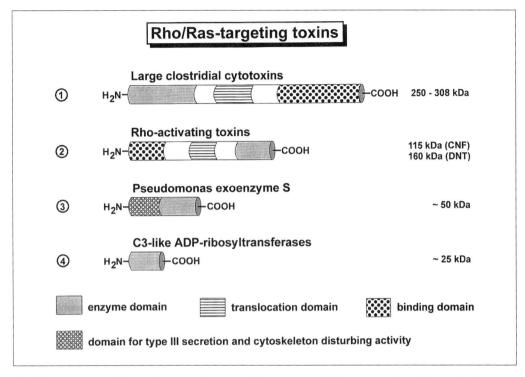

Fig. 3 The structure of Rho and Ras-targetting toxins. The large clostridial cytotoxins (1) and Rho-activating toxins (2) possess three functional domains involved in receptor binding, translocation, and enzyme activity. (3) *Pseudomonas* exoenzyme S consists of two domains. The C-terminal domain possesses ADP-ribosyltransferase activity, whereas the N-terminal domain is essential for type III secretion from the host bacteria. (4) C3-like ADP-ribosyltransferases consist of only an enzymatic component.

signal-transduction pathways (66). Exoenzyme S ADP-ribosylates Ras at Arg41 (66), although additional sites are also ribosylated *in vitro*. The functional consequences of these modifications are not clear. Early work revealed that ADP-ribosylation by exoenzyme S is strictly dependent on the presence of a cytosolic factor termed FAS (factor activating ExoS) (67), later identified as a member of the 14–3–3 protein family (68).

In vivo studies with exoenzyme S are complicated, because the exoenzyme is delivered to cells by the bacterial type III secretion system (69). Translocation of the exoenzyme depends, therefore on direct contact between bacteria and the host cell. The introduction of purified exoenzyme S into eukaryotic cells is not possible merely by adding the 'toxin' to cell culture medium.

Of special interest is the finding that exoenzyme S appears to harbour a biological activity in addition to its ADP-ribosyltransferase activity. Exoenzyme S is a 49 kDa protein that consists of at least two domains (70); the ADP-ribosyltransferase activity is located within the C-terminal half of the protein, while an activity located in the N-terminal fragment induces alterations of the actin cytoskeleton by an unknown mechanism. This N-terminal fragment of exoenzyme S reveals significant sequence similarity with YopE from *Yersinia*, which is also a type III secreted protein that causes effects on the actin cytoskeleton (60, 71). More recently, it has been suggested that both YopE and the N-terminal fragment of exoenzyme S affect the cytoskeleton by altering the Rho GTPase pathway (72, 73).

6. Large clostridial cytotoxins

Rho GTPases are the targets of large clostridial cytotoxins. Prototypes of this toxin family are *Clostridium difficile* toxins A and B (74), the major virulence factors involved in antibiotic-associated diarrhoea and pseudomembranous colitis (75–78). Other members of the family are the lethal and haemorrhagic toxins from *Clostridium sordellii* and the α-toxin from *Clostridium novyi*. *Clostridium sordellii* and *Clostridium novyi* toxins play a role as virulence factors in gas gangrene in man and are implicated in enterotoxaemia of animals (79–81). All these toxins are cytotoxic and cause destruction of the cytoskeleton and induce rounding up of cells. The cytotoxic effects, which can be prevented by addition of an antitoxin antibody, are used as a specific and sensitive method for the identification of these toxins in clinical isolates and for diagnosis of the toxin-related diseases. The underlying biochemical mechanism of the toxin-induced cytotoxic effects is the glucosylation of members of the Rho family.

6.1 Glucosylation of small GTPases

Clostridium difficile toxins A and B catalyse the monoglucosylation of Rho GTPases (82, 83) (Fig. 1), using UDP-glucose as a co-substrate. All Rho subfamily proteins, including Rho, Rac, and Cdc42, and their isoforms are substrates, but other small GTPases of the Ras, Rab, Ran or Arf subfamilies are not modified. *Clostridium sordellii* haemorrhagic toxin shows the same co-substrate and protein substrate specificity. However, the lethal toxin from *Clostridium sordellii*, which also uses UDP-glucose as a co-substrate, differs in its substrate specificity (84, 85) and glucosylates Rac and Cdc42 but not Rho. In addition, Ras subfamily proteins such as Ras, Rap, and Ral are modified and inactivated by this toxin. Thus, *Clostridium sordellii* lethal toxin, but not toxin A or B, inhibits the activation of the Ras-dependent MAP-kinase pathway (see Chapter 3). The α-toxin of *Clostridium novyi* differs by catalysing *N*-acetylglucosaminylation using UDP-GlcNAc as a co-substrate (86).

All these toxins modify Rho at threonine 37 (Rac and Cdc42 at Thr35). Thr37 is in the switch 1 region of the GTPase (87–90) and is directly involved in the coordination of the divalent magnesium cation, which is essential for guanine nucleotide binding (see Chapter 9). Crystallographic analysis of Rho proteins indicates that in the GTP-bound form the hydroxyl group of Thr37 is directed into the protein and participates in coordination of Mg^{2+} (89, 90). In the GDP-bound form of the GTPase, the hydroxyl group is oriented towards the solvent. Accordingly, Rho proteins are stoichiometrically glucosylated in the GDP-form but are resistant to modification when loaded with GTP (82, 83). Because Thr37 is located in the vicinity of Asn41, which is the acceptor amino-acid residue for C3-induced ADP-ribosylation, glucosylation inhibits subsequent ADP-ribosylation (91).

Recently the functional consequences of glucosylation of Rho and Ras proteins have been described (92). Glucosylation of Ras by *Clostridium sordellii* toxin inhibits Cdc25-stimulated nucleotide exchange by about tenfold, but the major effect of glucosylation is observed on its GTPase activity. The low intrinsic GTPase activity of Ras is even further reduced by glucosylation (by a factor of five), while the stimulation of GTPase activity by the Ras GTPase-activating protein p120 GAP is completely blocked. Perhaps more importantly, glucosylated Ras is no longer able to interact with Raf kinase, the best-characterized effector of Ras (see Chapter 3) and this would therefore explain the biological effects of the *C. sordellii* toxin (84). Similar studies with Rho have indicated that glucosylation induces more or less the same functional consequences as described for Ras. Toxin B-induced glucosylation of Rho results in a slight decrease in the affinity for nucleotides, inhibits p50RhoGAP-stimulated GTPase activity, and prevents its interaction with at least one effector, p65PAK (46).

Treatment of cultured cells with toxin B (or toxin A) causes destruction of the actin cytoskeleton and rounding up (93) (Fig. 4). Dependent on the toxin concentration, thin cell processes may remain, a phenomenon referred to as arborization (94). The toxin effect occurs not earlier than 30 minutes after addition of the toxins, the time required for entry into the cell and, in most cases, the full effects are observed after 2–3 hours. Because all Rho subfamily proteins (Rho, Rac, and Cdc42) are modified by these toxins, a complete breakdown of the actin cytoskeleton is observed, including the loss of stress fibres, the subcortical actin ring, lamellipodia, and filopodia.

6.2 Structure–activity relationship of large clostridial cytotoxins

Large clostridial cytotoxins are single-chain toxins of 250–308 kDa and are the largest bacterial toxins known (74). It has been suggested that the toxins consist of three functional domains (Fig. 3) required for (1) receptor-binding and endocytosis, (2) toxin translocation into the cytosol, and (3) modification of target GTPases.

The C-terminal part of the toxins is thought to be involved in cell-surface receptor binding (95) and is characterized by repetitive oligopeptides (96) that interact with carbohydrate moieties on the target cell. The nature of the toxin receptors is still unclear, although in rabbit ileum, toxin A binds to the brush border disaccharidase, sucrase–isomaltase (97). Although sucrase–isomaltase mediates the enterotoxicity in rabbits, this enzyme is not present in human colonic mucosa and cannot, therefore, be the human receptor. The central region of the toxins is characterized by a rather short hydrophobic domain (e.g. amino-acid residues 956–1128 in toxin B) that is thought to be involved in the translocation of the toxin from the endosome compartment into the cytosol. Finally, the N-terminal part of the toxins harbours the glucosyltransferase activity (98) and deletion analysis reveals that the 546 N-terminal amino-acid residues are sufficient for full enzymatic activity. Further deletion of 30

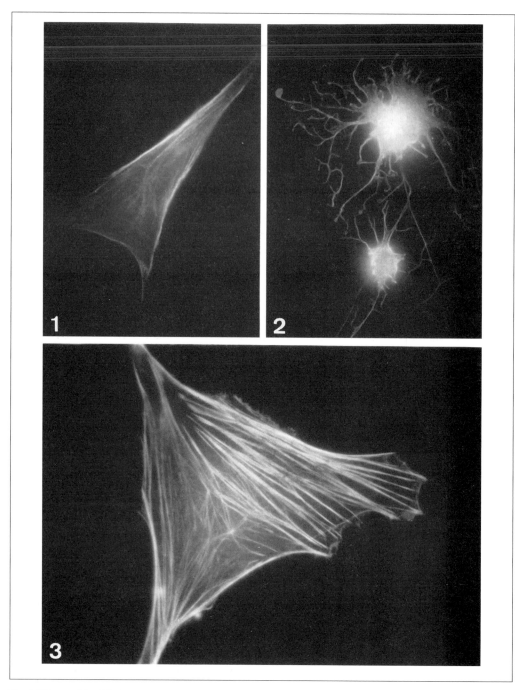

Fig. 4 Effects of *C. difficile* toxin B and cytotoxic necrotizing factor 1 (CNF1) on the actin cytoskeleton of NIH3T3 cells. NIH3T3 cells (non-treated, panel 1) were treated with toxin B (100 ng ml^{-1}) for 2.5 h (panel 2) or with CNF1 (200 ng/ml^{-1}) for 16 h (panel 3). The actin cytoskeleton was visualized with rhodamine-phalloidin. (From G. Schmidt and K. Aktories (1998) *Naturwissenschaften*, **85**, 253.)

amino-acid residues causes loss of transferase and of glucohydrolase activity. Moreover, microinjection of the enzymatically active fragment into cells results in the same phenotype observed after treatment of cells with the complete toxin. Because toxin B and the lethal toxin from *Clostridium sordellii* differ in their protein substrates, toxin chimeras have been constructed to identify the region defining substrate specificity. Substrate specificity of *Clostridium sordellii* and *Clostridium difficile* toxin is encoded between amino-acid residues 364 and 516 (99). Recently, we identified several amino-acid residues in *Clostridium sordellii* lethal toxin that are essential for its enzymatic activity (100). Exchange of the two aspartic acid residues at position 286 or 288 with alanine or asparagine decreases glucosyltransferase activity about 5000-fold and it has been suggested that this DXD motif is involved in the coordination of manganese ions, which participate in the binding of the UDP-glucose and are essential for the activity of all clostridial transferases (101). Interestingly, the DXD motif and the surrounding residues are conserved in a large number of glycosyltransferases, including eukaryotic and bacterial enzymes (101). Even more exciting, significant sequence similarities with large clostridial cytotoxins have been recognized in one of the open reading frames of the 92 kb plasmid of the enterohaemorrhagic *E. coli* (EHEC) O157:7, the predominant causative agent of haemorrhagic colitis (102). Finally, a gene found in *Chlamydia trachomatis* also encodes a protein showing sequence similarity to the large clostridial toxins (102). Thus, it appears that the large clostridial toxins may be a subgroup of a much larger family of glucosylating cytotoxins, widely distributed in various types of bacteria.

7. Activation of Rho GTPases by bacterial toxins

A group of bacterial toxins has recently been shown to activate Rho GTPases. Current members of this toxin family are the cytotoxic necrotizing factors (CNFs) 1 and 2 produced by certain *E. coli* strains and the dermonecrotic toxin (DNT) from *Bordetella* species.

7.1 Cytotoxic necrotizing factors, CNF1 and 2

In 1983, Caprioli and co-workers isolated a toxin, which they called CNF, from *E. coli* obtained from a patient with enteritis (103). CNF could induce necrotizing effects on rabbit skin and the formation of large multinucleated cells. Later the toxin was identified in various *E. coli* strains isolated from domestic animals and a second related toxin, CNF2, was also described (104). CNF1 and CNF2 are single-chain toxins with molecular masses of about 115 kDa, sharing more than 90% sequence similarity (105, 106). Whereas CNF1 is chromosomally encoded, CNF2 is encoded by plasmids.

CNFs are toxic for a wide variety of cells, including CHO, Vero, HeLa, and Caco2 cells. Treated cells are characterized by striking morphological changes that include flattening and spreading, as well as the transient formation of filopodia and membrane ruffles. Later a dense network of stress fibres is seen (Fig. 4) (105, 107).

Polymerization of actin induced by CNF was the first hint that Rho GTPases might be the targets of these cytotoxins. This hypothesis was supported by the observation that Rho proteins from CNF-treated cells migrate at an apparent higher molecular mass after SDS-gel electrophoresis, suggesting a covalent modification (105, 108). Subsequently, it was shown that the CNF can induce a modification of recombinant RhoA even in the absence of any additional factor (109) and mass spectroscopic analysis revealed an 1 Da was detected in a peptide spanning Gln52–Arg68 (110). Amino-acid sequencing of this peptide assigned the increase in mass to glutamine 63, which had in fact been deamidated to glutamic acid (109, 110). Gln63 is essential for the GTPase activity of RhoA and crystallographic structure analysis suggests that it participates in stabilizing the transition state during GTP hydrolysis (89, 111) (see Chapter 9). Biochemical analysis of CNF1-treated RhoA revealed an inhibition of both intrinsic and GAP-stimulated activity, indicating that the deamidation of Gln63 results in the formation of a constitutively activated Rho protein (Figs 1 and 2) (109, 110). Although many of the morphological changes observed in CNF-treated cells can be explained by activation of Rho, we have recently found that Rac and Cdc42 are also substrates for deamidation by CNF (112) at residue Gln61. In contrast to CNF-treated RhoA, deamidation of Rac or Cdc42 does not cause changes in their mobility on SDS gels (110), but it does inhibit GTPase activity. These recent observations account for the induction of filopodia and membrane ruffling after CNF treatment of cells (Fig. 4).

7.2 Structure–activity analysis of CNF

Lemichez and co-workers have postulated a structure of CNF consisting of an N-terminal cell-binding component, a translocation domain with hydrophobic sequences in the middle of the protein, and a catalytic domain located at the C terminus (Fig. 3) (113). In fact, a toxin fragment of amino-acid residues 720–1007 obtained from the C terminus of CNF possesses full deamidation activity, but is not able to enter cells. After microinjection this fragment induces the same morphological changes as those induced by intact CNF (114).

Recently, it has been observed that CNF (full length and the active fragment) possesses not only deamidating but also transglutamination activity (114). Thus, at least *in vitro*, CNF is able to incorporate primary amines (e.g. putrescine, ethylene-diamine, cadaverine) into RhoA at Gln63. Interestingly, other eukaryotic transglutaminases, such as tissue transglutaminase or factor XIII, can modify Rho GTPases, although no significant sequence similarity between CNF and transglutaminases has been observed. However, cysteine and histidine residues are located in the active site of eukaryotic tissue transglutaminases (115), and these have been looked for in CNF. Treatment of CNF with *N*-methylmaleimide or iodoacetamide blocks deamidase and transglutaminase activity, while substitution of cysteine 866 of CNF with serine results in a complete block of enzymatic activity. Similarly, exchange of histidine 881 with alanine blocks both enzyme activities of CNF. Thus, it appears that the active site of CNF includes essential cysteine and histidine residues, similar to the tissue transglutaminases (114).

7.3 Dermonecrotizing toxin (DNT)

The dermonecrotizing toxin is produced by *Bordetella bronchiseptica*, *B. pertussis*, and *B. parapertussis* (116, 117). The toxin causes dermonecrotic lesions in mice and other laboratory animals and is considered to be a virulence factor for turbinate atrophy in porcine atrophic rhinitis (118). In cell culture, DNT induces multinucleation and assembly of actin stress fibres. The toxin is a 160 kDa, heat-labile protein that shares significant sequence homology with CNF (116); the main regions of similarity being in a stretch of 260 amino-acid residues (residues 1160–1420 of DNT). In CNF, the corresponding residues are essential for enzyme activity and DNT induces a similar mobility shift of Rho after electrophoresis as reported for CNF (119). Both CNF and DNT induced Rho-dependent tyrosine phosphorylation of focal adhesion kinase and paxillin (120). Subsequently, it was shown that DNT induces the reorganization of actin stress fibres through deamidation of Gln63 of Rho (121). Moreover, by using an antibody that recognizes specifically deamidated Rho GTPases, it has been shown that Rac and Cdc42 are also substrates for DNT. In contrast to CNF, however, DNT treatment of intact cells causes the appearance of a faster-migrating Rho protein band after electrophoresis (121). Whether this is the result of a transglutamination reaction or is caused by another enzyme activity remains to be clarified.

8. Concluding remarks

GTPases are targets for numerous bacterial toxins and are modified by ADP-ribosylation, glucosylation, and deamidation. Recent work suggests that the number of bacterial toxins or exoenzymes that act on GTPases is still growing. Many pathogenic bacteria (e.g. *Listeria*, *Shigella*, *Edwardsiella*, and *Salmonella*) that invade eukaryotic host cells (122) also appear to subvert small GTPase signalling pathways for entry. For example, it has been demonstrated that the SopE protein from *Salmonella*, which is translocated into the host cell by type III secretion, acts as a GEF (guanine nucleotide exchange factor) and induces ruffling of the host cell by activating Rac and Cdc42 GTPases (123). Thus, GTPases, especially small GTPases of the Rho and Ras subfamilies are preferred targets of bacterial toxins and play an important role in the interactions of bacteria with their hosts. Advances in the understanding of these interactions will provide insights not only into the mode of action of virulence factors and pathogenic bacteria, but also into the cellular function and regulation of the eukaryotic toxin targets.

References

1. Aktories, K. and Just, I. (1993) GTPases and actin as targets for bacterial toxins. In *GTPases in biology,* (ed. B. F. Dickey and L. Birnbaumer), p. 87. Springer-Verlag, Berlin-Heidelberg.

2. Honjo, T., Nishizuka, Y., and Hayaishi, O. (1968) Diphtheria toxin-dependent adenosine diphosphate ribosylation of aminoacryl transferase II and inhibition of protein synthesis. *J. Biol. Chem.,* **243**, 3553.

3. Gill, D. M. (1977) Mechanism of action of cholera toxin. *Adv. Cyclic Nucleotide Res.,* **8**, 85.

4. Zhang, G.-F., Patton, W. A., Moss, J., and Vaughan, M. (1997) Cholera toxin: Mechanism of action and potential use in vaccine development. In *Bacterial toxins—tools in cell biology and pharmacology,* (ed. K. Aktories), p. 1. Chapman & Hall, Weinheim,

5. Hol, W. G. J., Sixma, T.K., and Merritt, E. A. (1995) Structure and function of *E. coli* heat-labile enterotoxin and cholera toxin B pentamer. In *Bacterial toxins and virulence factors in disease,* (ed. J. Moss, B. Iglewski, M. Vaughan, and A. T. Tu), p. 185. Marcel Dekker, New York.

6. Locht, C. and Antoine, R. (1997) Pertussis toxin. In *Bacterial toxins tools in cell biology and pharmacology,* (ed. K. Aktories), p. 33. Chapman & Hall, Weinheim.

7. Aktories, K. and Koch, G. (1997) *Clostridium botulinum* ADP-ribosyltransferase C3. In *Bacterial toxins: tools in cell biology and pharmacology,* (ed. K. Aktories), p. 61. Chapman & Hall, Weinheim.

8. Aktories, K., Braun, U., Habermann, B., and Rösener, S. (1990) Botulinum ADP-ribosyl-transferase C3. In *ADP-ribosylating toxins and G proteins,* (ed. J. Moss and M. Vaughan), p. 97. American Society for Microbiology, Washington.

9. Aktories, K., Mohr, C., and Koch, G. (1992) *Clostridium botulinum* C3 ADP-ribosyl-transferase. *Curr. Top. Microbiol. Immunol.,* **175**, 115.

10. Aktories, K. and Just, I. (1995) Monoglucosylation of low-molecular-mass GTP-binding Rho proteins by clostridial cytotoxins. *Trends Cell Biol.,* **5**, 441.

11. Aktories, K. (1997) Rho proteins: targets for bacterial toxins. *Trends Microbiol.,* **5**, 282.

12. Moldave, K. (1985) Eukaryotic protein synthesis. *Anun. Rev. Biochem.,* **54**, 1109.

13. Choe, S., Bennett, M. J., Fujii, G., Curmi, P. M. G., Kantardjieff, K. A., Collier, R. J., and Eisenberg, D. (1992) The crystal structure of diphtheria toxin. *Nature,* **357**, 216.

14. Naglich, J. G., Metherall, J. E., Russell, D. W., and Eidels, L. (1992) Expression cloning of a diphtheria toxin receptor: Identity with a heparin-binding EGF-like growth factor precursor. *Cell,* **69**, 1051.

15. Kounnas, M. Z., Morris, R. E., Thompson, M. R., FitzGerald, D. J., Strickland, D. K., and Saelinger, C. B. (1992) The alpha 2-macroglobulin receptor/low density lipoprotein receptor-related protein binds and internalises *Pseudomonas* exotoxin A. *J. Biol. Chem.,* **267**, 12420.

16. Kohno, K., Uchida, T., Ohkubo, H., Nakanishi, S., Nakanishi, T., Fukui, T., Ohtsuka, E., Ikehara, M., and Okada, Y. (1986) Amino acid sequence of mammalian elongation factor 2 deduced from the cDNA sequence: Homology with GTP-binding proteins. *Proc. Natl Acad. Sci. USA,* **83**, 4978.

17. Van Ness, B. G., Howard, J. B., and Bodley, J. W. (1980) ADP-ribosylation of elongation factor 2 by diphtheria toxin. *J. Biol. Chem.,* **255**, 10717.

18. Nygard, O. and Nilsson, L. (1990) Kinetic determination of the effects of ADP-ribosylation on the interaction of eukaryotic elongation factor 2 with ribosomes. *J. Biol. Chem.,* **265**, 6030.

19. Davydova, E. K. and Ovchinnikov, L. P. (1990) ADP-ribosylated elongation factor 2 (ADP-ribosyl-EF-2) is unable to promote translocation within the ribosome. *FEBS Lett.*, **261**, 350.

20. Collier, R .J. (1968) Effect of diphtheria toxin on protein synthesis: Inactivation of one of the transfer factors. *J. Mol. Biol.*, **25**, 83.

21. Gill, D. M. (1976) The arrangement of subunits of cholera toxin. *Biochemistry*, **15**, 1242.

22. Gill, D. M. and Rappaport, S. H. (1979) Origin of the enzymatically active A₁ fragment of cholera toxin. *J. Infect. Dis.*, **139**, 674.

23. van Heyningen, W. E., Carpenter, C. C. J., Pierce, N. F., and Greenough, W. B. (1971) Deactivation of cholera toxin by ganglioside. *J. Infect. Dis.*, **124**, 415.

24. Cassel, D. and Pfeuffer, T. (1978) Mechanism of cholera toxin action: Covalent modification of the guanyl nucleotide-binding protein of the adenylate cyclase system. *Proc. Natl Acad. Sci., USA*, **75**, 2669.

25. Abood, M. E., Hurley, J. B., Pappone, M.-C., Bourne, H. R., and Stryer, L. (1982) Functional homology between signal-coupling proteins: cholera toxin inactivates the GTPase activity of transducin. *J. Biol. Chem.*, **257**, 10540.

26. Van Dop, C., Tsubokawa, M., Bourne, H. R., and Ramachandran, J. (1984) Amino acid sequence of retinal transducin at the site ADP-ribosylated by cholera toxin. *J. Biol. Chem.*, **259**, 696.

27. Cassel, D. and Selinger, Z. (1976) Catecholamine-stimulated GTPase activity in turkey erythrocyte membranes. *Biochim. Biophys. Acta*, **452**, 538.

28. Gill, D. M. and Meren, R. (1978) ADP-ribosylation of membrane proteins catalyzed by cholera toxin: basis of the activation of adenylate cyclase. *Proc. Natl Acad. Sci., USA*, **75**, 3050.

29. Murayama, T. and Ui, M. (1984) [3H]GDP release from rat and hamster adipocyte membranes independently linked to receptors involved in activation or inhibition of adenylate cyclase. *J. Biol. Chem.*, **259**, 761.

30. Kahn, R. A. and Gilman, A. G. (1984) ADP-ribosylation of Gs promotes the dissociation of its α and β subunits. *J. Biol. Chem.*, **259**, 6235.

31. Tamura, M., Nogimuri, K., Murai, S., Yajima, M., Ito, K., Katada, T., Ui, M., and Ishii, S. (1982) Subunit structure of islet-activating protein, pertussis toxin, in conformity with the A-B model. *Biochemistry*, **21**, 5516.

32. West, R. E., Moss, J., Vaughan, M., Liu, T., and Liu, T.-Y. (1985) Pertussis toxin-catalyzed ADP-ribosylation of transducin. *J. Biol. Chem.*, **260**, 14428.

33. Gierschik, P. (1992) ADP-ribosylation of signal-transducing guanine nucleotide-binding proteins by pertussis toxin. *Curr. Top. Microbiol. Immunol.*, **175**, 69.

34. Nürnberg, B. (1997) Pertussis toxin as a cell biological tool. In *Bacterial toxins—tools in cell biology and pharmacology*, (ed. K. Aktories), p. 47. Chapman & Hall, Weinheim.

35. Kikuchi, A., Yamashita, T., Kawata, M., Yamamoto, K., Ikeda, K., Tanimoto, T., and Takai, Y. (1988) Purification and characterization of a novel GTP-binding protein with a molecular weight of 24,000 from bovine brain membranes. *J. Biol. Chem.*, **263**, 2897.

36. Just, I., Mohr, C., Schallehn, G., Menard, L., Didsbury, J. R., Vandekerckhove, J., van Damme, J., and Aktories, K. (1992) Purification and characterization of an ADP-ribosyl-transferase produced by *Clostridium limosum*. *J. Biol. Chem.*, **267**, 10274.

37. Just, I., Selzer, J., Jung, M., van Damme, J., Vandekerckhove, J., and Aktories, K. (1995) Rho-ADP-ribosylating exoenzyme from *Bacillus cereus* – purification, characterization and identification of the NAD-binding site. *Biochemistry*, **34**, 334.

38. Inoue, S., Sugai, M., Murooka, Y., Paik, S.-Y., Hong, Y.-M., Ohgai, H., and Suginaka, H. (1991) Molecular cloning and sequencing of the epidermal cell differentiation inhibitor gene from *Staphylococcus aureus*. *Biochem. Biophys. Res. Commun.*, **174**, 459.

39. Aktories, K., Weller, U., and Chhatwal, G. S. (1987) *Clostridium botulinum* type C produces a novel ADP-ribosyltransferase distinct from botulinum C2 toxin. *FEBS Lett.,* **212**, 109.

40. Aktories, K., Rösener, S., Blaschke, U., and Chhatwal, G. S. (1988) Botulinum ADP-ribosyltransferase C3. Purification of the enzyme and characterization of the ADP-ribosylation reaction in platelet membranes. *Eur. J. Biochem.,* **172**, 445.

41. Popoff, M. R., Boquet, P., Gill, D. M., and Eklund, M. W. (1990) DNA sequence of exo-enzyme C3, an ADP-ribosyltransferase encoded by *Clostridium botulinum* C and D phages. *Nucl. Acids Res.,* **18**, 1291.

42. Popoff, M. R., Hauser, D., Boquet, P., Eklund, M. W., and Gill, D. M. (1991) Character-ization of the C3 gene of *Clostridium botulinum* types C and D and its expression in *Escherichia coli. Infect. Immun.,* **59**, 3673.

43. Nemoto, Y., Namba, T., Kozaki, S., and Narumiya, S. (1991) *Clostridium botulinum* C3 ADP-ribosyltransferase gene. *J. Biol. Chem.,* **266**, 19312.

44. Sekine, A., Fujiwara, M., and Narumiya, S. (1989) Asparagine residue in the rho gene pro-duct is the modification site for botulinum ADP-ribosyltransferase. *J. Biol. Chem.,* **264**, 8602.

45. Aktories, K., Just, I., and Rosenthal, W. (1988) Different types of ADP-ribose protein bonds formed by botulinum C2 toxin, botulinum ADP-ribosyltransferase C3 and pertussis toxin. *Biochem. Biophys. Res. Commun.,* **156**, 361.

46. Sehr, P., Joseph, G., Genth, H., Just, I., Pick, E., and Aktories, K. (1998) Glucosylation and ADP-ribosylation of Rho proteins – effects on nucleotide binding, GTPase activity, and effector-coupling. *Biochemistry,* **37**, 5296.

47. Wiegers, W., Just, I., Müller, H., Hellwig, A., Traub, P., and Aktories, K. (1991) Alteration of the cytoskeleton of mammalian cells cultured *in vitro* by *Clostridium botulinum* C2 toxin and C3 ADP-ribosyltransferase. *Eur. J. Cell Biol.,* **54**, 237.

48. Paterson, H. F., Self, A. J., Garrett, M. D., Just, I., Aktories, K., and Hall, A. (1990) Micro-injection of recombinant p21[rho] induces rapid changes in cell morphology. *J. Cell Biol.,* **111**, 1001.

49. Chardin, P., Boquet, P., Madaule, P., Popoff, M. R., Rubin, E. J., and Gill, D. M. (1989) The mammalian G protein rho C is ADP-ribosylated by *Clostridium botulinum* exoenzyme C3 and affects actin microfilament in Vero cells. *EMBO J.,* **8**, 1087.

50. Ridley, A. J., Paterson, H. F., Johnston, C. L., Diekmann, D., and Hall, A. (1992) The small GTP-binding protein rac regulates growth factor-induced membrane ruffling. *Cell,* **70**, 401.

51. Nobes, C. D. and Hall, A. (1995) Rho, Rac, and Cdc42 GTPases regulate the assembly of multimolecular focal complexes associated with actin stress fibers, lamellipodia, and filopodia. *Cell,* **81**, 53.

52. Katoh, H., Aoki, J., Ichikawa, A., and Negishi, M. (1998) p160 RhoA-binding kinase ROKα induces neurite retractions. *J. Biol. Chem.,* **273**, 2489.

53. Jalink, K., Van Corven, E. J., Hengeveld, T., Morii, N., Narumiya, S., and Moolenaar, W. H. (1994) Inhibition of lysophosphatidate- and thrombin-induced neurite retraction and neuronal cell rounding by ADP ribosylation of the small GTP-binding protein Rho. *J. Cell Biol.,* **126**, 801.

54. Olson, M. F., Paterson, H. F., and Marshall, C. J. (1998) Signals from Ras and Rho GTPases interact to regulate expression of p21[Waf1/Cip1]. *Nature,* **394**, 295.

55. Stasia, M.-J., Jouan, A., Bourmeyster, N., Boquet, P., and Vignais, P. V. (1991) ADP-ribosylation of a small size GTP-binding protein in bovine neutrophils by the C3 exo-enzyme of *Clostridium botulinum* and effect on the cell motility. *Biochem. Biophys. Res. Commun.,* **180**, 615.

56. Fujisawa, K., Madaule, P., Ishizaki, T., Watanabe, G., Bito, H., Saito, Y., Hall, A., and Narumiya, S. (1998) Different regions of Rho determine Rho-selective binding of different classes of Rho target molecules. *J. Biol. Chem.,* **273**, 18943.

57. Aullo, P., Giry, M., Olsnes, S., Popoff, M. R., Kocks, C., and Boquet, P. (1993) A chimeric toxin to study the role of the 21 kDa GTP binding protein rho in the control of actin microfilament assembly. *EMBO J.,* **12**, 921.

58. Barth, H., Hofmann, F., Olenik, C., Just, I., and Aktories, K. (1998) The N-terminal part of the enzyme component (C2I) of the binary *Clostridium botulinum* C2 toxin interacts with the binding component C2II and functions as a carrier system for a Rho ADP-ribosylating C3-like fusion toxin. *Infect. Immun.,* **66**, 1364.

59. Olson, J. C., McGuffie, E. M., and Frank, D. W. (1997) Effects of differential expression of the 49-kilodalton exoenzyme S by *Pseudomonas aeruginosa* on cultured eukaryotic cells. *Infect. Immun.,* **65**, 248.

60. Frithz-Lindsten, E., Du, Y., Rosqvist, R., and Forsberg, A. (1997) Intracellular targeting of exoenzyme S of *Pseudomonas aeruginosa* via type II-dependent translocation induces phagocytosis resistance, cytotoxicity and disruption of actin microfilaments. *Mol. Microbiol.,* **25**, 1125.

61. Coburn, J., Dillon, S. T., Iglewski, B. H., and Gill, D. M. (1989) Exoenzyme S of *Pseudomonas aeruginosa* specifically ADP-ribosylates the intermediate filament protein vimentin. *Infect. Immun.,* **57**, 996.

62. Knight, D. A. and Barbieri, J. T. (1997) Ecto-ADP-ribosyltransferase activity of *Pseudomonas aeruginosa* exoenzyme S. *Infect. Immun.,* **65**, 3304.

63. Coburn, J. and Gill, D. M. (1991) ADP-ribosylation of p21^{ras} and related proteins by *Pseudomonas aeruginosa* Exoenzyme S. *Infect. Immun.,* **59**, (11), 4259.

64. Coburn, J., Wyatt, R. T., Iglewski, B. H., and Gill, D. M. (1989) Several GTP-binding proteins, including p21 c-H-ras, are preferred substrates of *Pseudomonas aeruginosa* exoenzyme S. *J. Biol. Chem.,* **264**, 9004.

65. McGuffie, E. M., Frank, D. W., Vincent, T. S., and Olson, J. C. (1998) Modification of Ras in eukaryotic cells by *Pseudomonas aeruginosa* exoenzyme S. *Infect. Immun.,* **66**, 2607.

66. Ganesan, A. K., Frank, D. W., Misra, R. P., Schmidt, G., and Barbieri, J. T. (1998) *Pseudomonas aeruginosa* exoenzyme S ADP-ribosylates Ras at multiple sites. *J. Biol. Chem.,* **273**, 7332.

67. Coburn, J., Kane, A. V., Feig, L., and Gill, D. M. (1991) *Pseudomonas aeruginosa* exoenzyme S requires a eukaryotic protein for ADP-ribosyltransferase activity. *J. Biol. Chem.,* **266**, 6438.

68. u, H., Coburn, J., and Collier, R. J. (1993) The eukaryotic host factor that activates exoenzyme S of *Pseudomonas aeruginosa* is a member of the 14–3–3 protein family. *Proc. Natl Acad. Sci., USA,* **90**, 2320.

69. Yahr, T. L., Goranson, J., and Frank, D. W. (1996) Exoenzyme S of *Pseudomonas aeruginosa* is secreted by a type III pathway. *Mol. Microbiol.,* **22**, 991.

70. Knight, D. A., Finck-Barbancon, V., Kulich, S. M., and Barbieri, J. T. (1998) Functional domains of *Pseudomonas aeruginosa* exoenzyme S. *Infect. Immun.,* **63**, 3182.

71. Rosqvist, R., Forsberg, A., and Wolf-Watz, H. (1998) Intracellular targeting of the *Yersinia* YopE cytotoxin in mammalian cells induces actin microfilament disruption. *Infect. Immun.,* **59**, 4562.

72. Mecsas, J., Raupach, B., and Falkow, S. (1998) The *Yersinia* yops inhibit invasion of *Listeria, Shigella* and *Edwardsiella* but not *Salmonella* into epithelial cells. *Mol. Microbiol.,* **28**, 1269.

73. Pederson, K. J., Vallis, A. J., Aktories, K., Frank, D. W., and Barbieri, J. T. (1998) The amino-terminal domain of ExoS, a bifunctional toxin, disrupts actin filaments via a Rho-dependent mechanism, *Mol. Microbiol.*, **32**, 393.

74. Von Eichel-Streiber, C., Boquet, P., Sauerborn, M., and Thelestam, M. (1996) Large clostridial cytotoxins—a family of glycosyltransferases modifying small GTP-binding proteins. *Trends Microbiol.*, **4**, 375.

75. Taylor, N. S., Thorne, G. M., and Bartlett, J. G. (1981) Comparison of two toxins produced by *Clostridium difficile*. *Infect. Immun.*, **34**, (3), 1036.

76. Bongaerts, G. P. A. and Lyerly, D. M. (1994) Role of toxins A and B in the pathogenesis of *Clostridium difficile* disease. *Microb. Pathog.*, **17**, 1.

77. Lyerly, D. M. and Wilkins, T. D. (1995) *Clostridium difficile*. In *Infections of the gastrointestinal tract*, (ed. M. J. Blaser, P. D. Smith, J. I. Ravdin, H. B. Greenberg, and R. L. Guerrant), p. 867. Raven Press, New York.

78. Kelly, C. P. and LaMont, J. T. (1998) *Clostridium difficile* infection. *Annu. Rev. Med.*, **49**, 375.

79. Bette, P., Oksche, A., Mauler, F., Eichel-Streiber, C., Popoff, M. R., and Habermann, E. (1991) A comparative biochemical, pharmacological and immunological study of clostridium novyi α-toxin, *C. difficile* toxin B and *C. sordellii* lethal toxin. *Toxicon*, **29**, 877.

80. Bette, P., Frevert, J., Mauler, F., Suttorp, N., and Habermann, E. (1989) Pharmacological and biochemical studies of cytotoxity of *Clostridium novyi* type A alpha-toxin. *Infect. Immun.*, **57**, 2507.

81. Martinez, R. D. and Wilkins, T. D. (1992) Comparison of *Clostridium sordellii* toxins HT and LT with toxins A and B of *C. difficile*. *J. Med. Microbiol.*, **36**, 30.

82. Just, I., Selzer, J., Wilm, M., Von Eichel-Streiber, C., Mann, M., and Aktories, K. (1995) Glucosylation of Rho proteins by *Clostridium difficile* toxin B. *Nature*, **375**, 500.

83. Just, I., Wilm, M., Selzer, J., Rex, G., Von Eichel-Streiber, C., Mann, M., and Aktories, K. (1995) The enterotoxin from *Clostridium difficile* (ToxA) monoglucosylates the Rho proteins. *J. Biol. Chem.*, **270**, 13932.

84. Just, I., Selzer, J., Hofmann, F., Green, G. A., and Aktories, K. (1996) Inactivation of Ras by *Clostridium sordellii* lethal toxin-catalyzed glucosylation. *J. Biol. Chem.*, **271**, 10149.

85. Popoff, M. R., Chaves, O. E., Lemichez, E., Von Eichel-Streiber, C., Thelestam, M., Chardin, P., Cussac, D., Chavrier, P., Flatau, G., Giry, M. (1996) Ras, Rap, and Rac small GTP-binding proteins are targets for *Clostridium sordellii* lethal toxin glucosylation. *J. Biol. Chem.*, **271**, 10217.

86. Selzer, J., Hofmann, F., Rex, G., Wilm, M., Mann, M., Just, I., and Aktories, K. (1996) *Clostridium novyi* α-toxin-catalyzed incorporation of GlcNAc into Rho subfamily proteins. *J. Biol. Chem.*, **271**, 25173.

87. Bourne, H. R., Sanders, D. A., and McCormick, F. (1991) The GTPase superfamily: conserved structure and molecular mechanism. *Nature*, **349**, 117.

88. Bourne, H. R., Sanders, D. A., and McCormick, F. (1990) The GTPase superfamily: a conserved switch for diverse cell functions. *Nature*, **348**, 125.

89. Ihara, K., Muraguchi, S., Kato, M., Shimizu, T., Shirakawa, M., Kuroda, S., Kaibuchi, K., and Hakoshima, T. (1998) Crystal structure of human RhoA in a dominantly active form complexed with a GTP analogue. *J. Biol. Chem.*, **273**, 9656.

90. Wei, Y., Zhang, Y., Derewenda, U., Liu, X., Minor, W., Nakamoto, R. K., Somlyo, A. V., Somlyo, A. P., and Derewenda, Z. S. (1997) Crystal structure of RhoA-GDP and its functional implications. *Nature Struct. Biol.*, **4**, 699.

91. Just, I., Fritz, G., Aktories, K., Giry, M., Popoff, M. R., Boquet, P., Hegenbarth, S., and Von Eichel-Streiber, C. (1994) *Clostridium difficile* toxin B acts on the GTP-binding protein Rho. *J. Biol. Chem.*, **269**, 10706.

92. Herrmann, C., Ahmadian, M. R., Hofmann, F., and Just, I. (1998) Functional consequences of monoglucosylation of H-Ras at effector domain amino acid threonine-35. *J. Biol. Chem.*, **273**, 16134.

93. Ottlinger, M. E. and Lin, S. (1988) *Clostridium difficile* toxin B induces reorganization of actin, vinculin, and talin in cultures cells. *Exp. Cell Res.*, **174**, 215.

94. Fiorentini, C. and Thelestam, M. (1991) *Clostridium difficile* toxin A and its effects on cells. *Toxicon*, **29**, 543.

95. Krivan, H. C., Clark, G. F., Smith, D. F., and Wilkins, T. D. (1986) Cell surface binding site for *Clostridium difficile* enterotoxin: evidence for a glycoconjugate containing the sequence Galα1–3Galβ1–4GlcNAc. *Infect. Immun.*, **53**, 573.

96. Eichel-Streiber, C. and Sauerborn, M. (1990) *Clostridium difficile* toxin A carries a C-terminal structure homologous to the carbohydrate binding region of streptococcal glycosyltransferase. *Gene*, **96**, 107.

97. Pothoulakis, C., Gilbert, R. J., Cladaras, C., Castagliuolo, I., Semenza, G., Hitti, Y., Montcrief, J. S., Linevsky, J., Kelly, C. P., Nikulasson, S. (1996) Rabbit sucrase–isomaltase contains a functional intestinal receptor for *Clostridium difficile* toxin A. *J. Clin. Invest.*, **98**, 641.

98. Hofmann, F., Busch, C., Prepens, U., Just, I., and Aktories, K. (1997) Localization of the glucosyltransferase activity of *Clostridium difficile* toxin B to the N-terminal part of the holotoxin. *J. Biol. Chem.*, **272**, 11074.

99. Hofmann, F., Busch, C., and Aktories, K. (1998) Chimeric clostridial cytotoxins: identification of the N-terminal region involved in protein substrate recognition. *Infect. Immun.*, **66**, 1076.

100. Busch, C., Hofmann, F., Selzer, J., Munro, J., Jeckel, D., and Aktories, K. (1998) A common motif of eukaryotic glycosyltransferases is essential for the enzyme activity of large clostridial cytotoxins. *J. Biol. Chem.*, **273**, 19566.

101. Wiggins, C. A. R. and Munro, S. (1998) Activity of the yeast *MNN1* α-1,3-mannosyl-transferase requires a motif conserved in many other families of glycosyltransferases. *Proc. Natl Acad. Sci., USA*, **95**, 7945.

102. Burland, V., Shao, Y., Perna, N. T., Plunkett, G., Sofia, H. J., and Blattner, F. R. (1998) The complete DNA sequence and analysis of the large virulence plasmid of *Escherichia coli* O157:H7. *Nucl. Acids Res.*, **26**, 4196.

103. Caprioli, A., Falbo, V., Roda, L. G., Ruggeri, F. M., and Zona, C. (1983) Partial purification and characterization of an *Escherichia coli* toxic factor that induces morphological cell alterations. *Infect. Immun.*, **39**, 1300.

104. de Rycke, J., González, E. A., Blanco, J., Oswald, E., Blanco, M., and Boivin, R. (1990) Evidence for two types of cytotoxic necrotizing factor in human and animal clinical isolates of *Escherichia coli*. *J. Clin. Microbiol.*, **28**, 694.

105. Oswald, E., Sugai, M., Labigne, A., Wu, H. C., Fiorentini, C., Boquet, P., and O'Brien, A. D. (1994) Cytotoxic necrotizing factor type 2 produced by virulent *Escherichia coli* modifies the small GTP-binding proteins Rho involved in assembly of actin stress fibers. *Proc. Natl Acad. Sci., USA*, **91**, 3814.

106. Falbo, V., Pace, T., Picci, L., Pizzi, E., and Caprioli, A. (1993) Isolation and nucleotide sequence of the gene encoding cytotoxic necrotizing factor 1 of *Escherichia coli*. *Infect. Immun.*, **61**, 4909.

107. Falzano, L., Fiorentini, C., Donelli, G., Michel, E., Kocks, C., Cossart, P., Cabanié, L., Oswald, E., and Boquet, P. (1993) Induction of phagocytic behaviour in human epithelial cells by *Escherichia coli* cytotoxic necrotizing factor type 1. *Mol. Microbiol.,* **9**, 1247.

108. Fiorentini, C., Donelli, G., Matarrese, P., Fabbri, A., Paradisi, S., and Boquet, P. (1995) *Escherichia coli* cytotoxic necrotizing factor 1: Evidence for induction of actin assembly by constitutive activation of the p21 Rho GTPase. *Infect. Immun.,* **63**, 3936.

109. Flatau, G., Lemichez, E., Gauthier, M., Chardin, P., Paris, S., Fiorentini, C., and Boquet, P. (1997) Toxin-induced activation of the G protein p21 Rho by deamidation of glutamine. *Nature,* **387**, 729.

110. Schmidt, G., Sehr, P., Wilm, M., Selzer, J., Mann, M., and Aktories, K. (1997) Deamidation of Gln63 of Rho induced by *Escherichia coli* cytotoxic necrotizing factor 1. *Nature,* **387**, 725.

111. Rittinger, K., Walker, P. A., Eccelston, J. F., Smerdon, S. J., and Gamblin, S. J. (1997) Structure at 1.65 Å of RhoA and its GTPase-activating protein in complex with a transition-state analogue. *Nature,* **389**, 758.

112. Lerm, M., Selzer, J., Hoffmeyer, A., Rapp, U. R., Aktories, K., and Schmidt, G. (1999) Deamidation of Cdc42 and Rac by *Escherichia coli* cytotoxic necrotizing factor 1 (CNF1)—activation of c-Jun-N-terminal kinase in HeLa cells. *Infect. Immun.,* **67**, 496.

113. Lemichez, E., Flatau, G., Bruzzone, M., Boquet, P., and Gauthier, M. (1997) Molecular localization of the *Escherichia coli* cytotoxic necrotizing factor CNF1 cell-binding and catalytic domains. *Mol. Microbiol.,* **24**, 1061.

114. Schmidt, G., Selzer, J., Lerm, M., and Aktories, K. (1998) The Rho-deamidating cytotoxic-necrotizing factor CNF1 from *Escherichia coli* possesses transglutaminase activity—cysteine-866 and histidine-881 are essential for enzyme activity. *J. Biol. Chem.,* **273**, 13669.

115. Pedersen, L. C., Yee, V. C., Bishop, P. D., Trong, I. L., Teller, D. C., and Stenkamp, R. E. (1994) Transglutaminase factor XIII uses proteinase-like catalytic triad to crosslink macromolecules. *Protein Sci.,* **3**, 1131.

116. Walker, K. E. and Weiss, A. A. (1994) Characterization of the dermonecrotic toxin in members of the genus *Bordetella. Infect. Immun.,* **62**, 3817.

117. Pullinger, G. D., Adams, T. E., Mullan, P. B., Garrod, T. I., and Lax, A. J. (1996) Cloning, expression, and molecular characterization of the dermonecrotic toxin gene of *Bordetella* spp. *Infect. Immun.,* **64**, 4163.

118. Roop, R. M. II., Veit, H. P., Sinsky, R. J., Veit, S. P., Hewlett, E. L., and Kornegay, E. T. (1987) Virulence factors of *Bordetella bronchiseptica* associated with the production of infectious atrophic rhinitis and penumonia in experimentally infected neonatal swine. *Infect. Immun.,* **55**, 217.

119. Horiguchi, Y., Senda, T., Sugimoto, N., Katahira, J., and Matsuda, M. (1995) *Bordetella bronchiseptica* dermonecrotizing toxin stimulates assembly of actin stress fibers and focal adhesions by modifying the small GTP-binding protein rho. *J. Cell Sci.,* **108**, 3243.

120. Lacerda, H. M., Pullinger, G. D., Lax, A. J., and Rozengurt, E. (1997) Cytotoxic necrotizing factor 1 from *Escherichia coli* and dermonecrotic toxin from *Bordetella bronchiseptica* induce p21^{rho}-dependent tyrosine phosphorylation of focal adhesion kinase and paxillin in swiss 3T3 cells. *J. Biol. Chem.,* **272**, 9587.

121. Horiguchi, Y., Inoue, N., Masuda, M., Kashimoto, T., Katahira, J., Sugimoto, N., and Matsuda, M. (1997) *Bordetella bronchiseptica* dermonecrotizing toxin induces reorganization of actin stress fibers through deamidation of Gln-63 of the GTP-binding protein Rho. *Proc. Natl Acad. Sci., USA,* **94**, 11623.

122. Rosenshine, I. and Finlay, B. B. (1993) Exploitation of host signal transduction pathways and cytoskeletal functions by invasive bacteria. *Bioessays*, **15**, 17.

123. Hardt, W.-D., Chen, L.-M., Schuebel, K. E., Bustelo, X. R., and Galán, J. E. (1998) *S. typhimurium* encodes an activator of Rho GTPases that induces membrane ruffling and nuclear responses in host cells. *Cell*, **93**, 815.

Index